Pest Management and the Environment in 2000

Edited by

Abdul Aziz S.A. Kadir,
Henry S. Barlow

C·A·B International
in association with the
Agricultural Institute of Malaysia

C · A · B International
Wallingford
Oxon OX10 8DE
UK

Tel: Wallingford (0491) 32111
Telex: 847964 (COMAGG G)
Telecom Gold/Dialcom: 84: CAU001
Fax: (0491) 33508

A catalogue record for this book is available from the British Library

ISBN 0 85198 775 3

Typeset, Printed and bound in Malaysia

Contents

PART 1
Introduction

Preface

The importance of pest management in raising agricultural productivity is well recognised. It has now evolved into a complex and involved management system. The widespread use of pesticides has affected not only pests but also human health, the environment and resistance of pests to many pesticides. As a result, interest in Integrated Pest Management has grown steadily over the years. In response to this interest the Agricultural Institute of Malaysia (AIM) and CAB International sponsored the international seminar on Pest Management and the Environment in the Year 2000. This seminar was held in Kuala Lumpur from 7 to 8 May 1991 with the objective of increasing awareness of the issues, dangers, trends and possible solutions in pest management.

A major challenge to plant protection specialists worldwide is the ability to integrate effectively pest control measures which must be selected on the basis of cost effectiveness, sustainability and environment friendliness. The seminar provided a forum for scientists, policy makers, agricultural extension workers and producers to share experiences and discuss issues and problems pertaining to pest management. In the developing countries several barriers, such as technological, institutional, economic and attitudinal, will have to be overcome before integrated pest management can be implemented widely and successfully. Their efforts to produce sufficient food both in terms of quantity and quality for the expanding populations can only be realised with effective and sustainable pest management systems.

The organising committee, on behalf of CAB International and AIM wish to thank all those who have helped towards the success of this conference. Special thanks are due to the Hon. Minister of Science, Technology and the Environment, the invited speakers; to the distinguished Chairmen of sessions, the Rubber Research Institute of Malaysia, the Asian Development Bank, International Rice Research Institute, Australian International Development Assistance Bureau, Overseas Development and Natural Resources Institute, UK, CIBA-Geigy Ltd, Basle, National Agricultural Research Centre, Japan, Royal Agricultural College, Cirencester, UK, Royal Botanic Gardens, Kew and FAO Indonesia.

Chapter one: Opening Speech by the Hon. Mr Law Hieng Ding, Minister of Science, Technology and the Environment

May I wish you all a "Good Morning" and bid a warm "Welcome to Malaysia" to our foreign guests in the Malaysian traditional way of "Selamat Datang".

I am pleased to be here this morning and would like to thank the organisers for inviting me to address you at this international gathering of eminent experts in pest management. The Agricultural Institute of Malaysia (AIM) and CAB International are to be congratulated for making this seminar a reality. I am sure that with the presence of scientists from many countries, the deliberations over the next two days will formulate strategies and programmes to enhance the agro-economic benefits of pest management.

The problem of pests in agriculture is not new, as you are familiar with crop devastations attributed to pest attacks. Rodent pests have caused post-harvest losses of disturbing proportions. Certain pest infestations lead to crop threatening diseases, while weedy pests deprive crops of their much needed nutrients. In other words, pests pose a serious economic burden on agriculture.

The traditional method has been to rely on chemical pesticides for seemingly effective control of pests. This may not necessarily be the best method. There have been different schools of thought. The proponents of chemical pesticide claim that if not for the use of such chemicals, there would not have been high productivity in agriculture. On the other hand, the plant breeders attribute higher productivity directly to the use of higher yielding varieties. However, the soil scientist will emphasise the role of fertilisers in raising productivity.

You, the scientists and experts from various disciplines, have to look into the different perspectives of this problem facing us today. While higher productivity in agriculture is the target, we need to look at the solution in terms of today's realities, so as not to avoid the issues of the environment and sustainability in agriculture, to ensure real improvement of the quality of life of mankind.

It is interesting to note that in Malaysia, many of our crops suffer little from pest problems. This means that farmers (who manage large and small holdings) need not spend too much money and effort in pest management. We are conscious of protecting our natural heritage and we are committed to ecologically sound pest management. As such, Malaysia has always adopted a sensible approach to pest management. This will also make our commodities more competitive in the world market.

Take for instance, the plantation sector. Plant protection specialists were quick to point out that excessive and calendar use of insecticides leads to outbreaks of oil palm bagworms. This led to a stop in indiscriminate use of insecticides in oil palms since the early sixties.

The rhinoceros beetle posed a serious threat to coconut in the early years of the coconut palm industry. Rather than resort to chemical insecticides, plant protection specialists found out that the outbreak of this pest infestation is caused by careless planters who left tree trunks to rot in the field. Legislation was introduced to ensure that all fallen trees were disposed of quickly.

This reduced the beetle to a non-pest status. The beetles breed in rotten logs. It was later found that by reducing these breeding sites, the beetle was effectively controlled by a disease which specifically kills the insect. This disease was subsequently exported to other countries in the Pacific to help them control this coconut pest. This seems to be a success story of biological control.

Rice is another crop where very low use of chemical pesticides helps farmers grow good quality rice without affecting the health of farmers. It has been shown that when farmers use insecticides regularly in their rice crops, the consequence has been serious pest problems. This is because indiscriminate use of the insecticides kills off predators and parasitoids of rice pests. The Malaysian authorities recognised that these predators and parasitoids are necessary for rice cultivation and have avoided attempts to introduce pesticide subsidies for rice farmers. This has been shown to be a very good move.

Despite the advantages of conserving existing natural enemies, there is a trend to increased usage of chemical pesticides in Malaysia. In 1989 alone, more than $300 million Malaysian ringgit were spent on chemical pesticides. We will have to review the wisdom of such a level of expenditure. It should be noted that control by chemical pesticides is only one of the pest management tools to be used when necessary. More money should be spent on research to develop ecologically-sound pest management.

There is greater awareness that chemical pesticides, besides being toxic to humans, can pollute the environment. By killing off natural enemies, pesticide use tends to set off more outbreaks. Often the organisms that the pesticides seek to kill, themselves become tolerant, leading to the phenomenon which scientists call "pesticide resistance".

A case in point was when the diamondback moth arrived from Europe together with cabbage cultivation in Malaysia. It soon become a limiting factor to cabbage cultivation in the Cameron Highlands because there were no effective predators and parasitoids. Farmers relied heavily on chemical insecticides. Very soon, the diamondback moth became resistant and was not easily killed by the insecticides used. Eventually, farmers had to use more insecticides, sprayed more frequently, before a crop could be harvested. The diamondback moth was not a problem in countries where effective parasitoids exist. Malaysian scientists then introduced "friendly" insects. Initially, these "friendly" insects did not have any impact on the pest because farmers continued with excessive use of insecticides. However, when the use of insecticides declined, the parasitoids were able to act and reduced the pest problem to a low level. Here we have a case where research triumphed and a sensible pest management programme was put in place.

Despite the success of "friendly" insects in keeping insect pests in check, farmers often failed to appreciate them. Hence, it is important that efforts be made to transmit this information for the benefit of farmers.

Chemical pesticide is fairly well recognised as an economic approach to control pests, but such chemicals are highly toxic to other species in the environment. There is now a growing concern worldwide over the indiscriminate use of such chemicals. Many arguments have been presented suggesting that over-dependence on chemical control is not in the long-term interest of mankind.

Many instances have been cited of the negative repercussions of indiscriminate use of chemical pesticides. Environmental pollution, especially of surface waters, is one serious consequence of pesticide run-off. The consequence is serious contamination of rivers, streams and even underground aquifers. Reports of fish-kill in rice fields have been a regular feature. Freshwater fish which used to thrive in abundance are now fast diminishing in rice fields. Chemical pesticide has been singled out as the most likely culprit.

Another issue concerns residual levels of toxic pesticide in fruits and vegetables. Cases have been mentioned of farmers who are over-generous in their pesticide application . The excess often ends up in the food consumed. It is a disturbing development and the Government views this very seriously and has adopted stringent monitoring controls on all farm produce. Such controls have minimised the incidence of toxic residues in our fruits and vegetables.

An equally hazardous consequence of pesticide use is in its health implications. Workers who are in constant contact with pesticides have reported negative health effects, not to mention deliberate recourse to pesticide in suicide attempts. In some circles, pesticide misuse has evolved into a worrying social problem.

It is clear that chemical pesticides have helped boost agricultural output. However, it is also obvious that pesticides require proper handling. Indiscriminate use can precipitate undesirable consequences such as environmental contamination and toxic residues in food. Experiences in the past have confirmed that chemical pesticides can be abused to the detriment of mankind.

Educating farmers about environment-friendly insects should be further emphasised. Such activities have been carried out in the northern rice states of Kedah and Perlis and should be done in the other rice growing states. Educating the users is definitely a useful approach to pursue. Farmers should be educated on correct dosages. Workers should be informed about protective clothing and other precautions.

I understand that scientists are giving serious attention to looking for acceptable control strategies, particularly biological means of control. R & D efforts in such areas are showing encouraging results which should pave the way for further reduction in pesticide use.

The Malaysian Government is fully aware of the benefits of investment in R & D. We are supportive of all research efforts to develop technologies for the continued progress of mankind. It would be ideal if we can develop technologies which can harmonise economic pursuits with sustenance of our ecological well-being. In the Fifth Malaysia Plan, a sum of $400 million Malaysian ringgit was allocated for R & D under the Intensification of Research in Priority Areas (or IRPA) mechanism. For the forthcoming Sixth Malaysia Plan, an even bigger allocation is envisaged. This is a clear demonstration of the Government's commitment to R & D efforts in this country. In this endeavour, we hope the private sector will also allocate more funds for their R & D programmes.

In relation to studies on pest control, I have been told that a number of sectors have made commendable progress. Rodent control in oil palm plantations is now achieved effectively using barn owls. This should definitely reduce the use of chemical rodenticides. The rearing of sheep under rubber has, to a significant extent, reduced use of chemical weedicides. These are developments that scientists should be proud of.

However, we must never be complacent. Instead, scientists should continue to explore other avenues to further improve pest mangement. You need to be bold to look at alternative solutions to our problems not only until the year 2000, but beyond into the twenty-first century.

I am pleased that this seminar will promote further exchange of ideas and information on the subject of pest management. The location of the CAB International Regional Office for Asia in Kuala Lumpur provides for better interaction among scientists in Asia and other regions in general and Malaysia in particular.

We in the Ministry of Science, Technology and the Environment look forward to receiving a copy of your deliberations as it could provide valuable inputs to our own R & D planning.

In closing, I would like our foreign participants not only to meet their local counterparts, but to take some time out to enjoy the rich cultures and traditions of our multi-ethnic country. I wish you every success in your seminar. On this note, I now have pleasure in declaring this Seminar officially open.

Chapter two:
Pest Management in 2000: What We Might Learn from the Twentieth Century

M.J. Whitten
CSIRO, Division of Entomology, Canberra, Australia

Summary

The twentieth century has witnessed the greatest advances — and some substantial retreats — in mankind's capacity to deal with pests, embracing vertebrates, invertebrates, pathogens and weeds. Extraordinary advances have occurred in the physical and especially the biological sciences, and these have been translated into novel forms of pest control, some that were not even imaginable last century, e.g. genetic control and sterile males. One hundred years of biological control was celebrated with the centenary Vedalia conference at the University of California, Riverside Campus in 1989, and this traditional discipline has experienced a renaissance of late. Several generations of pesticides have come and gone, curtailed either through resistance in the target species or because environmental or health effects have became unacceptable. The abysmal failure of pesticides to contain key pests of rice and vegetables in much of Asia — indeed pesticide-induced pests are a feature of the non-sustainable agriculture practised in parts of Asia today — poses one of the great problems for the future.

Improved physical containment, especially for post-harvest storage of grain crops, is becoming an increasingly important option for much of the developing world. Behaviour modifying chemicals such as pheromones represent a significant feature of modern pest management options for monitoring presence and abundance of pest species, and in some instances for disrupting their behaviour.

Over the past decade, scientific and technical revolutions have occurred in information technology and molecular biology. The promise of these new technologies has yet to be fully realised but improved modelling, genuine expert and forecasting systems, genetically engineered crop plants and improved natural enemies will be a conspicuous feature of pest management in the twenty-first century. But the distinguishing feature of pest control for decades to come will be the integration of cultural, chemical, biological, and physical approaches to pest management, combined with genetically superior plants and animals.

Presently, plant and animal protection in much of the developed and developing world is marred by misapplication of traditional technology. Current initiatives in biological control and IPM by international agencies such as FAO, WHO and CABI are largely driven by the need to overcome misuse of pesticides which can induce pests, create severe environmental and health problems, squander valuable but finite resources, and generally lead to unsustainable agricultural systems.

Part of the address will be devoted to detailing the arsenal of pest management tools that should be available in the year 2000 and beyond, and how we might integrate these in a cost-effective, environmentally friendly and sustainable manner. However, the evidence is mounting to suggest that the overriding challenge will not be the creation of additional pest management options, but the application of those we already have, in a sound political, social and economic context. In other words, technology itself is unlikely to be the limiting or critical factor. We will need to continue to produce high quality, residue-free commodities as cheaply as possible; and the techniques we employ will need to be as effective 20 years down the track as when first introduced. In other words, they will need to be robust and sustainable.

We will need to produce highly qualified technologists in the disciplines of taxonomy, ecology, toxicology, pathology, molecular biology and physiology, and information technology. These will need to be complemented by social and political scientists and resource economists who can ensure that the solutions we apply go well beyond the technical. An ability and willingness to talk with and win the support of political and community leaders about the critical importance of plant and animal protection will also need to be an accepted skill of our leading pest managers in the years ahead.

In addition, new methods of non-formal adult education will need to be devised to make the millions of farmers in the next century literate in pest management practice. Future generations will require pest management options that are genuinely suitable for the circumstances and communities who are supposed to benefit, and not the commercial interests of some board room in another place. In this sense, the technologies must serve the producer and consumer, not the converse.

Finally, we will only be buying time, if the increased carrying capacity for mankind is not coupled with a truly effective human population management strategy which does not threaten key aspects of sustainability such as the quality of our water and air, the fertility of our soils, and the biodiversity of terrestrial and aquatic ecosystems.

Introduction

Many agricultural systems in both developed and developing countries have become dependent on regular usage of fertilisers and pesticides, and are characterised by soil degradation, water contamination and loss of biological diversity (Anon, 1988, 1989; Langeweg, 1989). Thus, otherwise valuable agricultural production technologies can prove to be harmful to mankind at large: especially if our analysis encompasses the indirect impacts, the so-called externalities. When all the sums are done, some technologies, especially high input/output systems, can be unprofitable — at least to producers, consumers or more often taxpayers because of indirect subsidies; they can be environmentally damaging; and they sometimes score low on sustainability. Quite frequently we witness an erosion of non-renewable resources such that the productive capacity cannot be maintained over decades, let alone centuries. Accordingly, to give some credibility to our predictions, it is necessary to address two issues.

First, we need to anticipate the range of technologies potentially available to the pest manager in 2000 and beyond. Second, we need to appreciate the social, political, economic and environmental contexts in which the pest manager will be choosing among these technical options. For example, pesticide residues in foodstuffs may be deemed irrelevant; or they may have a health dimension, real or perceived; or the residues may be exploited as a trade barrier. Pesticide contamination of the environment could be a critical factor, depending on what intrinsic value is placed on individual species or biodiversity generally. The level of importance placed on sustainable production systems, and the rights of future generations to eke an existence from the same non-renewable resources, will also impact on the choice of technologies. Even the emphasis we place on research itself to identify and develop novel options, will be influenced by these broader and non-technical issues.

Let us take a specific example from events that have occurred over the last two decades in the Sahel of West Africa. Following a major outbreak of grasshoppers in 1974, probably due to short-term fluctuations in climate (Fishpool & Popov, 1981), the FAO undertook the first ever large-scale campaign to control insect pests in subsistence crops by aerial and ground spraying. Over one million hectares were treated. Although short-term control was achieved, the grasshopper situation rapidly worsened during the eighties and infestations of grasshoppers were routinely controlled in most seasons in the Sahel with a range of broad spectrum insecticides. Between 1986 and 1990 about 16 million litres of concentrated insecticides and 14 million kilograms of insecticidal dusts were applied to nine million hectares of subsistence crops, but the grasshopper problem continues to worsen.

Earlier research had shown that large numbers of locust eggs and nymphs are consumed by predators and parasites in the Sahel (Farrow, 1975). Recent pilot studies by the FAO on the environmental effects of grasshopper spraying showed that fenitrothion and chlorpyrifos killed up to 80% of one group of beneficial insects. Some of these, such as the relatively sedentary tenebrionid beetles, which are important predators of grasshopper egg pods, did not recover, suggesting that the increasing severity of outbreaks is, in part, a pest resurgence phenomenon. Because of these negatives, biological control and a range of novel methods are now being pursued. Thus the options that will become available in the future to deal with the grasshopper problems in the Sahel are certain to be influenced by interconnected questions of technical feasibility, sustainable productivity, and the value we place on biodiversity and the environment. The efficacy of these options will be governed by the emphasis placed on research topics over the next 5–10 years. No doubt, pesticides will continue to play a key role in such crises for some time to come, but they are increasingly being viewed as a 'necessary evil'.

To give a further example; genetic engineering provides, for the first time, an ability to create novel organisms that had little or no prospect of arising through the natural evolutionary process. The availability of a previously benign virus rendered virulent by the insertion of a gene coding for a protein toxin or anti-feedant, or the insertion of similar genetic information into a crop plant, could lower the pest status of plant herbivores like *Heliothis armigera,* the brown plant hopper, or the diamondback moth. To the scientist, some of these products of genetic engineering would appear absolutely safe and pose a novel, cost-effective and environmentally friendly way of managing pest species. Exactly what was the genetic manipulation in a given instance, will often remain a complete puzzle to many in the community; their assessment will necessarily be made in partial or total ignorance of the technical aspects of the option. The lay person may have a philosophical or religious view — it's unnatural, it's playing God — or hold an unsound ecological perception of what is entailed. But it is likely to be these perceptions and ideologies that will dictate whether the option is ever exercised. The current hesitation in Europe and North America, and the withdrawal of financial support by commercial interests around the world, for genetic engineering and biotechnology is very much influenced by this uncertainty.

The Technical Options

An Overview

A simple way to tackle the scope of technical options that should be available to the pest manager around the year 2000 and beyond is to give a brief outline of entomological research conducted by the CSIRO's Division of Entomology. The Division has the principal responsibility for undertaking research on insects and related arthropods to provide knowledge relevant to issues of economic and social importance to Australia and the surrounding Asian and Pacific region. If the Division is doing its job with any degree of competence, and with an appropriate level of forward thinking, then the research and development portfolio should give some inkling of future options for pest management.

The Division has some 350 scientists and support staff, headquartered in Canberra, whose research and development responsibilities span a range from economic entomology of plant and animal pests, to fundamental studies on insect taxonomy, behaviour, development, ecology and evolution, through to defining the place of insects in the environment. We need always to be mindful of the fact that the vast majority of insects are not harmful to the interests of mankind. Indeed, without them, most ecosystems would crash. Insects are increasingly useful as models in basic research and can give us valuable insights into phenomena such as learning, short-term and long-term memory, development and behaviour (Whitten, 1989). Thus there is an overwhelming case, on economic, cultural and scientific grounds for maintaining strong centres for entomological research in every country regardless of its level of development.

Despite the above, the pragmatic person knows that the opportunity for entomological research would be severely limited, were it not for the fact that some species of insects have the temerity to compete with mankind for food and fibre, act as important disease vectors, or be valuable biological control agents. Accordingly, applied entomology has to occupy a conspicuous place on the research agenda. In consequence, much effort is devoted to seeking more effective ways of using chemical pesticides and fumigants and integrating these with other methods of control. However, our overriding philosophy is to reduce mankind's dependence on chemical pesticides by identifying other methods which are economic, environmentally sound and lasting in effectiveness. Seeking these alternatives is the exciting challenge.

Often the search takes us into the "hi-tech" arena — but it need not be so. Knowing how an insect behaves and understanding its niche in the ecosystem can sometimes suffice to allow the resourceful scientist to develop a "low-tech" but effective and durable method of control. One such example is the development of a simple trap, by Dr Robert Sutherst and his colleagues (pers. comm.), for removing buffalo fly *Haematobia exigua*, from its host, the dairy cow. The trap is strategically placed at the entrance and exit from the milking yard and, as the cows pass through, the flies are disturbed by a set of dangling straps. The flies are attracted to the sunlit roof of the trap where they are killed by solar heat. No chemicals are required, and it is improbable that the fly will evolve behavioural resistance to the trap. The cattle are so appreciative of the relief, that they actually queue for the privilege of moving through the trap!

The burial of vast amounts of cattle dung following the introduction of exotic dung beetles from Africa and Europe further removes larvae feeding resources for the buffalo fly, the bush fly and other nuisance dung breeding dipterans over much of Australia (Tyndale-Biscoe, 1990). Such classical biological control, or more strictly, control by ecological manipulation of an insect's habitat, will continue to play a prominent role in pest management, well into the twenty-first century, as entomologists identify and attempt to remedy ecological disturbances caused by the activities of mankind during the twentieth century. As the work of Tyndale-Biscoe illustrates, information technology through computer-based modelling and decision-support systems, also has an important complementary part in facilitating classical biological control.

Introduction technology also has a major role to play in developing decision-support and expert systems where we wish to manage pests, pesticide resistance and integrate these with other crop management practices such as irrigation or fertiliser regimes. CLIMEX is one such package which permits prediction of ultimate distribution and pest status of an exotic pest (Maywald & Sutherst, 1989). CLIMEX is also readily applicable to forecasting changed pest status and distribution under climate change as global warming proceeds.

A major effort within the Division centres around ecology, especially in terms of defining parameters which determine abundance and distribution of pests and their natural enemies. Such studies focus on life cycles; local and long-distance dispersal using radar, kites and other traps; insect/plant interactions; case studies on individual major pests like red-legged earth mite, spider mite, salvinia and its natural enemy, and the sheep blowfly (for review see Whitten, 1991). Without detailed ecological studies, we are generally unable to determine why a particular control strategy worked or failed, or to evaluate the efficacy of a novel control approach, such as the genetic control of sheep blowfly or the mating disruption of codling moth with sex pheromones. The major water weed, salvinia, was considered to be refractory to biological control. However, some 70 research papers later (Room, 1990), and after 10 years of sound entomological field research, salvinia has been controlled in Australia, PNG, and in some Asian and African countries. Kevin Gallagher, later in the symposium, should be able to draw the same conclusion about IPM of insect pests of rice in Asia.

The record demonstrates that effective applied entomology is invariably underpinned by sound basic research. Incidentally, the salvinia and rice projects also demonstrate that much of the research needs to be done or validated in the country where the problem is occurring. Certainly much valuable information can be acquired from developed countries, but the unquestioning transfer of ideas and technologies from temperate to tropical countries can sometimes smack of "scientific colonialism".

Classical biological control continues to play a major role for insect and weed control in Australia, Asia and the Pacific, and this traditional approach to pest management will retain an important place in the decades to come (Waterhouse & Norris, 1987). I will return later to the place of international agencies such as FAO and CABI in ensuring that these opportunities are effectively exploited.

Work in CSIRO, and elsewhere, continues on the development of improved natural enemies (Whitten & Hoy, in press), and micro-organisms, such as the fungus, *Metarrhizium anisopliae* (Beckman, 1991), various bacteria, and mycoherbicides, especially for grassy weeds where specificity has proved a major obstacle in the search for suitable natural enemies (Wapshere, 1990). Entomopathogenic fungi and nematodes are being field tested in Australia, China and Europe for inundative periodic treatment of a range of insect pests. CSIRO has pioneered the use of nematodes in classical and inundative biological control, and has a strong interest in the use of sex pheromones for monitoring pest abundance or for their control by disrupting mating behaviour. Some of these topics are discussed in more detail below.

We have over 20 scientists studying ways to insert foreign genes into insects, the development of general systems for genetic transformation of insects. They, and colleagues around the world, study the molecular basis for pesticide resistance so that we can design novel pesticides, or extend the useful life of existing pesticides (Soderlund & Bloomquist, 1990). These latter studies are done in conjunction with ecological geneticists who endeavour to determine the precise circumstances in the field that favour resistant genotypes when they are exposed to specific pesticide regimes (Roush & Daly, 1990). Molecular biologists are identifying and isolating strategic genes, those which code for proteins that are toxic, e.g. wasp venom proteins, or proteins that act as anti-feedants, or which otherwise disrupt some vital behaviour (Binnington & Baule, 1991). In conjunction with plant molecular biologists elsewhere, who are developing the generic capability to transform crop plants, our scientists are identifying the strategic genes, and also developing the delivery systems. We need to have a range of suitable genes well beyond *Bacillus thuringiensis* (*Bt*) toxins, and we need a wider selection of delivery systems. These will range from virulent, but self-destruct viruses of the baculovirus and entomopox virus groups, to RNA viruses small enough to be integrated *in toto* into the plant genome. In passing, much of the work on genetic engineering of insect pathogens and crop plants is deemed to have commercial value and often the work is done by R & D agencies under secrecy agreements. Thus, no clear overview of what is happening around the world is possible though the broad intention is fairly obvious.

We will now turn to taxonomy, a discipline central to pest management. It is particularly interesting to see how the traditional field of taxonomy is forging linkages with new technologies such as with information technology, scanning electron microscopy and molecular biology in such a manner that taxonomy is enjoying a true renaissance. Only a few years ago science administrators in Australia viewed taxonomy as a soft science, a hobby of the privileged few! This unenlightened view has not been restricted to Australia.

Taxonomy is the science of handling vast amounts of biological data, and it is these biological data that are proving central to informed decisions concerning economic entomology and the environment. On reflection, taxonomists are properly viewed as "knowledge experts", and it is information technology that will transform our taxonomists from the status of cloistered monks, and our major insect collections from the status of inaccessible medieval libraries, into accessible databases of valuable and often critical information that will assist a wide range of scholars and applied scientists.

To give one example, Dr John Lawrence of the Australian National Insect Collection ran a workshop in Hawaii in June 1991 showing how he used the DELTA software package, developed by Michael Dallwitz, to assemble a comprehensive database covering the larval stages of the 350,000 known species of beetles — at least to family, sub-family or tribe level. DELTA is coupled with another software package, INTKEY, which permits the user to interact in a friendly manner with the database, to identify and access known information about specimens of interest (Dallwitz, in press). This package, together with glossary of terms, and appropriate graphics, when eventually available, will be encoded in several floppy disks or CD-ROM.

There are two rate-limiting processes to the consummation of this remarkable marriage between taxonomy and information technology. First, in Australia and Asia, fewer than half the species of invertebrates already known to science, have been named, and we have useful biological information on far less. A larger army of taxonomists is required to address this problem, yet their numbers are diminishing. Second, much of this material and information is relatively inaccessible, being lodged in major collections in Europe, the United Kingdom and North America. It will be a major effort to database electronically this vast body of information contained in the physical records of the major entomological collections around the world, and make it readily available to a wider range of entomologists and other biologists. However, there is a move by several of the major collections (*viz.* the British Museum (Natural History), the Smithsonian in Washington, the Australian National Insect Collection and the Canadian Biosystematics Research Centre) to begin constructing an inter-connected global database.

More rapid, comprehensive and reliable access to the world's storehouse of taxonomic information will represent a major advance for applied entomologists, and will be of enormous value to environmental scientists and those responsible for managing the world's biodiversity. CABI is a well-placed institution to act as a leading information-broker in facilitating information technology to assume a central role in pest management by the year 2000 and thereafter. Let me now speak in more detail about four areas of entomological research to illustrate the shape of technologies to come:

- storage of dry food commodities
- molecular biology
- insect-killing nematodes, and
- pheromones

Storage of Dry Food Commodities

Control of insect pests in storage facilities has centred around physical and chemical techniques. Absence of biological methods is likely to continue for the foreseeable future, though genetically-engineered micro-organisms such as *Bt* may find a narrow niche. Information technology, with computer-based decision support and expert systems, will certainly become more prominent, offering remote, automatic and sensitive control of storage conditions (Heong, 1990). Tomorrow's technologies, both for bulk and bag storage, are likely to include one or more of the following options:

1. ***Sealed storage***, ranging from sealed packages (Mitsuda & Yamamato, 1980), bag stacks (Annis, 1990), plastic covered bunkers (Yates & Sticka, 1984) through to sealed permanent structures (Newman, 1990).
2. ***Controlled atmospheres***, using supplied gas (CO_2, N_2), on-site gas generation (CO_2, N_2), or flow-through fumigants such as phosphine (PH_3) using processes such as SIROFLO (Champ *et al.*, 1990).
3. ***Persistent pesticide***, adhered to dusts which are removed prior to processing the grain. Such applications promise to be cost-effective, leave little residue in the commodity, and have reduced probability of resistance developing in the insect pest (Desmarchelier & Ahern, 1988).
4. ***Heat drying and disinfestation***. Most grains can be treated as fluids when air is passed through under appropriate conditions. Pre-heating the air provides the option of grain drying or pest disinfestation. High volume (200 tonnes per hour) fixed heat disinfestators are available, as are low volume (20 tonnes per hour) mobile dryers (Sutherland, 1988).
5. ***Removable inert dusts***, such as dryacide, a silica-based product, can be applied to grain to disrupt normal behaviour and development (Ebeling, 1976). The husks of paddy rice are comprised mainly of silica — so this solution has a distinguished history.

The range of technologies available today, or at an advanced stage of development, seem to suggest that by the year 2000, losses of dry food commodities, whether in bulk or small container storage, should be largely preventable.

Molecular Biology

By the turn of the century, we can expect molecular biology to have contributed to advances in both the diagnosis and control of pest populations. From a diagnostic perspective, there are at least three areas where molecular techniques are likely to prove invaluable: the identification, quantification and genetic monitoring of pest populations. Unequivocal identification of pests is a necessary prerequisite for implementation of rational control strategies. While morphological and cytological techniques are generally employed for this purpose, the separation of sibling species in a pest complex can be beyond the capabilities of such methods. In these cases, appropriate molecular techniques can provide the necessary diagnosis of species composition of a pest population. Nevertheless, identification of pests is not necessarily through separation at species level. Particular strains, races or pathotypes of species can have very different biological characteristics, and thus very different status as pest organisms. For example, while crown gall disease of orchard trees is caused by infection of a plant with a virulent strain of *Agrobacterium tumefaciens*, it is possible to protect against the disease by inoculating plants with a non-virulent strain of the same species (Kerr, 1980). Identification of bacterial strains within a growing area could thus determine whether such treatment was warranted; in different situations determination of potential crop exposure to microbial pathotypes can form a rational base for choice between resistant and susceptible varieties or cultivars.

A further potential for molecular biology is the quantification of pest populations which cannot be directly or easily visualised. Serologically-based tests could provide rapid estimates of, for example, biomass of mite pests in the soil, again aiding in rational application of pesticide to crops. Another important application of molecular diagnostic techniques lies in the monitoring of both presence and frequency of genes of interest within a pest population. Thus, for example, genes for resistance to specific classes of pesticide can be detected, their frequency periodically assessed, and this information used in the design and continued implementation of rational pest management strategies.

A variety of molecular techniques can be and are used for the purposes described above. These are based on analyses of either the proteins or nucleic acids of the organisms under study. Protein-based tests include studies of isoenzyme patterns, and the more recently developed immunological assays. These latter make use of interspecific (or intervarietal, etc.) antigen variation, and, depending on the organisms involved and their degree of relatedness, are based on polyclonal or monoclonal antisera. The most critical part of these tests is the laboratory production of discriminatory antisera; once achieved, it is relatively simple to incorporate individual components into self-contained test kits (Miller & Williams, 1990). Socio-economic factors are then the major barrier to their successful use. It is hoped that the availability of such kits will make an important contribution to the future of pest diagnosis in the field. Tests based on nucleic acid methodologies are at present largely confined to laboratory settings; while very powerful, they require specialised equipment and a high level of expertise. In their current forms, these tests generally rely on binding of short fragments of DNA to nucleic acid extracts of the organism, and subsequent detection of the interaction through labelling or specific amplification. Advances which allow field-adaptability of such techniques will make a great contribution to pest management.

Techniques of molecular biology can also be utilised in the control aspects of pest management. This work relies largely on improvement of the efficiency with which natural pathogens of pests kill or debilitate their hosts, and on improvement in formulations used for their delivery (Whitten & Oakeshott, 1990). It is possible to enhance the virulence of microbial pathogens through artificial selection processes, but both theory and experience predict that field releases will then be followed by reversion to a lower virulence level. Of greater potential interest, therefore, is the possibility of using disease agents as vectors for the delivery of group-specific toxins. This work generally means manipulation of the pathogen genome through "genetic engineering" methodologies, and specifically, the insertion of heterologous genes which encode toxic or otherwise detrimental products.

This approach must be supported by a comprehensive understanding of the characteristics of the pathogen genome, in order to ensure the regulated expression of the toxin-encoding genes. Perhaps the most advanced example of this approach to pest control is the study of the occluded viral pathogens of insect pests, and in particular, that of the nuclear polyhedrosis virus sub-group of the baculovirus family (Cochran *et al.*, 1986). Genetic manipulation of these viruses, and the expression of heterologous inserted genes at predictable times and levels have been repeatedly demonstrated (Maeda, 1989). A number of groups around the world are now actively engaged in the development of recombinant microbial pesticides using these viruses and this strategy. The efficacy of these products in the field, and thus their contribution to pest management in the year 2000, has not yet been demonstrated. While it would now appear to be high, realisation of their full potential rests on three further contingencies. These can be summarised: first, choice of appropriate heterologous genes for expression in the recombinants; secondly, development of methods for cheap large-scale manufacture of the recombinants; and thirdly, permission to release such products into the environment.

Gene products detrimental to pests can be of two types—either proteins which are specifically toxic to the organism, or proteins which are generally present in the organism, but which when misexpressed interfere with homeostatic or developmental mechanisms. Again, using insect pests as an example, both types of genes are under active consideration. A number of toxins which kill insects but not vertebrates have been identified in the venoms of scorpions, itch mites and wasps; examples of each have been cloned and sequenced, and these are now being used to construct recombinant viral insecticides (Binnington & Baule, 1991; Carbonell *et al.*, 1988). Alternatively, a variety of insect genes encoding the production or degradation of insect hormonal products have been used in a similar fashion and, in laboratory-scale experiments, some of these have shown dramatic effects. While recombinant products of the type described here can be routinely grown in special conditions in the laboratory, this is an expensive, and therefore unsatisfactory method for mass production. Major research effort thus needs to be applied to what may be perceived as the less glamorous aspects of large-scale and/or cottage industry production, if this work is to have genuine applicability to pest control throughout the world.

Finally, but by no means least, following satisfactory safety testing, the recombinant products must be cleared for release into the environment. While it is wise to approach this matter with appropriate caution, unwarranted mistrust of all products generated through genetic manipulation should not be allowed to hinder their use. This must be especially emphasised when it is realised that in cases such as these, the products are likely to represent a safe and target-specific alternative to the many wide-spectrum, environmentally-damaging pesticides currently in use.

The distribution of all types of pesticides through the environment can be reduced if appropriate formulation and delivery strategies can be developed. Arguably the most important contribution of molecular biology to this area is the capacity to express pesticidal protein moieties within transgenic plants (Hilder *et al.*, 1990). Using this strategy, molecules with gut-acting pesticidal activity are delivered at the precise site of feeding of herbivorous pests, but are not free to circulate within the environment at large. A salient example of this strategy is the development of transgenic plants which express the *Bt* delta-endotoxin; it has been proven that a variety of plant species of commercial importance can be protected from caterpillar attack by this method, and it is likely that some of these species will be field-tested in the near future.

There is also potential to control a wide variety of other plant pests and pathogens using transgenic plants, with limits imposed only by the nature of the proteins which can be tolerated by the plant itself, and by the discovery and/or development of molecules with appropriate activity. Proteins which can be used to cross-protect against viral infection have also been successfully expressed in transgenic plants, and it is likely that a variety of plant proteins with feeding-deterrent activity are waiting to be discovered. Transgenic plants can also produce antibodies of the type naturally restricted to vertebrates, and the current level of interest in this technology suggests that it is likely to be at an advanced stage by the year 2000. In addition to expression of protective proteins, plants can express the catalytic nucleases known as ribozymes, with which RNAs can be specifically cleaved. These have already been shown to be highly active against plant viral diseases (Young *et al.*, in press), and efforts are now in progress to target their activity to other groups of plant pests.

As, and when, the generic capability to transform insects become available, it should be possible to develop a new range of options for improving genetic or autocidal methods of control, especially for dipteran pests which are often refractory to biological methods of control, and prone to develop resistance to chemical insecticides (Whitten & Oakeshott, 1990). Conditional lethal genes, sex transforming genes, etc., can be developed for one species and "counterpart" genes screened from a range of distantly related species because of nucleic acid homologies, such that work done on model species like *Drosophila melanogaster* or *Lucilia cuprina* would find ready application in unrelated species.

Finally, it should be restated that the impact of molecular biology on pest control in the year 2000 is unlikely to be limited by shortcomings in technological capabilities. A far more serious impediment to the successful application of these methodologies to world agriculture will be the recurrent problems of product development into forms acceptable to growers and consumers and the community at large, and product availability at prices which growers can afford to pay.

Insect Killing Nematodes

A novel means of insect control that has developed rapidly in the last two decades and is expected to show considerable promise by 2000 is the use of parasitic nematodes. Whereas a majority of these parasites either have little effect on their hosts or give very low levels of infection, there are some that sterilise or kill and are able to produce high levels of infection; these can, particularly with human intervention, make a significant impact on insect pest populations.

The first nematode to be used on a large scale for insect control, *Deladenus siricidicola*, is an example of a classical biological control agent. It completely sterilises female *Sirex noctilio* which nonetheless disperse the nematode throughout infested pine forests during pseudoviposition. Because of the nematode's density dependence and extraordinary life cycle, populations of the pest crash rapidly. The parasitic cycle is followed by a series of free-living cycles which enable the nematodes to breed up in vast numbers while feeding on the symbiotic fungus of sirex as it spreads throughout the tree (Bedding, 1972, 1984a). The pest manager usually only needs to intervene to liberate the nematode in forests newly infested by sirex. However, it has required intensive screening of hundreds of isolates and a thorough understanding of the nematodes' biology and development of special techniques to arrive at this situation (Bedding & Akhurst, 1974; Bedding, 1979; Haugen *et al.*, 1990). The European *Sirex noctilio*, the only species out of some 40 related insects that is able to kill relatively healthy pine trees, is a major pest of *Pinus radiata* forests in Australia and New Zealand and has recently entered Uruguay and Brazil where it is already causing severe damage and is likely to spread before the year 2000 to the huge *P. radiata* forests of Chile. In Australia alone, it has been calculated that damage from sirex to the nearly one million hectares planted to pine trees, in the absence of control, could range from $A1-$A4 billion during each rotation. Eventually it is to be expected that sirex will spread to all regions of the world that grow susceptible pine species. Provided that adequate control measures are introduced, and in particular nematode control, economic losses should be limited.

Unfortunately, the tylenchid nematode, *D. siricidicola* is highly specific to sirex and its associated symbiotic fungus and could not be used in the control of other insects. Most other tylenchids could not be manipulated so readily since they do not have both parasitic and free-living cycles. Hopefully, however, other tylenchids will be discovered that can be used in pest control, and techniques for their in *vivo* mass production and dissemination will be developed.

Mermithids are certainly the most obvious nematode parasites of insects, occasionally reaching 0.5 metre or more in length in large insects, and eventually filling the entire body cavity of their host which dies soon after the nematodes emerge as preadults. A major problem with manipulating them for insect control, apart from their usually very long life cycles, is that like the obligate parasitic tylenchids, it has not been possible to culture them satisfactorily outside an insect host, in spite of numerous attempts. With *Filipjevimermis leipsandra*, which unlike many mermithids has a relatively short life cycle, considerable progress has been made towards in-vitro culture (Fassuliotis & Creighton, 1982) and by the year 2000 this nematode could conceivably be used against the very important corn root worm, *Diabrotica undecimpunctata*. Various species of mermithids are known to heavily parasitise populations of short winged grasshoppers on the tablelands of south east Australia (Baker, 1986). While it is not possible to manipulate these directly, simple procedures such as controlling the timing of stock grazing to avoid destruction of *Mermis quirindiensis* eggs, may in the future be a means of increasing rates of parasitisation (Baker, pers. comm.). In addition, planting perennial grasses or reafforestation in areas unsuitable for mermithids would make these areas also unsuitable for grasshopper breeding so that no major migrations would occur into regions where grasshopper control by nematodes is otherwise satisfactory (Farrow & Baker, in press).

It is the entomopathogenic nematodes, *Heterorhabditis* and *Steinernema* species, that are currently providing the most exciting possibilities for controlling a wide range of different insect pest species and by 2000 they should be established as among the most important insect pathogens used for this purpose. There are at least 30 species of heterorhabditid and steinernematid nematodes, comprising several hundred isolates from around the world (Bedding, 1990). Infective juveniles of these nematodes seek out insects, penetrate natural orifices (Poinar, 1979) or interskeletal membranes (Bedding & Molyneux, 1982) and enter the haemocoele. The nematodes release specific symbiotic bacteria, *Xenorhabdus* species, into the insect's blood which causes septicaemia followed by death of the insect within a day or so (Poinar & Thomas, 1966). The bacteria are highly pathogenic to insects, provide essential nutrients for the nematodes and produce antibiotics that preserve the insect cadaver while the nematodes multiply (Akhurst, 1982). Within about two weeks, infective juveniles (about one million per gram of insect) migrate from the host cadaver and may survive many months, or infect new hosts immediately.

Most species of insect can be infected by any species of entomopathogenic nematode, but insect susceptibility and nematode infectivity vary considerably, with no single nematode species being most infective for a majority of insect species, and, there can be a several hundred-fold difference in LD50 between two species (Bedding *et al.*, 1983). Currently *Steinernema carpocapsae* and *Heterorhabditis bacteriophora* are the most commonly available species, but by the year 2000 it is to be hoped that the best species of nematode to use against any of a range of important pests will be known and these nematodes will be available commercially. Continued use will also have been made of the simple *Galleria* trapping technique (Bedding & Akhurst, 1975) to isolate many more species and strains, and all of these will be stored under liquid nitrogen to maintain genetic diversity. At the same time there will doubtless be ample room for genetic improvement by artificial selection and even by genetic engineering, as has been the case for domesticated animals. Further improvements are also likely from manipulations of the symbiotic relationship between nematodes and *Xenorhabdus* with new combinations, perhaps arrived at by genetic engineering, being more effective than those now existing.

At first sight it might seem that, because nematodes require a moisture film in which to migrate, their use to control insects could be severely limited despite their broad host range. However, over 90% of insects spend at least part of their life in soil and there are many plant boring insects; in these situations, nematodes have a distinct advantage over other microbials and insecticides in that they can seek out their hosts. There are also possibilities for microenvironmental manipulation which could allow their use in other situations (Begley, 1990). On foliage there is considerable potential for use of water thickeners which not only allow excess water to be retained on the plant (Webster & Bronskil, 1968) but also provide the necessary viscosity for nematode migration and penetration (Bedding, unpublished); exsheathment and chemical activation of nematodes will also speed up the infection process which, together with artificial selection for more rapidly acting strains and improved application technology, may enable high levels of infection before the nematodes desiccate.

Baiting is another means of avoiding the problem of nematode desiccation. Already it is proving important in controlling the banana weevil, *Cosmopolites sordidus* in Australia (Treverrow & Bedding, 1990) and this will doubtless extend to other countries, and it also has possibilities for controlling the termite *Mastotermes darwiniensis* (Miller & Bedding, pers. comm.) and cockroaches (Deseo, pers. comm.).

It is only recently that methods for the mass production (Bedding, 1984b, 1990) and large-scale storage (Bedding, 1988, 1990) of entomopathogenic nematodes have made possible their use for commercial control of some insect pests. To date commercial utilisation has been on a relatively small scale and on high-value crops. Nevertheless, economic control with *Steinernema* or *Heterorhabditis* species has been achieved in some countries against pests such as the black vine weevil *Otiorhynchus sulcatus* (Bedding & Miller, 1981), beet army worm *Spodoptera exigua*, artichoke plume moth *Platyptilia carduidactyala*, blackcurrant borer *Synanthedon tipuliformis* (Miller & Bedding, 1982), fruit tree borer *Synanthedon myopaeformis*, and fig carpenter worm *Prionoxystus robiniae* (Bari & Kaya, 1984; Deseo, 1986).

Use of entomopathogenic nematodes to control insect pests is ideally suited to developing countries. Whereas the cost of imported insecticides is high, nematode-based control can be relatively inexpensive and production can be adapted as a local cottage industry (Bedding, 1990). Small-scale control has been achieved against the live wood termite *Glyptotermes dilatatus* in Sri Lanka (Dantharayana & Vitarana, 1987) and injections of *Steinernema carpocapsae* are now the standard means of eliminating carpenter worm *Holcocerus insularis*, from shade trees in some of China's major northern cities (Qin *et al.*, 1988). Of even greater significance, it is now clear that China's major apple pest, the peach borer, *Carposina nipponensis* could be effectively and economically controlled using soil applications of *Steinernema carpocapsae* instead of insecticides (Wang, 1990; Wang & Li, 1987; Doeleman 1990). For this to be accomplished, nearly three quarters of a million hectares would require treatment using some 300 tonnes of nematodes (6 x 1014) over 6–8 weeks each year. Currently CSIRO Australia, the Biological Control Institute, Beijing and the Guangdong Entomological Institute, Guangzhou are co-operating, under the auspices of the

Australian Centre for International Agricultural Research (ACIAR), to make this possible, hopefully by 2000.

For developing countries, mass rearing using monoxenic, solid culture, in self-aerating trays (Bedding, 1988) may at the moment be most economical because of low labour cost. However, commercial interests in Australia, the United States of America and the United Kingdom have been investigating the potential of liquid culture to produce huge batches of entomopathogenic nematodes with limited labour input (Pace *et al.*, 1986; Friedman *et al.*, 1989) and significantly better economies of scale (Friedman, 1990). Whether such methods will soon be suitable for developing countries or flexible enough to cope with a range of different species remains to be seen.

So far, there have been successes with entomopathogenic nematodes against several insect pests even though this means of control is yet in its infancy. By the year 2000 there will be further developments in the technology for producing, storing and using them; there will be a better understanding of factors affecting field efficacy; various adjuvants will be used to enhance their infectivity; they will be integrated with other means of control; a larger armoury of them will be assembled and assessed. As a result, it will be possible to control many more important pest species using these important organisms.

Pheromones in Pest Management

The potential of sex pheromones as a tool in insect pest management was recognised some 50 years ago by the German entomologist, Gotz, who wrote that "the use of attractants to create artificial sources of stimulation, unknown to nature in the same strength, could fundamentally change pest control, which

Table 1. Uses listed by suppliers in 1988 for commercially available pheromones

	Numbers of arthropod species		
Order	Monitoring	Mating disruption	Lure and kill
Blattoidea	3		1
Coleoptera	27		4
Diptera	7		5
Homoptera	4		
Hymenoptera			1*
Lepidoptera	189	18	15
Total	230	18	26

* Swarm trapping (After Inscoe *et al.*, 1990).

today is based on toxic chemicals' (Gotz, 1940). The assessment of Gotz is even more valid, half a century later!

Gotz envisaged mass trapping as the means by which control could be achieved, and although this particular technique has since met with limited success, the role of pheromones in pest management is now well established. In a survey of the uses to which commercially available pheromones were being put, Inscoe *et al.* (1990) revealed that monitoring was the most common application, followed by lure-and-kill and mating disruption (see Table 1).

Monitoring

Pheromone traps are, generally speaking, species-specific. They are biologically sensitive, require little maintenance and can be operated by non-entomologists, all of which make them ideally suited for monitoring purposes. Their main application is to determine the seasonal abundance of pest species, making possible the judicious application of control measures, most commonly pesticides, by careful interpretation of trap catch. Because of the many variables that influence trap catch, such interpretation is often difficult and in industrialised countries, where damage thresholds may be as low as 1%, there is often so little margin for error that monitoring traps have limited value. In developing countries, however, the threshold is sometimes higher and monitoring traps play a valuable role in the timing of insecticide applications. Examples here include the bollworm *Heliothis armigera* in India, the potato moth *Pthorimaea operculella* in South America and the African army worm *Spodoptera exempta* in eastern and central Africa (Campion & Nesbitt, 1981).

Monitoring traps also serve a useful purpose as quarantine tools, where they are used to indicate whether or not a particular species is present. Examples include their use to detect infestations of fruit flies (*Dacus* and *Ceratitis* species, Bateman, 1982), the gypsy moth (*Lymantria dispar*) and the Japanese beetle (*Popillia japonica*) in the USA (Kennedy, 1981) and the codling moth (*Cydia pomonella*) in Western Australia.

In a relatively recent development, pheromone traps have been used to monitor the incidence of insecticide resistance. Small quantities of the insecticide are incorporated in the sticky surface of the trap and mortality of the trapped males compared with those caught at insecticide-free traps. Alternatively, topical applications of insecticide are given to males caught at pheromone traps. The incidence of resistance of azinphos methyl has been determined for the tufted apple bud moth *Platynota idaeusalis* (Knight & Hull, 1989) and for the light brown apple moth *Epiphyas postvittana* (Suckling *et al.*, 1987) using these methods.

In New Zealand, pheromone traps have been used since 1985 to guide insecticide use at trial sites in Nelson and Hawkes Bay. These traps monitored the population density of three major leafroller pests of apples, *Epiphyas postvittana*, *Ctenopseustis obliquana*, and *Planotortrix excessana*, and insecticide sprays withheld during periods of low moth activity. This has enabled two or three sprays to be omitted each season which represents a reduction of up to 38% while achieving the same level of control as the recommended programme. This IPM strategy is being adopted by orchardists in New Zealand (Walker *et al.*, 1989).

Mating disruption

Mating disruption is a technique that involves release into the atmosphere of large amounts (relative to those produced by the pest population) of synthetic pheromone. Communication between the sexes is disrupted, mating is prevented or at least severely curtailed and the population suppressed. For some species the method has worked successfully as a control method in its own right, e.g. the oriental fruit moth *Cydia molesta* (Vickers *et al.*, 1985) and the pink bollworm *Pectinophora gossypiella* (Critchley *et al.*, 1985). For others, such as the bollworms *Earias insulana* and *E. vitella*, reductions in the number of insecticide applications were made possible by releasing the main component of *Earias* pheromone as a mating disruptant to assist in reducing population levels (McVeigh *et al.*, 1990).

Attract and kill

This method relies upon the removal of sufficient adults to limit mating to such an extent that the population can no longer maintain itself. Removal may be achieved through trapping ("mass trapping") or by combining insecticide with the pheromone ("attracticide") so that adults contacting the lure are either killed by the insecticide or are so debilitated by it that they are unable to mate.

Removal by trapping is more likely to succeed in species where females or both sexes are attracted to the lure rather than males alone which, at least for most lepidopteran pest species, are capable of mating several times. Campion and McVeigh (1984) considered that for species where only males are trapped, over 95% would have to be removed to have a significant impact. Even at low population densities this may require the deployment and maintenance of so many traps that labour costs alone make the method uneconomic unless the crop is of particularly high value, as is the case of the citrus flower moth *Prays citri* in Israel (Sternlicht *et al.*, 1990).

Where the attractant is extremely potent, mass trapping can be economical and effective, even if only males are removed (Lanier, 1990). Methyl eugenol, an empirical fruit fly attractant, is more potent than any known natural source and has been used, for example, to eliminate occasional introductions of the Medfly *Ceratitis capitata* and other Tephritidae from Florida and California (Steiner *et al.*, 1961). While the future of methyl eugenol is in some doubt because of its possible carcinogenic properties, several non-carcinogenic alternatives have now been identified (cited in Lanier, 1990).

The mass trapping of both sexes has been employed mainly as an aid in the control of coleopteran pests. There is circumstantial evidence to suggest that an outbreak of the spruce bark beetle *Ips typographus* in Norway and Sweden was curtailed in a massive trapping programme that removed an estimated 7.4 billion beetles from Norway alone (Bakke, 1982). In another example, tree mortality caused by the western pine beetle *Dendroctonus brevicomus* in central California fell from 227 per year to less than 73 per year in plots of 2.6 sq km, following a mass trapping programme (Bedard & Wood, 1974).

Prospects for pheromones

By 1982 compounds acting as sex attractants had been identified for over 900 species, the majority of them being lepidopteran (Klassen *et al.*, 1982). The total now is almost certainly well over 1000, yet from Table 1 it can be seen that pheromones are commercially available for the management of fewer than 280 species.

One obvious reason for there being such a discrepancy between the number of identifications and commercially available pheromones is that many of the former relate to species of little or no economic importance. However, it must also be said that a major factor mitigating against the more widespread use of pheromones is the effectiveness, relatively cheap price and ease of use of insecticides.

While pheromones have an obvious role to play in pest management, their full potential is still to be realised. Improvements in the techniques used to isolate, identify and synthesise pheromone components have made what was once considered one of the most difficult steps in the processes leading to pheromone use in pest management, into one of the easiest.

Similar advances are now required in our understanding of insect behaviour, particularly in the context of mating disruption. Despite several often-repeated hypotheses about likely mating disruption mechanisms, we still cannot point to any particular one as being operative against species for which the method is known to work. A sound understanding of the way in which behaviour is manipulated by the release of synthetic pheromones is likely to lead to further successes with other species. This provides a further example of the value of sound basic research underpinning novel methods of pest control.

In the context of mating disruption, further improvements are required in the technology used to dispense pheromones. Most current dispensing systems are at least a decade old, and while they may well be adequate for the few species for which they have been successfully employed, they are often unsuitable for releasing pheromone components such as aldehydes and dienes that are particularly susceptible to oxidation and ultra-violet degradation. The development of a dispenser that provides this sort of protection will facilitate mating disruption of a wider range of pest species.

There are a number of other essentially technical problems which, if overcome, will greatly improve the prospects of pheromone application in pest management. However, there are also obstacles of a non-technical nature which need to be confronted. Standardising and simplifying the requirements to register pheromones for commercial use will encourage manufacturers to bring to market new products by reducing the costs associated with their registration. A willingness on the part of governments to promote alternative pest control strategies will also encourage their use. In the case of pheromones, this may require education programmes to make growers aware of the benefits to be gained by using fewer pesticides, the provision of expertise to assist them in establishing and maintaining such strategies and perhaps encouragement in the form of subsidies to make adoption of the new strategies financially viable, bearing in mind that it is not just the grower that will benefit, but the whole community.

Although reasonable progress has been made so far, the potential of pheromones to make a significant contribution to pest management has still to be fully realised. In this context their use should not be considered in isolation from other control measures but, where appropriate, as part of the arsenal in integrated control programmes. If only some of the problems outlined here can be overcome by the turn of the century we will have gone a long way towards realising that potential.

The Social Issues

While being a science manager for 10 years has been a major factor in shaping my views on pest management in the year 2000, some other experiences have had a sobering influence. Recent ones include:

- membership of the FAO/UNEP Panel of Experts on IPM through the 80s
- Mission leader for mid-term review of "FAO Intercountry program for the development and application of integrated control in rice in south and south east Asia" (October–November 1990)
- participant at a FAO/UNEP/USSR Workshop in Kishinev, Moldavia on Integrated Pest Management, June 1990
- participant in a Workshop, sponsored by the Stockholm Environment Institute in Stockholm, December 1990, to consider the possibility of establishing an independent international biosafety panel to provide advice on request regarding the release of transgenic organisms into the environment
- membership of the Australian Government's genetic manipulation advisory committee.

These wider experiences have caused me to appreciate the need for pest managers to have full access to the range of technologies available if they hope to develop cost-effective and sustainable pest management systems; but, equally important, these options must be developed and implemented in a sound socio-economic, political and environmental context. The social experiences provide a real-world balance to the exhilarating enthusiasm of science-driven technology which can characterise an organisation peopled with scientists.

Before we approach pest management in 2000 and beyond, it is prudent to reflect on what we have achieved and are prepared to learn from our pest management experiences in the twentieth century where mankind entered the century with 1.6 billion individuals and leaves with an additional 4.5 billion. Otherwise the endless pursuit of new technologies becomes a fruitless "displacement activity", science driven with little enduring joy for those who are supposed to benefit — the producers and the consumers. First, let us consider how the so-called developed world is faring with sustainable pest management, beginning with Europe.

High input/high output agriculture in highly developed agricultural communities such as the Netherlands is economically possible in the short term, principally because the taxpayer subsidises many aspects of production, storage and export of major commodities such as wheat (Zadoks, 1989). However, recent recognition of soil and water contamination resulting from intensive use of agricultural chemicals, together with the more comprehensive assessment of economic costs of agricultural production, has forced the Dutch Government to seek a major reappraisal of agricultural practices in that country (Langeweg, 1989). In consequence, it has moved towards low input/low output farming, but maintaining the economic returns to growers. The Dutch Government, in its Multi-Year Crop Protection Plan, has set the target of reducing pesticide usage by 50% before the year 2000. A similar policy is being contemplated in other EEC countries.

Recent publications such as *Alternative Agriculture* (National Research Council (Anon, 1989)) suggest that the position in the USA is not dissimilar to EEC countries. Both direct and indirect subsidies by taxpayers and unsustainable production systems characterise late-twentieth-century North American agriculture. Once again, the USA has access to, and indeed is a key initiator of, the new technologies of the day, whether they be classical biological control, improved natural enemies, broad spectrum pesticides, behaviour modifying chemicals, cultural practices, improved crop varieties, biotechnologies or the adroit combinations of two or more of these into some sophisticated integrated package. Yet, despite access to sophisticated pest management technology, the USA is beginning to recognise the need for a radical rethink of its approach to agricultural production with increasing emphasis on environmentally friendly and sustainable systems. The NRC (Anon, 1989) in the USA reported that there was a 46% reduction in insecticide usage on field crops in the country between 1976 and 1982 following uptake of IPM on cotton, alfalfa, apples and peanuts. What is largely lacking to bring this revolution to reality in a sensible time frame is a preparedness by the consumer to pay higher direct prices for commodities and foodstuffs produced under such systems.

A first glance at agricultural technology in Eastern Bloc countries, such as

Poland and East Germany, has revealed a situation far worse than in the EEC or other developed regions of the world. There, pollution and lack of sustainable production systems are not even partly offset by temporarily enhanced efficiency on production, transport or storage. More reliable information was provided on "socialist" agriculture during the FAO/UNDP/USSR Workshop in Kishinev, Moldavia, USSR on Integrated Pest Management in June 1990. This workshop, attended by members of the FAO Panel of Experts on IPM, compared experiences with Soviet scientists from a range of research institutions, hosted by the All-Union Institute of Biological Methods of Plant Protection (AUIBMPP) in Kishinev.

According to N. Filippov, Director of AUIBMPP, during 1989 some 250,000–300,000 tons of more than three hundred pesticides were applied over some one hundred and seventy million hectares in the USSR. This accounted for 5.3% of pesticide usage worldwide. Some two billion roubles were spent on pesticide production and application during 1989, adding about eight billion roubles in value to crops including grain (seventeen to twenty million tons), potatoes (eight million tons), sugar beet (nineteen million tons), raw cotton (1.8 million tons), and fruit and vegetables (6.5 million tons). The 4:1 benefit to cost ratio is remarkably similar to estimates in western countries (Morley *et al.*, 1989). However, Filippov also acknowledged the serious negative effects: environmental pollution and destruction of natural ecosystems, pollinators, natural enemies, wild animals and birds.

According to K.V. Novozhilov, Director of the All-Union Research Institute for Plant Protection (VIZR), forty species of arthropod and fungi were registered as resistant to one or more pesticides and, in 1989, 5.7% of foodstuffs were contaminated with pesticides, a three-fold decrease since 1985. On the credit side, the USSR operates some 1700 commercial factories designed to mass produce 20 species of parasite, predator or microbial preparation. More than 750 mechanised production lines are engaged in mass rearing *Trichogramma* spp. for application to 17 million ha. One production facility produced four–five million *Trichogramma* individuals every 24 hours whose release controlled pests over 350,000–450,000 ha during the 1989 growing season. Table 2 gives an overview of chemical and biological control of some major pest groups in the USSR in 1989. According to Filippov, some 26.7 million ha were treated in 1988 with biocontrol agents at a cost of 569 million roubles adding 970 million roubles of value by way of extra production. This figure, at first glance, is not as favourable as the 4:1 benefit : cost ratio for pesticides, but the negative externalities are far lower than for pesticide usage, and the level of sustainability is clearly much higher. The Kishinev Workshop was also told that 4% of hectarage devoted to fruit and vegetable production in the USSR accounted for 40% of total output. The 4% was in the hands of individuals who had an incentive to succeed, while the remaining 96% was managed by collectives and State farms where there was little or no incentive to succeed.

The need for the producer of goods to share ownership of the production

Table 2. Chemical and biological control of a number of agricultural pests in the USSR (Filippov, pers. comm.)

Species of pests	Volume of work done in 1989 (1000 ha)		
	Chemical control	Biological control	Total control
Rodents — *Apodemus* *Meriones* *Microtus* spp.	7,500.6	4,271.8	11,772.4
Pyrausta sticticalis D	5,329.4	2,107.9	7,437.3
Agrotis spp.	132.6	4,472.2	4,604.8
Mamestra, Spodoptera spp., etc.	175.1	3,965.3	4,140.4
Ostrinia nubilalis Hb	17.8	1,950.1	1,967.9
Heliothis armigera Hb	453.0	4,968.5	5,421.5
Leptinotarsa decemlineata	6,187.0	305.1	6,492.1
Complex of pests of vegetables	1,285.1	803.5	2,088.6
Pests of fruits complex	3,699.0	620.8	4,319.8
Pests of vine complex	1,830.3	142.4	1,972.7
Pests of cotton complex	2,687.0	6,305.1	8,992.1

process is almost trivially obvious, yet the twentieth century has been characterised in many countries, under quite diverse political systems, by an alienation of the worker from the production process: ironically much of the alienation has occurred in socialist countries. Orthodox Marxist philosophy predicted precisely this unfavourable outcome.

The second experience which influenced my perspective on pest management in the year 2000 occurred in late 1990 when I led an FAO mission conducting a mid-term review of FAO's project entitled "Integrated Pest Management of Rice Pests in South and Southeast Asia". The rice IPM project also has important take-home messages which should be addressed before we launch ourselves with great enthusiasm towards the technological opportunities of the twenty-first century. The rice project, under the leadership of Dr Peter Kenmore, aims to imbue a philosophy of pest management in some 200 million rice growers and their families in south and southeast Asia, covering nine countries in the region, including Malaysia. The findings and recommendations of the Review Mission are contained in a Report cleared by FAO in March 1991 (Anon, 1990). Some insects, but particularly the brown plant hopper *Nilaparvata lugens* (BPH), emerged as significant pests of rice after the introduction of high yielding dwarf varieties was made widely available in Asia. Kenmore and colleagues had shown that regular use of pesticides depleted the abundant and diverse natural fauna of predators and parasites, allowing for periodic outbreaks of invertebrate herbivores such as BPH.

In China the "cargo-cult" mentality that pesticides represented some "intrinsic good", and should be applied regularly along with fertilisers, led to early recognition by Chinese scientists in the 1970s that pesticides can add to production costs, lead to reduced outputs by depleting natural enemy populations, generate environmental and health concerns and finally, contaminate foodstuffs (Anon, 1988, 1990). The collective farms were conducive to calendar application of pesticides. China was among the first countries to collect scientific evidence on the detrimental aspects of misapplication of pesticides. It was certainly the first country in the world to formulate a national policy favouring IPM, doing so in the late 1970s (Anon, 1988).

Two factors have favoured successful uptake of IPM philosophy for rice production in China: first, the Chinese Government, and not foreign owned multinational companies, control pesticide production levels which, according to government officials, is dictated by farmer needs rather than commercial interests; and secondly, the 1980s have witnessed the abandonment of collective farms in China, with agricultural production returning to individual farmer control.

The FAO project is helping the Chinese Government and extension workers to assist peasant farmers become expert pest managers by the application of sophisticated non-formal adult education technology. It is a unique experience to talk with otherwise illiterate farmers and their families who are now back in control of the production process, are genuine pest management experts, and have demonstrated a high capacity to transfer laterally this philosophy to neighbouring untrained farmers. Pesticides are a tool used as a last resort by these farmers, only when the wealth of natural enemies proves to be inadequate to keep rice herbivores below economic threshold levels.

Vietnam is following China's example, more slowly in the North than the South, in abandoning collective farming, but the Vietnamese Government so far is less committed to IPM philosophy than China. There was also evidence of influence of chemical companies based in Japan and elsewhere on government departments in Vietnam to maintain pesticide dependence for rice production in the country (Anon, 1990).

In Indonesia, a strong central government had been persuaded by external experts in the 1960s to switch from traditional rice farming to the use of high yielding varieties dependent on high inputs of agricultural chemicals. Rapid yield increases and a move to self-sufficiency in rice production justified the Government's policy. To encourage enhanced fertiliser application, the Government introduced hefty subsidies and the army was used to encourage farmers to avail themselves of the cheap farm chemicals. Pesticides were included under the fertiliser subsidy. This 85% subsidy of pesticide production costs grew from $US seven million in 1976 to around $US 160 million by 1986 (Anon, 1990).

Although the circumstances were different between China and Indonesia, the outcome was similar; external forces and interests were dictating farmer practices. The farmers themselves had been alienated from the production process,

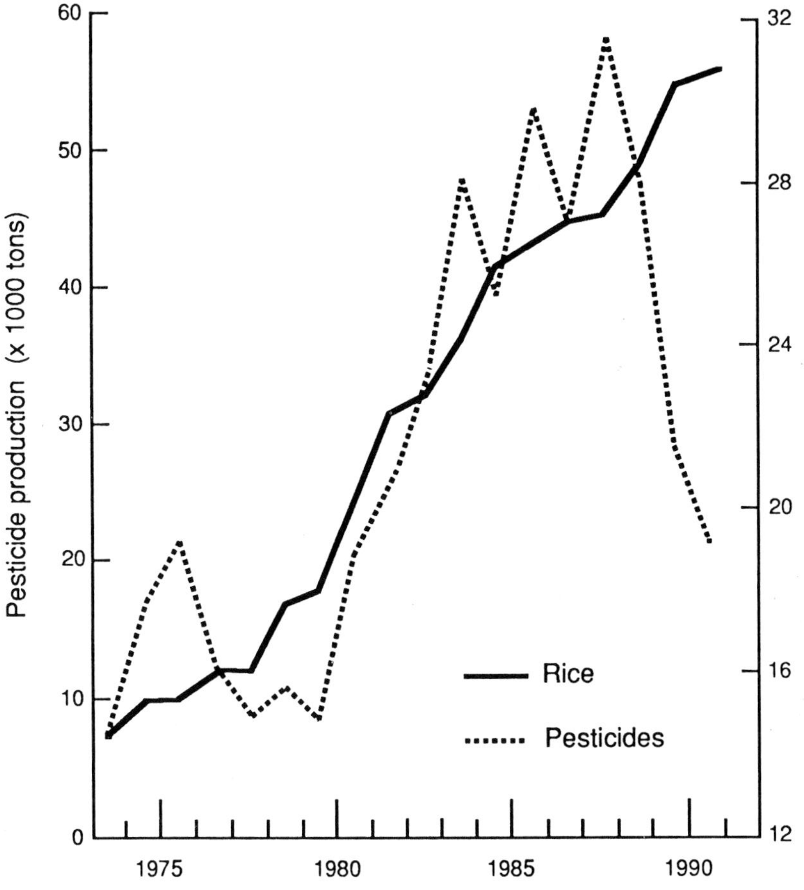

Fig. 1. Pesticide usage and rice production, Indonesia (after Anon, 1990)

with disastrous consequences (Figs. 1, 2 and 3). The estimated production losses to Indonesia between 1974 and 1986 due to pesticide-induced BPH outbreaks have been estimated at $US 1.75 billion (Anon, 1990). Combined with unnecessary pesticide subsidies, Indonesia was therefore some $US 3.35 billion the poorer because of inappropriate pest control measures in rice alone. To this can be added the considerable environmental and health problems due to the overuse of pesticides. The FAO Project has assisted the Indonesian Government move towards IPM of rice herbivores as firm government policy, the counter-productive pesticide subsidy has been abolished, pesticide usage for rice has declined 60% and the country has once again moved towards self-sufficiency in

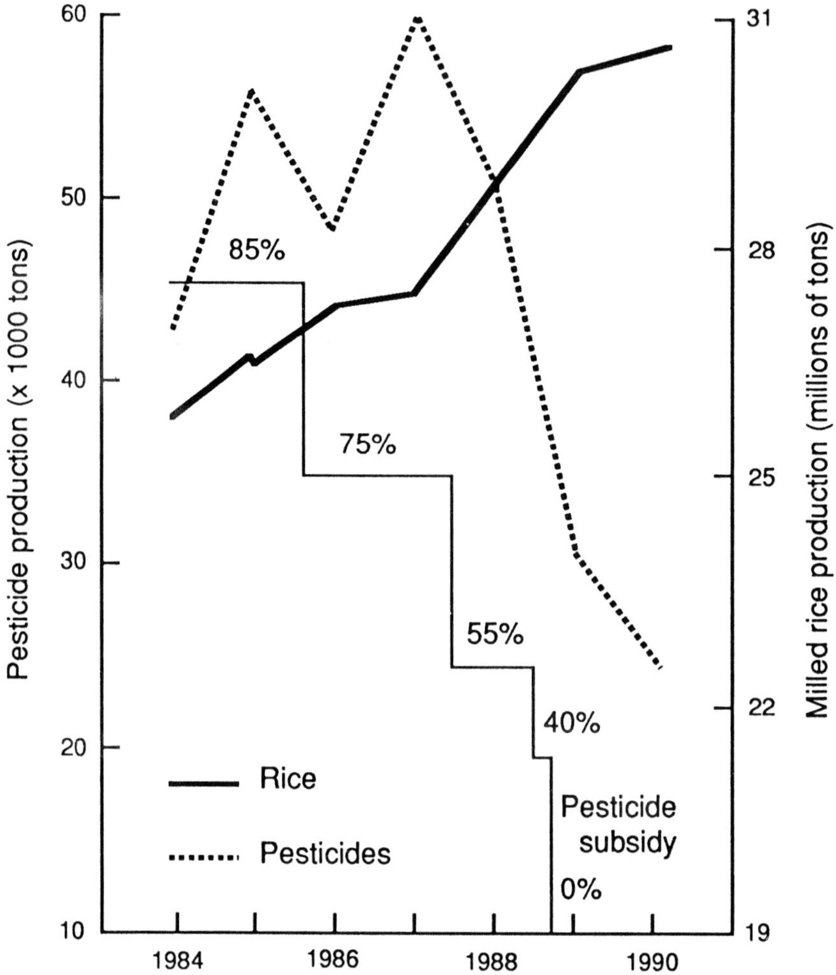

Fig. 2. Rice pesticides in Indonesia (after Anon, 1990)

rice production. Indonesia's response to the BPH problem has become a model for neighbouring countries in south-east Asia.

Probably the most significant achievement of the rice IPM for Indonesia has been the fact that rice farmers and families are now much more in control of the production process. This remarkable revolution was reported by Professor Michael Useem and Louis Setti in the 11 December 1990 issue of the *Boston Globe* which stated, "*Given that millions of Third World rice farmers lack even the rudiments of literacy and numeracy, let alone an understanding of entomology*

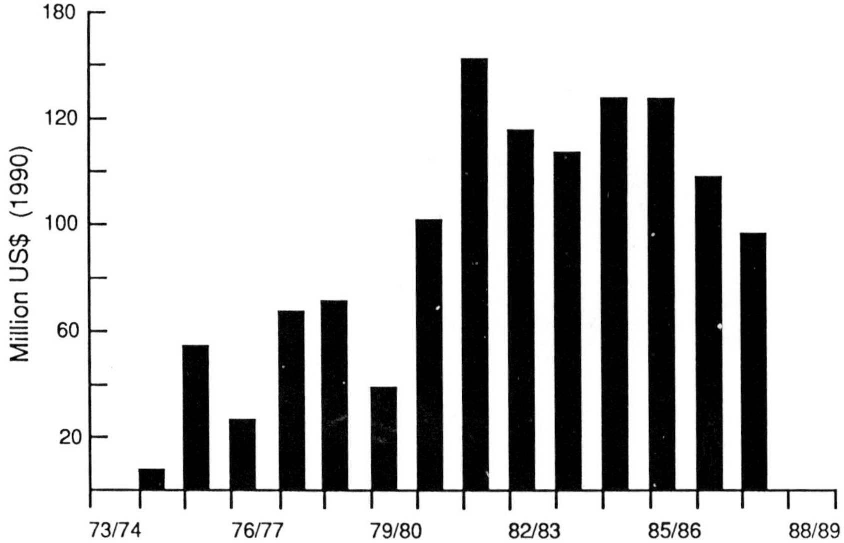

Fig. 3. Pesticide subsidies for rice in Indonesia (after Anon, 1990)

and hydrology, the task was daunting. But since the ecology of each rice field was different, there was no other way. Farmers had to become their own experts. ... Integrated pest management distributed information to the grass roots, decentralised power to use the information, and promoted local farming networks to ensure effective application. Combined with tangible and immediate gains, you have an organisational fuel that has proven irresistible to even the most sceptical. ... American organisations would also do well to watch the new art of Javanese management."

The USSR and the rice IPM experiences stress the fact that modern technologies, if pursued outside a sound socio-economic and environmental context, can exacerbate the problem of providing food and fibre of acceptable quality at affordable prices. At best, they can lead to a temporary increase in the carrying capacity of the environment (the main impact of broad spectrum pesticides between 1950 and 1990); at worst, they can reduce agricultural output, and lead to significant negative externalities.

Central policy changes by governments in China, the Philippines, Indonesia, India and Bangladesh promoting IPM practices for managing rice pests have proved essential for farmers to receive the necessary signals that pesticides are not an "intrinsic good". The FAO project has established beyond doubt that farmers can become their own experts in pest management and can successfully transfer the capability and appropriate attitudes to neighbouring farmers. But even this successful programme has a long way to go before the majority of the 200 million plus rice farmers in Asia control the technologies of crop protection, rather than the converse. Governments and bureaucrats in other participating countries in the rice IPM project are moving in the same direction, but more slowly towards farmer uptake of IPM practices for management of rice herbivores. And then we need to look at pest management technology for rice beyond Asia, and beyond rice to other commodities such as vegetables and other crops which an increasingly affluent society in the twenty-first century will demand.

The FAO project has demonstrated the need for scientists to operate beyond the technical sphere, to be prepared to discuss issues and practices with senior politicians and bureaucrats, and to be equally at home with extension workers and peasant farmers. Of course, the technical options need to be developed and available, but such efforts will come to nought, even be counter-productive if they are not subservient to the legitimate social and economic demands of the producer and consumer; and with a measure of commitment to the well-being of future generations.

Pesticide-induced pest problems for rice production in tropical Asia should not be generalised to rice production outside the tropics and sub-tropics where a rich and diverse fauna of natural enemies has evolved alongside the rice herbivores. In temperate rice-growing regions such as Japan, and especially Australia where the crop itself is a recent introduction, it may well be the case that some pests such as BPH would frequently exceed economic thresholds and periodic treatment with pesticides would be necessary. In Japan, another unique factor applies. Rice imports are prohibited and domestic prices are 6–10 times world prices. Consequently, the total cost of crop protection — surveillance, information distribution and pesticide treatments — exceeds $US 2 billion per year for an annual harvest of around 20 million tonnes (Anon, 1990), running into hundreds of dollars per hectare for insecticides per year.

Certain other key crop pests in Asia, such as the diamondback moth *Plutella xylostella*, are exotic, and may therefore encounter few, if any, effective endemic natural enemies in Asia. Thus, there is no reason to believe that endemic natural enemies are decimated by the intensive pesticide applications. For species such as the diamondback moth, neither pesticides nor biological control would appear to pose a solution, and in such cases we may need to resort to novel technologies. Other speakers at this conference are likely to argue that classical biological control of the diamondback moth remains to be given a fair trial because of the prevalence of pesticides in the environment.

I have deliberately chosen to dwell on the deficiencies of the pesticide technologies, especially in the manner that they are sometimes transferred by the developed to the developing world, because until we face the human, economic and social shortcomings that inappropriate application of these powerful technologies has revealed, we shall repeatedly revisit these problems as and when we attempt to apply the two new technologies that will predominate the twenty-first century: information technology and biotechnology.

The Role of Scientists and Research Institutions

We have discussed the technical options and the social framework in which these might be implemented in setting the agenda for pest management in the year 2000. Among changes in attitudes of scientists, interactions between existing institutions and the establishment of new types of organisations, the following considerations seem relevant.

Scientists themselves will have to become more active in communication. They will need to explain themselves more fully to those who fund their research and those who stand to benefit. The skills required will need to go beyond an ability to convey information concisely and comprehensively, and at a level appropriate to the particular audience. Success in communication will be measured not just by the level of knowledge and understanding transferred, but by the degree to which attitudes have been altered. Scientists have become increasingly able to transfer the information, but are generally less prepared to grapple with the interactive dimension of communication. In the past, their task was deemed to be done once the message was delivered. This will no longer be adequate; attitudes of politicians, bureaucrats and farmers need to be changed so that a new conviction brings with it a firm commitment to a new order. This pro-active dimension to communication is what distinguished the rice IPM project in Asia from many other potentially successful research and development ventures. Peter Kenmore's success as leader of the rice project is in part due to his calibre as a scientist, but principally due to his charisma and commitment to influence decision makers. It will be the Kenmores of the world that we can expect to have the major impact in shaping pest management in the year 2000.

At the institutional level, we will also witness significant reforms over the coming decades. I have already referred to the moves to establish a global database for insect taxonomy. This month the heads, or their representatives, of 10 major entomological collections are meeting in Washington, to consider, among other topics, the global database and ways in which the vast body of information contained in the collections can be shared. More of a challenge will be the way in which the small numbers of taxonomists around the world, charged with the massive task of cataloguing, describing, naming and acquiring some understanding of the invertebrate biodiversity of the world, can tackle the task in a meaningful timeframe.

It is not necessary to assume that some higher order institutional arrangement needs to be put in place to capture these benefits. However, international agencies such as CABI could play a valuable facilitating role as information brokers as the national institutions forge more tangible collaborative links.

If the IPM project for rice is to achieve its objectives in Asia, and perhaps encompass other rice growing areas and extend to other commodities such as vegetables, possible institutional arrangements will need to be considered carefully. FAO has played a pivotal role in the establishment and successful operation of the rice IPM project. However, it may be timely to discuss whether an international agriculture centre focusing on IPM should be established. Given that IPM might be considered a philosophy, an attitudinal approach rather than a technical package, it could be counter-productive to found an institution similar to the consultative group centres such as the International Rice Research Institute (IRRI) in the Philippines, but dedicated to IPM theory and practice. It therefore raises the spectre of a new type of facility, not devoted to research *per se* but which identifies opportunities, secures funds and manages IPM implementation programmes.

I have already stated that classical biological control has considerable potential well into the twenty-first century. Indeed, it is an under-resourced field of applied entomology. However, biological control is costly to execute at an appropriate level of activity. While there are various *ad hoc* arrangements between countries to reduce exploration and evaluation costs, there are two initiatives that warrant comment. A grouping of biological laboratories for weed and insect pests is occurring in Montpellier, France. The French, American and Australian Governments are constructing neighbouring laboratories under the aegis of AGROPOLIS, a French consortium of international biological research agencies. Second, the former Commonwealth Institute of Biological Control, CIBC, has now gone international, and is known as International Institute of Biological Control (IIBC). IIBC, along with its entomological and mycological sister institutes, offers considerable economies for countries that share common pests, such as water weeds, leucaena psyllid and orthopteran pests in Africa, to share the costs of implementing projects; but without the need to share the profits of successful biological control!

Conclusion

Pest management in the year 2000 and beyond will be characterised by a mixture of old, proven technologies, and a sprinkling of novel applications of biotechnology and information technology. It will feature an integration of control measures. We must discard the tendency to view IPM as some alternative to chemical, biological or cultural control measures. IPM is more obvious than that — it really represents a "no-holds barred" approach; it is devoid of the luxury of any specific ideology but with a prudent eye to cost-effectiveness, durability and environmental friendliness.

For their part, the scientists will need to leave their laboratories and experimental plots more often, and rub shoulders in a real way with decision makers in government, the community and industry. The community needs to raise its level of understanding of the newer technologies, or else it risks discarding valuable options, like genetic engineering, through fear and ignorance. Industry will need to take a hard look at corporate policies which pressure their sales staff, especially in developing countries, to maximise pesticide usage. It is not sufficient for a company to espouse IPM philosophy without addressing the financial implications for staff on commision wages. And finally, for the farmer and producer, the technologies of the next century must empower, not enslave.

Acknowledgements

I would like to thank the following people who have assisted in the preparation of this paper, and in some instances provided text: Peter Annis, Robin Bedding, Tom Bellas, David Dall, Lynn Maxwell, Ebbe Nielsen and Richard Vickers. Figures were prepared by Chris Hunt and Sandy Smith, and keyboard assistance was provided by Cheryl Gay.

References

Akhurst, R.J. (1982) Antiobiotic activity of *Xenorhabdus* spp., bacteria symbiotically associated with insect pathogenic nematodes of the families Heterorhabditidae and Steinernematidae. *Journal of General Microbiology* 128, 3061–3065.

Annis, P.C. (1990) Sealed storage of bag stacks: status of the technology. In: *Fumigation and Controlled Atmosphere Storage of Grain: Proceedings of an international conference* (Eds. Champ, B.R., Highley, E. and Banks, H.J.), pp. 180–187. *ACIAR Proceedings* No 25.

Anon (1988) *The Developments in Integrated Control of Rice Diseases and Insect Pests in China.* General Station of Plant Protection (GSPP), Ministry of Agriculture, Animal Husbandry and Fishery, PRC., Zhezhiang Scientific and Technological Publishing Company, 354 pp.

Anon (1989) *Alternative Agriculture.* National Research Council, National Academy Press, Washington D.C., 448 pp.

Anon (1990) *Mid-term Review of FAO Intercountry Program for the Development and Application of Integrated Pest Control in Rice in South & South East Asia.* Mission Report, FAO, Rome, pp. 141 plus pp. 40 appendices.

Baker, G.L. (1986) The ecology of mermithid nematode parasites of grasshoppers and locusts in south eastern Australia. In: *Fundamental and Applied Aspects of Invertebrate Pathology* (Eds. Samson, A.R., Vlak, J.M. and Peters, D.), pp. 277–280. Foundation of the 4th International Colloquium of Invertebrate Pathology, Wageningen, The Netherlands.

Bakke, A. (1982) Utilisation of aggregation pheromones for control of the spruce bark beetle. In: *Insect Pheromone Technology: Chemistry and Applications* (Eds. Leonhardt, B.A., and Berosa, M.), pp. 219–227. American Chemical Society, Washington, D.C.

Bari, M.A. and Kaya, H.K. (1984) Evaluation of the entomogenous nematode *Neoaplectana carpocapsae* (=*Steinernema feltiae*) Weiser (Rhabdita: Steinernematidae) and the bacterium *Bacillus thuringiensis* Berliner var. kustaki for suppression of the artichoke plume moth (Lepidoptera: Pterophoridae). *Journal of Economic Entomology* 77, 225–229.

Bateman, M.A. (1982) Chemical methods for suppression or eradication of fruit fly populations. In: *Economic Fruit Flies of the South Pacific Region* (Eds. Drew, R.A.I., Hooper, G.H.S. and Bateman, M.A.), pp. 115–128. Queensland Department of Primary Industries, Brisbane.

Beckman, R. (1991) Fungi to control pests in the soil. *ECOS* 68 (Winter 1991), 25–27.

Bedard, W.D. and Wood, D.L. (1974) Programs utilizing pheromones in survey and control. Bark beetles — the western pine beetle. In: *Pheromones* (Ed. Birch, M.D.), pp. 441–449. North-Holland, Amsterdam.

Bedding, R.A. (1972) Biology of *Deladenus siricidicola* (Neotylenchidae) an entomophagous nematode parasitic in siricid woodwasps. *Nematologica* 18, 482–493.

Bedding, R.A. (1979) Manipulating the entomophagous-mycetophagous nematode, *Deladenus siricidicola* for the biological control of the woodwasp *Sirex noctilio* in Australia. In: *Current Topics in Forest Entomology* (Ed. Waters, W.E.), pp. 144–147. U.S.D.A. Forest Service, Gen. Tech, Pap. WO-8.

Bedding, R.A. (1984a) Nematode parasites of Hymenoptera. In: *Plant and Insect Parasitic Nematodes* (Ed. Nickle, W.R.), pp. 755–795. Marcel Dekker, New York.

Bedding, R.A. (1984b) Large scale production, storage and transport of the insect parasitic nematodes *Neoaplectana* spp. and *Heterorhabditis* spp. *Ann. Appl. Biol.* 104, 117–120.

Bedding, R.A. (1988) Apparatus and Method for Rearing and Harvesting Nematodes. Australian Patent Application No. PJ 0630/88.

Bedding, R.A. (1990) Logistics and strategies for introducing entomopathogenic nematode technology into developing countries. In: *Entomopathogenic Nematodes in Biological Control,* pp. 233–246. CRC Press, Boca Raton, Florida.

Bedding, R.A. and Akhurst, R.J. (1974) Use of the nematode *Deladenus siricidicola* in the biological control of *Sirex noctilio* in Australia. *J. Aust. Entomol. Soc.* 13, 129–135.

Bedding, R.A. and Akhurst, R.J. (1975) A simple technique for the detection of insect parasitic rhabditid nematodes in soil. *Nematologica* 21, 109–110.

Bedding, R.A. and Miller, L.A. (1981) Use of a nematode, *Heterorhabditis heliothidis*, to control black vine weevil, *Otiorhynchus sulcatus,* in potted plants. *Ann. Appl. Biol.* 99, 211–216.

Bedding, R.A. and Molyneux, A.S. (1982) Penetration of insect cuticle by infective juveniles of *Heterorhabditis* (spp.) (Heterorhabditidae: Nematoda). *Nematologica* 28, 354–359.

Bedding, R.A., Molyneux, A.S. and Akhurst, R.J. (1983) *Heterorhabditis* spp., *Neoaplectana* spp. and *Steinernema kraussei*: Interspecific and intraspecific differences in infectivity for insects. *Experimental Parasitology* 55, 249–257.

Begley, J.W. (1990) Efficacy against insects in habitats other than soil. In: *Entomopathogenic Nematodes in Biological Control* (Eds. Gaugler, R. and Kaya, H.K.), pp. 215–227. CRC Press, Boca Raton, Florida.

Binnington, K.C. and Baule, V.J. (1991) Naturally occurring insecticidal molecules as candidates for genetic engineering. In: *Molecular Approaches to Pure and Applied Entomology* (Eds. Whitten, M.J. and Oakeshott, J.G.), Springer-Verlag, New York.

Campion, D.G. and McVeigh, E.M. (1984) Controlling insect pests with pheromones. *Span* 27, 100–102.

Campion, D.G. and Nesbitt, B.F. (1981) Lepidopteran sex pheromones and pest management in developing countries. *Tropical Pest Management* 27, 53–61.

Carbonell, L.F., Hodge, M.R., Tomalski, M.D. and Miller, L.K. (1988) Synthesis of a gene coding for an insect-specific scorpion neurotoxin and attempts to express it using baculovirus vectors. *Gene*. 73, 409–418.

Champ, B.R., Highley, E. and Banks, H.J. (Eds.) 1990 *Fumigation and Controlled Atmosphere Storage of Grain: Proceedings of an international conference*. Singapore, 14–18 February 1989. *ACIAR Proceedings No 25*, pp. 301.

Cochran, M.A., Brown, S.E. and Knudson, D.L. (1986) Organisation and expression of the baculovirus genome. In: *The Biology of Baculoviruses* (Eds. Granados, R.R. and Federici, B.A.), pp. 239–258. CRC Press, Boca Raton, Florida.

Critchley, B.R., Campion, D.G., McVeigh, L.J., Cavanagh, G.C., Hosny, M.M., Nasr, R.S.A., Khidr, A.A. and Naguib, M. (1985) Control of pink bollworm, *Pectinophora gossypiella* (Saunders) (Lepidoptera: Gelechiidae), in Egypt by mating disruption using hollow-fibre, laminate-flake and microencapsulated formulations of synthetic pheromone. *Bull. Entomol. Res.* 75, 329–345.

Dallwitz, M.J. (in press) DELTA and INTKEY. In: *Proceedings of Workshop on Artificial Intelligence and Modern Computer Methods for Systematic Studies in Biology*. (Ed. Fortuner, R.), University of California, Davis.

Dantharayana, W, and Vitarana, S.I. (1987) Control of the live-wood termite *Glyptotermes dilatatus* using *Heterorhabditis* sp. (Nemat.). *Agriculture, Ecosystems and Environment* 19, 332–342.

Deseo, K.V. (1986) Control of tree and stem-borer pests with insect-parasitic nematodes. In: *Fundamental and Applied aspects of Invertebrate Pathology: International Colloquium on Invertebrate Pathology* (Eds. Samson, R.A., Vlak, J.M. and Peters, D.), pp. 271–274. Wageningen, The Netherlands.

Desmarchelier, J.M. and Ahern, C.M. (1988) Insecticide-retentive carriers 2. Fenitrothian-impregnated clays. *Australian Journal of Experimental Agriculture* 28, 271–278.

Doeleman, J.A. (1990) Benefits and costs of entomopathogenic nematodes: two biological control applications in China. *ACIAR Economic Assessment Series* No 4, 15.

Ebeling, W. (1976) Insect integument: a vulnerable organ system. In: *The Insect Integument* (Ed. Hepburn, H.R.), pp. 383–400. Elsevier Amsterdam.

Farrow, R.A. (1975) The African migratory locust in its main outbreak area of the Middle Niger: quantitative studies of solitary populations in relation to environmental factors. *Locusta* 11, 1–198.

Farrow, R.A. and Baker, G.L. (in press) The reduction of wingless grasshopper infestations by sound farm management practices. *Aust. Wool Corp. Res. Rev. Confr.* 28 May – 1 June, 1989, Ballarat, Australia.

Fassuliotis, G. and Creighton, C.S. (1982) In vitro cultivation of the entomogenous nematode *Filipjevimermis leipsandra*. *Journal of Nematology* 14 (1), 126–131.

Fishpool, L.D.C. and Popov G.B. (1981) The grasshopper faunas of the savannas of Mali, Niger, Benin and Togo. *Bulletin de l'I.F.A.N.* 43, 275–410.

Friedman, J.J. (1990) Commercialization and Application Technology. In: *Entomopathogenic Nematodes in Biological Control* (Eds. Gaugler, R. and Kaya, H.K.), pp 153–170. CRC Press, Boca Raton, Florida.

Friedman, J.J., Langston, S.L. and Pollitt, S. (1989) Mass production in liquid culture of insect-killing nematodes. International Patent No. WO 89/04602.

Gotz, B. (1940) Sexualduftstoffe als Lockmittel in der Schädlingsbekämpfung. *Umschau* 44, 794–796.

Haugen, D.A., Bedding, R.A., Underdown, M.G. and Neumann, F.G. (1990) National strategy for control of *Sirex noctilio* in Australia. *Australian Forest Grower* 13, No 2 (liftout).

Heong, K.L. (1990) Computer expert systems for improving insect pest management. *Rev. Agricultural Entomology* 78, 1–11.

Hilder, V.A., Gatehouse, A.M.R. and Boulter, D. (1990) Genetic engineering of crops for insect resistance using genes of plant origin. In: *Genetic Engineering of Crop Plants* (Eds. Lycett, G.W. and Grierson, D.), pp. 51–66. Butterworths, London.

Inscoe, M.N., Leonhardt, B.A. and Ridgway, R.L. (1990) Commercial availability of insect pheromones and other attractants. In: *Behaviour-modifying chemicals for insect pest management* (Eds. Ridgway, R.L., Silverstein, R.M. and Inscoe, M.N.), pp. 631–715. Marcel Dekker Inc., New York.

Kennedy, J.W. (1981) Practical application of pheromones in regulatory pest management programs. In: *Management of Insect Pests with Semiochemicals* (Ed. Mitchell, E.R.), pp. 1–11. Plenum Press, New York.

Kerr, A. (1980) Biological control of crown gall through production of agrocin 84. *Plant Disease* 64, 24–30.

Klassen, W., Ridgway, R.L. and Inscoe, M. (1982) Chemical attractants in integrated pest management programs. In: *Insect Suppression with Controlled Release Pheromone Systems,* vol 1 (Eds. Kydonieus, A.F. and Beroza, M.), pp. 13–130. CRC Press, Boca Raton, Florida.

Knight, A.L. and Hull, L.A. (1989) Use of sex pheromone traps to monitor azinphosmethyl resistance in tufted apple bud moth (Lepidoptera: Tortricidae). *J. Econ. Entomol.* 82(4), 1019–1926.

Langeweg, Ir.F. (Ed.) (1989) *Concern for Tomorrow: A National Environmental Survey 1985–2010.* National Institute of Public Health and Environmental Protection, Bilthoven, The Netherlands, pp. 350.

Lanier, G.N. (1990) Principles of attraction-annihilation. In: *Behaviour-modifying Chemicals for Insect Pest Management* (Eds. Ridgway, R.L., Silverstein, R.M. and Inscoe, M.N.), pp. 25–45. Marcel Dekker Inc., New York.

Maeda, S. (1989) Expression of foreign genes in insects using baculovirus vectors. *Ann. Rev. Entomol.* 34, 351–372.

Maywald, G.F. and Sutherst, R.W. (1989) CLIMEX: Recent developments in a computer program for comparing climates in ecology. *Proceedings from the Simulation Society of Australia Inc.,* Eighth biennial conference 25–27 September 1989, pp. 134–140.

McVeigh, L.J., Campion, D.G. and Critchley, B.R. (1990) The use of pheromones for the control of cotton bollworms and *Spodoptera* spp. in Africa and Asia. In: *Behaviour-modifying Chemicals for Insect Pest Management* (Eds. Ridgway, R.L., Silverstein, R.M. and Inscoe, M.N.), pp. 407–416. Marcel Dekker Inc., New York.

Miller, L.A. and Bedding, R.A. (1982) Field testing of the insect parasitic nematode, *Neoaplectana bibionis* (Nematoda: Steinernematidae) against currant borer, *Synanthedon tipuliformis* (Lep.: Sesiidae) in blackcurrants. *Entomophaga* 27, 109–114.

Miller, S.A. and Williams, R.J. (1990) Agricultural diagnostics. In: *Agricultural Biotechnology: Opportunities for International Development* (Ed. Persley, G.J.), pp. 87–107. CAB International, UK.

Mitsuda, H. and Yamamato, A. (1980) Advances in grain storage in a CO_2 atmosphere in Japan. In: *Controlled Atmosphere Storage of Grains* (Ed. Shejbal, J.), pp. 235–246. Elsevier, Amsterdam.

Morley, A.W., van Rijswijk, G. and Reis, R.G. (1989) Is there life after DDT? *Proceedings of the 5th Australasian Conference on Grassland Invertebrate Ecology*, pp. 102–108.

Newman, C.J.E. (1990) Specification and design of enclosures for gas treatments. In: *Fumigation and Controlled Atmosphere Storage of Grains: Proceedings of an international conference* (Eds. Champ, B.R., Highley, E. and Banks, H.J.), pp. 108–130. *ACIAR Proceedings* No 25.

Pace, G.W., Grote, W. Pitt, D.E. and Pitt, J.M. (1986) Liquid culture of nematodes. International Patent No. WO 86/10174.

Poinar, G.O. (1979) *Nematodes for Control of Insect Pests*. CRC Press, Boca Raton, Florida, pp. 277.

Poinar, G.O. and Thomas, G.M. (1966) Significance of *Achromobacter nematophilus* Poinar and Thomas (Achromobacteraceae: Eubacteriales) in the development of the nematode, DD136 (*Neoaplectana* sp.: Steinernematidae). *Parasitology* 56, 385–390.

Qin, X., Kao, R., Yang, H. and Zhang, G. (1988) Study on application of entomopathogenic nematodes of *Steinernema bibionis* and *Steinernema feltiae* to control wood borers. *Forest Research* 1, 179–185.

Room, P.M. (1990) Ecology of a simple plant-herbivore system: biological control of salvinia. *Trends in Ecology and Evolution* 5, 74–79.

Roush, R.T. and Daly, J.C. (1990) The role of population genetics in resistance research and management. In: *Pesticide Resistance in Arthropods* (Eds. Roush, R.T. and Tabashnik, B.E.), pp. 97–152. Chapman and Hall, USA.

Soderlund, D.M. and Bloomquist, J.R. (1990) Molecular mechanisms of insecticide resistance. In: *Pesticide Resistance in Arthropods* (Eds. Roush, R.T. and Tabashnik, B.E.), pp. 58–95. Chapman and Hall, USA.

Steiner, L.F., Rohwer, G.G., Ayers, E.L. and Christenson, L.D. (1961). The role of attractants in the recent Mediterranean fruit fly eradication program in Florida. *J. Econ. Entomol.* 54, 20–35.

Sternlicht, M., Barzakay, I. and Tamim, M. (1990) Management of *Prays citri* in lemon orchards by mass trapping of males. *Entomol. Exp. Appl.* 55, 59–67.

Suckling, D.M., Rodgers, D.J., Shaw, P.W., Wearing, C.H., Penman, D.R. and Chapman, R.B. (1987) Monitoring azinphosmethyl resistance in the light brown apple moth (Lepidoptera: Tortricidae) in New Zealand. *J. Econ. Entomol.* 80(4), 733–738.

Sutherland, J.W. (1988) The drying of grain with heated air. *Australian Refrigeration, Air Conditioning and Heating* Jan. 1988, 35–38.

Treverrow, N. and Bedding, R.A. (1990) Control of the banana weevil borer, *Cosmopolites sordidus* (Germar) with entomopathogenic nematodes. *Proceedings and Abstracts of Vth International Colloquium on Invertebrate Pathology and Microbial Control*, Adelaide, Australia, pp. 233.

Tyndale-Biscoe, M. (1990) *Common Dung Beetles in Pastures of South-eastern Australia*, CSIRO Australia, p. 71.

Vickers, R.A., Rothschild, G.H.L. and Jones, E.L. (1985) Control of the oriental fruit moth, *Cydia molesta* (Busck) (Lepidoptera: Tortricidae), at a district level by mating disruption with synthetic sex pheromone. *Bull. Entomol. Res.* 75, 625–634.

Walker, T.S., Suckling, D.M., Shaw, P.W. and White, V. (1989) Evaluation of pheromone traps to reduce insecticide sprays in New Zealand apple orchards. In: *Applications of Pheromones to Pest Control, Proc. Joint CSIRO-DSIR Workshop* July 1988, pp. 105–114.

Wang, J.X. (1990) Use of the nematode *Steinernema carpocapsae*, to control the major apple pest *Carposina nipponensis* in China. *Proceedings and Abstracts of 5th International Colloquium on Invertebrate Pathology and Microbial Control,* Adelaide, Australia, p. 392.

Wang, J.X. and Li, L.Y. (1987) Entomopathogenic nematode research in China. *Rev. Nematol.* 10, 483–489.

Wapshere, A.J. (1990) Biological control of grass weeds in Australia: an appraisal. *Plant Protection Quarterly* 5 (2), 62–75.

Waterhouse, D.F. and Norris, K.R. (1987) *Biological Control — Pacific Prospects.* Australian Centre for International Agricultural Research, Inkata Press, Melbourne, p. 454.

Webster, J.M. and Bronskil, J.F. (1968) Use of Gelgard M and an evaporation retardant to facilitate control of larch sawfly by a nematode-bacterium complex. *Journal of Economic Entomology* 61, 1370–1373.

Whitten, M.J. (1989) The relevance of molecular biology to pure and applied entomology. *Entomol. Exp. Appl.* 53, 1–16.

Whitten, M.J. (1991) Australian insects in scientific research. In: *The Insects of Australia,* 2nd edn. (Ed. Naumann, I.D.), pp. 236–251. Melbourne University Press, Carlton, Vic.

Whitten, M.J. and Hoy, M.A. (in press) Genetic improvement and other genetic considerations for improving the efficacy and success rate of biological control. In: *Principles and Application of Biological Control* (Ed. Fisher, E.). U.C. Press Riverside, California.

Whitten, M.J. and Oakeshott, J.G. (1990) Biocontrol of insects and weeds. In: *Agricultural Biotechnology: Opportunities for international development* (Ed. Persley, G.J.), pp. 123–142. CAB International, UK.

Yates, C.J. and Sticka, R. (1984) Development and future trends in bunkers storage. In: *Controlled Atmosphere and Fumigation in Grain Storage* (Ed. Ripp, B.E. *et al.*), pp. 589–600. Elsevier, Amsterdam.

Young, M.J., De Feyter, R.C. and Gerlach, W.L. (in press) Transgenic tobacco plants expressing a ribozyme have increased resistance to virus expression. *Nature.*

Zadoks, J.C. (1989) Development of farming systems. Evaluation of the five-year period 1980–1984, Wageningen, Pudoc., 90 pp.

PART 2
Integrated Pest Management

Chapter three: **IPM in 2000**

T.J. Perfect
Natural Resources Institute, Overseas Development Administration, Chatham, UK

Introduction

Integrated Pest Management has as many definitions as practitioners, but there is consensus on its major feature: minimal use of synthetic pesticides and maximum reliance on natural regulatory mechanisms to maintain pests below the level at which they cause economic damage. The approach is rooted in ecological thinking and, despite its attractions, has been slow to move from theory to practice. This failure has generated increasing concern in the international community and resulted in a number of initiatives to take stock of the constraints to implementation of IPM. The success of such undertakings and the means to deploy their findings will have a critical impact on progress in the next decade. This paper examines key characteristics of IPM, their present status and the likely influence of future developments.

IPM and Ecology

IPM applies ecological theory to the dynamics of pest populations in agroecosystems and seeks to identify appropriate points of intervention that will depress the equilibrium level or size of population peaks. Knowledge of the system is an essential prerequisite for such an approach if unexpected effects, often running counter to the overall objective of reducing loss of production, are to be avoided. Far too few studies of agroecosystems have collected data comparable to those a classical ecologist would require to derive a model of a natural habitat; on the other hand, agricultural systems are usually much simplified and such models may not be relevant.

In the tropics, indigenous mixed cropping systems are closer to natural habitats and credited with robustness to pest damage; the agricultural intensification associated with pressures for higher yield is a common cause of increased pest problems. A real opportunity exists for rigorous study of indigenous cropping systems and their response to intensification as a means of retaining important characteristics conferring sustainability.

IPM Technologies

Pesticides

Pesticides will remain a key means of intervention and an essential component of most IPM strategies. Paradoxically, their injudicious use represents the greatest threat to IPM and yet has provided the catalyst for virtually all IPM programmes. Just one recent example among many is the growing resistance to pyrethroids of the American bollworm (*Heliothis armigera*) in India, due to pesticide misuse. The bollworm now poses the biggest single threat to India's agriculture at a cost of $500 million per year (Sharma, 1991). In response, the Indian Council of Agriculture has instigated major production of natural enemies of *H. armigera,* including nuclear polyhedrosis virus (NPV) and parasitic wasps of the *Trichogramma* genus.

We are currently facing a major challenge: how can the use of pesticides be controlled and optimised such that they remain an effective tool in the face of growing problems of pesticide resistance, natural enemy mortality and pest resurgence? Over 500 insect pests are now resistant to chemical control (Jones, 1989). The use of chemicals, such as insect growth regulators, which affect morphogenesis or pest physiology, provides an ancillary approach but not an alternative to pesticides (Barton *et al.*, 1989). In the long term, it is possible that pesticide resistance may be used to advantage: most of the genes involved in resistance are single mutants of large effect, and it may become possible to manipulate these genes in insect pest populations (Jones, 1989).

However, this future technology aims to facilitate the use of synthetic pesticides, and fails to address the environmental problems and effects on neighbouring populations caused by the over-use of synthetic pesticides. A more practical aim is to reduce the amount of pesticide applied by the use of appropriate timing and/or low-volume application techniques, or by finding alternatives to synthetic pesticides.

Economic Thresholds

Population surveys and thresholds for insecticide applications have been successfully used within many IPM programmes, including control of soybean pests in Brazil (Moscardi & Ferreira, 1985; Gazzoni, 1991), and control of cassava mealybug in Africa (Alphen *et al.*, 1989). There is a need for researchers to overcome the problems farmers face in population assessment. For example, pheromone traps can be used at a simple level for the assessment of activity as a proxy for population levels (Whitten, 1991).

An aspect of IPM that is much underplayed is pest recognition: at the gross level, failure to distinguish pest from non-pest (by no means always straightforward) can result in the adoption of inappropriate and counter-productive control measures, leading farmers to apply the wrong pesticide and to the development of pesticide-resistant strains. One example of a new and potentially simple technology is an egg-squash field identification kit to distinguish the related species *H. peltigera* and *H. punctigera,* which works on the basis of antibodies and staining techniques (Whitten, 1991).

Biorationals

The term "biorationals" has been coined to cover a range of alternative approaches to synthetic pesticides. They are generally of greatly increased specificity and therefore avoid non-target mortality and the associated problems. Pheromones have been used to good effect in mating disruption (see for example Critchley *et al.,* 1987; Lohuis, 1989); sterile male releases have been in use for some time (Barclay & van den Driessche, 1989), and sterile female releases have been shown to be effective (Whitten, 1991).

Insect pathogens are playing an increasingly important role in crop protection. The bacterium *Bacillus thuringiensis* (*Bt*) , which produces insecticidal crystal proteins, has been used extensively for control of lepidopteran, dipteran and coleopteran pests: so far 22 different crystal proteins and their corresponding genes have been identified (De Waele, 1991). However, resistance has been recorded in *Heliothis* and *Plutella* spp., and is an increasing problem (Stone *et al.,* 1989). An international *Bt*-Resistance Working Group has been established to co-ordinate a speedy response to this potential problem.

Viruses, bacteria and nematodes all have great potential as pest control agents: viruses have been used successfully in the field, for example NPV for control of mealybug (Alphen *et al.,* 1989), and methods for their efficient and low-cost production in developing countries have been outlined (Jones, 1988). The fungus *Metarrhizium* has been used successfully in South America; nematodes have been used to control plant pests in China (Whitten, 1991).

The biological control of plant pathogenic fungi is still in its early stages in the field, but there is potential for the use of species such as *Bacillus subtilis, Agrobacterium* spp., *Trichoderma* spp., and some success has been achieved in the control of forest and orchard pathogens (Rishbeth, 1988). A number of commercial mycoherbicides are available in the USA, including "Collego" a formulation of the plant pathogen *Colletotrichum gloeosporioides* for the control of the weed species *Aeschynomene virginica.*

In the light of rapid progress over the past decade, the use of such methods can be expected to grow dramatically. In particular there is a need for more sophisticated delivery systems derived through advances in biotechnology.

Biological control has contributed much to pest management and has resulted in spectacular successes in the developed and developing worlds. In the context of IPM, the focus has increasingly been on promoting practices that husband existing populations of predators and parasitoids within cropping systems. This is usually closely linked with controlling pesticide use and with identifying appropriate agronomic practices. Population enhancement and the use of exotics ("classical" biological control) may play an increasing role in the future . For example, the maintenance of pathogen reservoirs, and conservation of potential pathogen vectors, can increase the efficacy and stability of indigenous pathogens of insect pests (Hochberg, 1989).

Resistant Cultivars

Perhaps the most important advance in facilitating IPM has been the breeding of pest-resistant crop cultivars. This seed-borne technology has produced immense direct benefits through increased yields and reduced expenditure on pesticides. Almost equally important are the indirect benefits gained through reducing the secondary effects of pesticides. Host-plant resistance has often been deployed as a unitary approach to pest management, with predictable problems. Resistance has often been short lived and many difficulties have arisen through pest interactions. Biotechnology is already drastically altering our approach to plant resistance and this trend can be expected to accelerate. Rapid progress has been made over the past ten years in molecular biology: transgenic plants were first reported in 1983, and the insertion of single genes of known function into crop plants can now be considered as routine (Schell, 1987). Virus resistance can be induced by the insertion of coat protein, satellite sequences or antisense sequences (Hull, 1990). Insect resistance has been obtained by insertion of single *Bt* genes, so far in tobacco, tomato, potato and cotton (De Waele, 1991). Trypsin inhibitor genes have been inserted in cowpea (Boulter *et al.,* 1989). Herbicide/pesticide resistance has been introduced into plants by the insertion of single enzyme genes, for example a project to engineer into cotton plants resistance to the biodegradable herbicide bromoxynil, which would reduce dependence on more environmentally damaging herbicides (Shulman, 1990). We must look to better strategies for resistance management in the future through cultivars with multiple resistance.

The cloning of plant resistance genes is an important goal which may soon be achieved. Workers on *Arabidopsis thaliana* (which provides a useful model system) are concentrating on plants with deletion mutations in resistance genes; other groups, working on tomato, are mapping resistance genes using restriction fragment length polymorphisms (RFLPs) (North, 1990). Projects are now under way in the USA, UK and Japan to map the entire genomes of various crop plants, including rice, wheat and barley (Ferrell, 1991).

Resistance Management

It is worth emphasising that resistance management is likely to be key to successful IPM in the future; the concept must embrace both the battle against resistance of pests to pesticides (of all kinds), and enhancement of resistance of crop plants to pests. The use of information technology for data management and simulation modelling must increase in importance in this regard. "Integrated resistance management" (IRM) involves prediction of the risks of resistance build-up and continued monitoring of variations in sensitivity, along with the reduction of and introduction of alternatives to pesticides. Recently, an International Organization for Resistant Pest Management has been established by the US Environmental Protection Agency as a co-ordinating body to facilitate the spread of knowledge and resources (Smale, 1991). The increasing importance of variability in pest populations based on susceptibility to control measures (primarily pesticides) and/or virulence to crop varieties is creating problems that are more dependent on biotechnology for their solution. Information technology for access to and management of more traditional indicators of variability will become more important, as will simple and robust methods derived from sophisticated biotechnology.

Socio-economic Aspects of IPM

Pest control has in the past been largely unimodal and has attempted to be prescriptive; it has met with considerable success in the short term, but IPM has grown out of the failure of such approaches to provide sustained control. Pest management implies the ability to identify and define the problem, to review and assess the options available and to make informed decisions given the prevailing agronomic and economic circumstances. Locale specificity is high and within-site factors are often critical.

Adaptive Technology

Effective IPM must centre on the farmer. Pest management technologies must be both affordable and comprehensible to provide farmer access and facilitate decision making; they must also be robust. Mechanical and cultural methods, so long as they are not too demanding on manpower and time, can be extremely valuable and in some cases offer the only real alternative, for example for the control of *Striga* spp. in millet and sorghum crops in semi-arid Africa. Crop varieties with simple architectural changes can give protection to natural enemies of pests: for example, leafless pea plants have been shown to increase the predation of aphids by ladybird beetles, which tend to fall from the slippery leaves of conventional plants (Kareiva & Sahakian, 1990). While it is easy for a farmer to plant a different variety (assuming such varieties are available and affordable), these simple inputs are backed up by detailed ecological research on the complex interactions at several trophic levels, and on neighbouring communities. An extremely robust and surprisingly simple control technique is suggested for the control of bean weevils in stored seeds: 'tumbling' the weevils by rotating seed containers twice a day has been shown to reduce populations by around 97%.

The weevil larva requires leverage against a hard surface to penetrate the seedcoat, a process which takes 24–48 hours and, if interrupted, has to be commenced afresh (Spencer *et al.,* 1991)

Many more such simple applications of research are needed, along with innovations in communications and education. The CNPSO-EMBRAPA project on soybean in Brazil has benefited enormously from good farmer education and communications; in particular a weekly TV programme, Alert System, has been set up by the extension services to inform farmers on IPM technology (Moscardi & Ferreira, 1985). A major task for scientists, educators and extension services will be to create an operating environment for the farmer in which he is able to make informed and effective decisions at the field level.

IPM and Policy

Given the growing importance attached by all nations to the sustainable use of renewable natural resources, it is difficult to see that anything other than IPM will provide an acceptable or affordable basis for pest control in the future. It will be of the utmost importance that governments give explicit recognition to this principle and create a policy environment in which IPM can take root. Five countries in Asia have already declared IPM as a national policy; by the year 2000 we may hope it will be an international one.

References

Alphen, J.J.M. *et al.* (1989) Insect invasions: the case of the cassava mealybug and its natural enemies evaluated. *Entomologist* 108, 38–55.

Barclay, H.J. and Driessche, P. van den (1989) Pest control models of combinations of sterile releases and trapping. *Insect Science and its Application* 10, 107–116.

Barton, A. E., Wing, K. D., Le, D. P., Slaweckir, A. and Feyereisen, R. (1989) Arylpyridylthiosemicarbazones - a new class of anti-juvenile hormones active against Lepidoptera. *Experientia* 45, 580–583.

Baulcomb, D. and Hull, R. (1989) Plant viruses: Spreading the word. *Nature* 341, 189.

Boulter, D., Gatehouse, A.M.R. and Hilder, V. (1989) Use of cowpea trypsin inhibition (CpTI) to protect plants against insect predation. *Biotechnology Advances* 7, 489–497.

Critchley, B.R., Campion, D.G., Cavanagh, G.G. and Chamberlain, D.J. (1987) Control of three major bollworm pests of cotton in Pakistan by a single application of their combined sex pheromones. *Tropical Pest Management* 33, 374.

De Waele, D. (1991) Insect control through plant engineering with *Bacillus thuringiensis* crystal protein genes. Presented at XII International Plant Protection Congress, August 1991, Rio de Janeiro.

Ferrell, J. (1991) Rice genome project. *Nature* 353, 99.

Gazzoni, D.L. (1991) Economic and social aspects of integrated pest management programmes in Brazil. Presented at XII International Plant Protection Congress, August 1991, Rio de Janeiro.

Hochberg, M.E. (1989) The potential role of pathogens in biological control. *Nature* 337, 262.

Hull, R. (1990) Non-conventional resistance to viruses in plants - concepts and risk. In: *Gene Manipulation in Plant Improvement II*. Proceedings of the 19th Stadler Genetics Symposium, Missouri, March 1989 (Ed. Gustafson, J.P.), New York: Plenum.

Jones, J.S. (1989) Evolution: Responses to chemical warfare. *Nature* 337, 690.

Jones, K. (1988) The use of insect viruses for pest control in developing countries. *Aspects of Applied Biology* 17, 425-433.

Kareiva, P. and Sahakian, R. (1990) Tritrophic effects of a simple architectural mutation in pea plants. *Nature* 345, 433–434.

Lohuis, H. (1989) Attractants against pests - what can be used? *PSP Pflanzenschultz Praxis* 4, 30–32.

Moscardi, F. and Ferreira, B.S.C. (1985) Biological control of soybean caterpillars. In: *World Soybean Research Conference III* (Ed. R. Shibles), London: Westview Press. pp. 703–711.

North, G. (1990) Plant pathology: The race for resistance genes. *Nature* 347, 517.

Rishbeth, J. (1988) Biological control of air-borne pathogens. In: *Biological Control of Pests, Pathogens and Weeds: Developments and Prospects*. Proceedings of a Royal Society Discussion Meeting, February 1987, London (Eds. Wood, R.K.S. and Way, M.J.), London: Royal Society.

Schell, J.S. (1987) Transgenic plants as tools to study the molecular organization of plant genes. *Science* 237, 1176–1183.

Sharma, D. (1991) India battles to eradicate major crop pest. *New Scientist* 10 August, p. 15.

Shulman, S. (1990) Agricultural biotechnology: One man's pesticide. *Nature* 347, 577.

Smale, P. (1991) The International Organization for Resistant Pest Management (IORPM). Presented at XII International Plant Protection Congress, August 1991, Rio de Janeiro.

Spencer, J.L., Quentin, M.E. and Miller, J.R. (1991) Tumbling controls bean weevils. *Nature* 352, 387.

Stone, T.B., Sims, S.R. and Marrone, P.G. (1989) Selection of tobacco budworm for resistance to a genetically engineered *Pseudomonas fluorescens* containing the A-endotoxin of *Bacillus thuringiensis* subsp. *kurstaki*. *Journal of Invertebrate Pathology* 53, 228–234.

Whitten, M.J. (1991) The appropriate applications of biotechnology in third world countries. Presented at XII International Plant Protection Congress, August 1991, Rio de Janeiro.

Chapter four: Management of Rice Diseases — A Future Perspective

T.W. Mew
The International Rice Research Institute
P.O. Box 933, Manila, Philippines

Summary

Rice (*Oryza sativa*) is attacked by many plant pathogens, including bacteria, fungi, nematodes, mycoplasma-like organisms and viruses. In general, with a few exceptions of the virus diseases in the tropics which are vector-borne, most of the other pathogens, especially those of fungi, nematodes and bacteria are widely distributed throughout the rice growing countries in the world. In the past three decades we have experienced the shifty nature of plant diseases. New diseases emerge when agriculture moves towards higher productivity. With a change of rice ecology, some formerly less important diseases appear to become more important. As rice production has moved to a new scenario, rice disease management has also changed to cope with new and different problems over the years. This paper analyses the changes and identifies areas where research is needed and takes into account new developments in computer technology and biotechnology.

Rice Production and Crop Protection

To predict rice disease management needs towards the year 2000, it is essential to assess the crop production and the crop growing environments. The triangular relationship between pathogen–host–environment is well known to all students of plant pathology. An examination of this relationship partially explains why we face the problems as they have emerged and how and what new information is needed in order to push further current research and development to cope with the new challenges. Cultivated rice has a recorded history of 7000 years in which it has evolved in response to changing land, water, soil and weather conditions. The rice environments include the water regime (deficit, excess, or optimum), drainage (poor or good), temperature (optimum, high or low,) soils (normal or adverse), and topography (flat or undulating). For convenience we group them according to water regimes in four broad ecosystems: rice under irrigation, rainfed lowland, deepwater and tidal wetland and upland conditions. On top of these rice growing environments, production is affected by crop season, crop intensity and crop management. These activities affect the crop (host) environments and thus the outcome or development of a disease epidemic.

Based on yield estimates, rice producing countries may be classified into four categories (IRRI, 1984, 1985): Group 1, the yield is greater than 5 t/ha, e.g. Korea, Japan; Group 2, less than 5 t/ha but greater than 3 t/ha, e.g. countries like China, Indonesia, Burma; Group 3, less than 3 t/ha but greater than 2 t/ha. Countries like Malaysia, India, Bangladesh, Philippines fall into this category; Group 4, less than 2 t/ha. Countries like Madagascar, Thailand, Nepal belong to this category. This classification was made more than five years ago and obviously many production conditions have changed, and even within a 5 t/ha country, there are districts producing over 5 t/ha and others less. Similarly, within countries in the lower yield brackets, there are farmers producing higher yields than the average. No matter how crude the estimate may be, the classifications help to underline the disease management needs. Disease management is part of crop management and, in turn, crop management cannot be dissociated from crop productivity in the various rice growing regions.

The real goal for countries or districts in a country producing more than 3 t/ha is to sustain the gain. Crop protection thus becomes an integral part of crop production in these groups of countries that have higher yields. The major effort is to prevent the development of disease epidemics. The use of agricultural chemicals to prevent the loss is essential. Crop protection not only aims at high epidemic-potential diseases such as blast in Korea or Japan, but also addresses chronic loss due to endemic diseases such as sheath blight and others. In Group 3 countries, crop protection relies heavily on host plant resistance to combat major diseases. If there is any practice or measure of disease control or management, it is minimal. In Group 4 countries, while crop loss due to plant diseases occurs, concomitantly there is only very little, if any, input invested in crop protection. Crop production is influenced by crop intensity that subsequently affects the crop and disease occurrence environments. Crops intensity involves continuous cropping of rice in the same field, either a double or single rice crop and a rice-based cropping system. In the tropics, when water supply is adequate and assured, continuous rice cropping with at least three crops a year is possible (De Datta, 1981). It is increasingly evident that continuous year-round rice cropping is often undesirable even though the water supply may be adequate. One of the undesirable factors is that it provides a breeding ground for rice pathogens and pests. The grouping of other crops in rotation with rice will lessen the pressure of the inoculum-carrying capacity from one crop to other. Cropping intensity not only relates to water supply but also climatic conditions of which temperature is the dominant factor. Differences in latitudes and growing seasons among localities produce differences in conditions under which rice is grown. The degree-growth days and rice growing periods in the different localities differ. In turn, this also affects crop intensity. Crop intensity favours disease development for two reasons: plant density and crop frequency changes the micro-climate and crop intensity perpetuates the disease inoculum.

The Host — Modern Cultivars

Rice production in Asia has changed in a revolutionary manner in the past two decades. The traditional cultivars which are tall and produce low tillers are gradually being replaced by cultivars with improved plant types. The cultural practices also change. This includes the change of cultivars from long to short duration, from photoperiod sensitive to insensitive, from wide spacing to close spacing, from low tiller to high tiller, and most recently from transplanting to direct seeding. The planting date is amended to dates when irrigation water is available. Such crop intensity changes the spectrum of rice diseases. Many of the bacterial diseases of rice reported recently are due to the change of agronomic practices. All of them must have been associated with the cultivation of rice for centuries. Their importance however, has been recognised only recently because of changes in cultural practices which are a result of change from traditional to modern cultivars.

The development of modern semi-dwarf cultivars of rice has been a key factor for the increase in rice yield in modern times. The increase in yield with modern cultivars does not come without cost. As production intensifies through higher fertiliser use and more rice crops per year, disease problems become one of the most important destablising factors to affect rice production. The continuous planting of uniform genotypes over large areas for longer periods has exerted a high selection pressure on pathogen populations to which the cultivars are resistant, and the photoperiod insensitivity enhances infection in favourable environmental conditions of otherwise less important diseases. In the monsoon climate of the rice growing seasons in the Asian tropics, the grain filling period from panicle initiation to harvest during the rainy season, is curtailed by high temperatures and humidity, making it vulnerable to pathogen attack. This has predisposed cultivars which are photoperiod insensitive to infection, resulting in grain/sheath discoloration in recent years.

The pathogen populations are not genetically uniform and stable. Rice cultivars developed for resistance to current pathogen populations may not be resistant to future ones and, therefore, breeding for resistance to disease is a continuing process. Diseases previously regarded as less important may become devastating. Thus, many diseases can be regarded as potential threats to rice production even though they may currently be unimportant or only regionally important. Modern agriculture often is also monoculture and this greatly favours epidemics. The epidemiology of each important disease should be researched so that we can predict epidemics through a thorough knowledge of the environmental conditions which favour epidemic outbreaks in rice. Impressive progress has been achieved in breeding for resistance to lessen the damage caused by some of the most prevalent diseases at that time such as blast; however, because of unexpected epidemics of rice virus diseases and bacterial blight, experienced shortly after the introduction of these modern cultivars, we are still far from the goal of limiting losses due to diseases to below the economic threshold.

Changes in plant types and agronomic practices — the two major features of modern rice production using these improved cultivars — have in turn changed the spectrum of diseases attacking rice. Sheath blight, sheath rot and stem rot and grain discoloration caused by the pseudomonad pathogens, for instance, are important diseases when modern technology is used, but were not considered important in traditional agriculture. As traditional agriculture gives way to modern agriculture, further changes in the relative importance of various diseases will occur. The wisdom of crop improvement and management relies on our understanding the balance of the pathosystems and the agroecosystems in order to provide a sustainable agricultural system which is both productive and safe. Obviously, there is no single tactic for effective management of all diseases of rice. Thus, a holistic approach which should combine an ecological and an epidemiological basis is to be recommended.

Changing Scene of Rice Diseases

During the past two decades, we have achieved rice disease management through varietal resistance. The effort is limited to a very few high epidemic potential diseases. The truth is that pathogens are shifty and we have experienced over the years a changing scenario of the diseases in certain rice growing environments. Such change is obviously related to the replacement of traditional with improved cultivars. While the Bengal famine due to brown spot (Padmanabhan, 1973) is unlikely to repeat itself, other rice diseases, whether their epidemic potential is high or low are unlikely to create a similar scenario. However, despite the optimism, diseases remain, a production constraint that is inevitably a reality. The blast epidemics in Korea, Egypt and elsewhere; the bacterial blight and virus diseases (grassy stunt, ragged stunt and tungro) in south and southeast Asia in the past decade never could match in scale the catastrophe of the Bengal famine, but have caused enough panic and hardship to farmers in areas where the epidemics occurred.

While blast and brown spot are diseases which have been recognised for centuries, their importance is only minimised in time and in specific locations. In recent years in tropical Asia where an intensive rice improvement programme has shown its impact, we have witnessed the emergence of rice virus diseases as one of the major production limiting factors. Tungro was found in the 1960s, grassy stunt in early 70s and ragged stunt in late 70s (Ou, 1985). Each has become epidemic and has threatened rice production across several countries at one or different times. Through a joint effort, however, grassy stunt and ragged stunt have been effectively controlled for the time being while tungro is still a nuisance. Bacterial blight was recognised as a serious disease only after improved but susceptible cultivars were introduced into production. When resistance is incorporated in the breeding programme and new cultivars are released to rice farmers, the disease is under control. While the single gene source of resistance (*Xa-4* derived from TKM 6) remains effective in some areas, the emergence of new virulent races appears in other areas where cultivars carrying this gene are continuously planted. Because of the use of photoperiod-insensitive cultivars in rice production, the rice sets flowers and is also harvested in the midst of the rainy reason. Grain and sheath diseases have become more prevalent than in the past. The etiology of some of the sheath and grain diseases, however, is not yet fully understood. There is evidence that the problems involve both fungal and bacterial pathogens. Among the bacterial pathogens, *Pseudomonas* spp. are predominant.

The change of disease spectrum in rice relates directly to the change of agronomic practices. In Japan, after rice production had become mechanised and seedlings were raised indoors in seedboxes, many seedling and grain diseases caused by bacterial pathogens were reported (Goto *et al.*, 1988). Most of these pathogens are *Pseudomonas* spp. including *P. glumae, P. panici, P. plantarii* and *P. syringae* cv. *syringae*. They are seed-borne and seed-transmitted. Because of the conditions in nurseries, they have become economically important. The bacterial sheath brown rot (*P. fuscovaginae*) first reported in Japan was recently found to be distributed worldwide. It is responsible for sheath and grain discoloration and has caused severe damage to actual crop production in the field and also to the quality of the grain. Another "new disease" named as "bacterial red stripe", first reported in Indonesia in 1988 and later found to occur in many other parts of southeast Asia is apparently widespread. Its etiology has not yet been established. Both bacteria and fungi were isolated from lesions on the leaf blade which exhibited a typical brown spot with an extended "tail" pointing toward the tips of the leaf blade. While the causal agent is still awaiting identification, its widespread occurrence on certain rice cultivars under intensive crop management suggests that certain diseases become prevalent and severe because of the strong influence of, and interactions between, genotypes, plant nutrients and crop management.

Disease Management

Many countries in the Third World in Asia have achieved self-sufficiency in food. However, much of the self-suffiency is marginal. There is a tendency, that is, once self-sufficiency is reached for resource allocation to support crop protection research to be transferred to other areas. This risk will arise when a major epidemic occurs. Then it is too late to recover from lack of research in a short span of time. In recent years, this has happened in many countries in Asia. It is vital that research support for crop protection is continued. It is equally important to develop a critical mass of plant pathologists to continue the research effort to find economically and ecologically feasible disease-control technology. Only if this is recognised, food security in the nations in Groups 1, 2, 3, or 4, can then be safeguarded against unpredictable epidemics of new diseases or old diseases in new forms.

When agricultural production is poor, crop protection tends to be least recognised by policy makers. The emphasis of national rice research programmes will depend on the financial health and resource allocation priorities of individual countries and on the success of these rice programmes. The major need to increase rice production and research capacity among nations that have the least resources must be recognised. Any gain will contribute to the livelihood of the people of the rice dependent populations, which are often poor. To prevent the chronic loss caused by diseases will contribute partly to such gain.

Rice is liable to be attacked by a range of diseases of the panicle (grains), leaf, stem and roots, which at epidemic levels have the potential to reduce yield by 50% or more and seriously affect grain quality. Varietal resistance is the most effective method of control, and even fungicides are available. This is often the most economically feasible technology for small rice farmers in Asia in general, and tropical Asia in particular. More importantly there is evidence that even where the fungicides are available, and economically feasible, ecological concern within the rice farming environments for pollution by pesticides and for resistance can and has minimised the use, both in diversity and rates needed for effective disease management.

But effective and efficient approaches for rice disease management must take into account both the nature of the interactions and also the life cycles of the pathogens. Rice diseases may be further classified as those with high epidemic and those with low epidemic potential. Diseases with high epidemic potential are those that produce abundant inoculum with many production cycles. They are characterised by a high infection rate and are capable of causing catastrophic crop loss on a large scale. A disease with high epidemic potential is often of high economic importance, even if it does not cause frequent epidemics. The frequency of epidemic occurrence is determined by the triangular relationship between the pathogen, the host and the environment. On the other hand, it is possible that a disease with low epidemic potential may be of high economic importance because of chronic losses that it causes.

If disease resistance could be maintained (durable resistance) it could remove and reduce the use of other forms and methods of disease management. However, the pathogens themselves continue to evolve and give rise spontaneously to new races able to overcome current resistance. More realistically, current breeding for resistance focuses only on a very few "major" diseases mentioned earlier. The emergence of new diseases previously considered "minor" has been well documented in rice (Ou, 1985; Mew, 1990) in response to crop intensity, and there is an urgent need for new plant types, and alternative methods. Likewise the effectiveness of some fungicides has been met by the resistance of new pathogen strains. It is unlikely that these problems can be solved through current resistance breeding programmes.

Resistance to blast and tungro has been a serious problem in specific ecosystems, with the occurrence of new races. In the case of rice tungro, which is one of the most destructive virus diseases in tropical Asia, the useful resistance donor has yet to be identified and efficiently incorporated in the resistance breeding programme. Varietal resistance will remain for many years to come the single most important and effective technology for rice disease management in developing countries. Until recently major genes were used because they were easy to recognise and to incorporate into breeding lines. Many rice cultivars with major genes have been developed and released to farmers. Like their counterparts in other food crop improvement programmes, rice scientists search for long lasting resistance to the high-epidemic potential diseases. To identify sources of durable resistance and to utilise these sources in resistance breeding is the goal set for rice improvement. Durable resistance as Johnson (1981) defined it, is a retrospective concept. Resistance of a cultivar can only be called durable if the resistance of one cultivar remains effective while another succumbs to an epidemic after being planted to a large hectarage for many years. One process which is always required in breeding for durable resistance is continuous field testing for extended periods at diverse geographical locations. From the available information, it is clear that many rice cultivars appear to possess durable resistance, either across wide geographical regions or in specific regions. The main problems encountered with these cultivars in breeding for durable resistance is that the mechanism of durable resistance is not yet fully understood, particularly its epidemiological and genetical aspects. Lee *et al.* (1989) indicated that a cultivar might have short-term resistance to one disease and durable resistance to another. The level of durability is different for different diseases and for the same disease under different environments. Partial resistance to blast is durable in tropical lowland irrigated rice; in temperate regions, there is no clear-cut evidence available. The major gene *Xa-4* for bacterial blight resistance seems to be durable. The gene from *Oryza nivara* for grassy stunt resistance also seems to be durable.

Disease management experts in general need to understand the pathogen–host interaction and the life cycle of a pathogen. There are dual characters of a plant pathogen: as a parasite and as a saprophyte. The degree of a pathogen's parasitism increases when its saprophytism decreases. This is also true under reverse conditions (Fig. 1). Following this concept of plant pathology, a pathogen's parasitism bears a close relationship to the degree to which it has become specialised in the infection of its host plant. Rice pathogens are therefore classified into two broad categories: specialised and unspecialised. Specialised pathogens usually have a narrow host range and are more host-dependent for survival. In disease management research, whether we consider a strategy using host resistance, or the chance of success in the selection of a disease management approach, this character of a pathogen is an important consideration. The disadvantage of a highly specialised pathogen in terms of host resistance is that

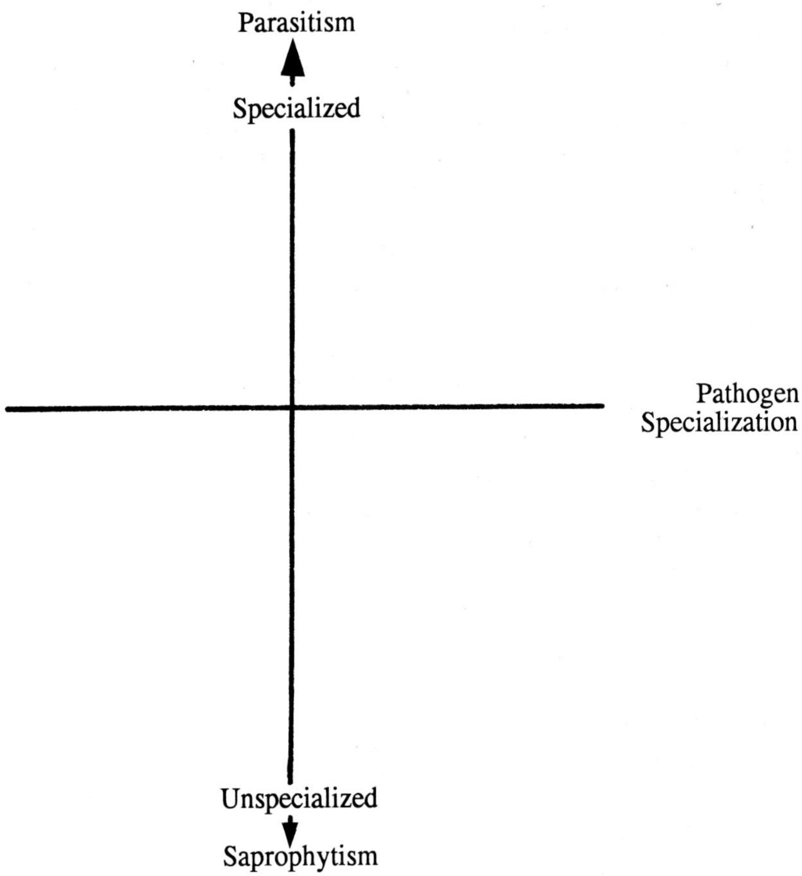

Fig. 1. The dual characters of plant pathogens

the pathogen is normally highly variable, and the diversity of virulence is closely related to the resistance, such as *Pyricularia oryzae*. On the other hand, when a pathogen's saprophytism increases, it also becomes unspecialised. Thus it has a wider host range and is less variable in terms of virulence. Its survival is not dependent on any particular one of its hosts. The disadvantage of an unspecialised pathogen in terms of host resistance is that it is difficult to search for the resistance, or if the resistance is there, it is not high enough to provide adequate protection to the crop. *Rhizoctonia solani* which causes sheath blight of rice is such a pathogen.

Other disease management strategies are needed. Chemical control has, in addition to genetic resistance, provided an immediate solution to protect the crop from infection. However, chemical control may be expensive and environmentally hazardous, and may have effects on non-target organisms. While fungicides will remain the last resort a farmer may have, they should be used only as needed, and their effects should be assessed in the context of integrated pest management. While many of the traditional cultural management practices used by rice farmers, such as straw burning, removal, mixture of cultivars, etc., may be used for disease management, the efficiency of these practices in crop protection has seldom been evaluated. Careful study and integration will enhance the disease management outcome of the practices.

This leads to another important aspect of a plant pathogen, its life cycle. In rice, with the exception of the viruses, and most of the nematodes, essentially all the fungal, and perhaps most of the bacterial pathogens can be seed-borne (Fig. 2). Although the seed-borne inoculum in terms of primary sources of infection varies, the fact that many rice pathogens are seed-borne is a reality. In the past, seed health has been considered only in the context of plant quarantine regulations. As plant quarantine objects for germplasm exchange, their importance cannot and should not be ignored. There are several rice pathogens which are of plant quarantine importance in rice germplasm exchange. Some infect seed and seed is a very important vehicle for their distribution, such as *Tilletia barclayana* (kernel smut) and *Aphelenchoides besseyi* (white tip), just to mention two. Although the diseases caused by these pathogens are seldom visible in rice fields

LIFE CYCLE OF RICE PATHOGENS

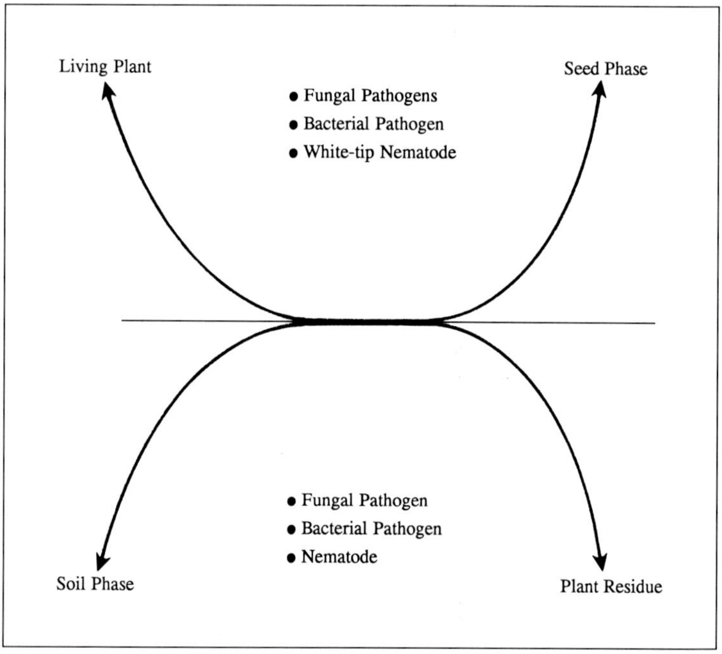

Fig. 2. The life cycle of rice pathogens based on major medium of survival in rice ecosystems

in the tropics, their detection from rice seeds is frequent. The infected seed becomes deformed or loses viability. Based on the frequency of their detection and the levels of seed infection, the seed-borne pathogens of rice may be categorised into four groups (Fig. 3). The high frequency of detection and level of infection appears to bear no relationship to their epidemic potential. Again, one has to take the ecosystems into consideration in actual disease management. Seed health, however, is more realistically associated with crop management. Quality seed produces a quality crop. The health of the seed safeguards the realisation of its genetic potential. Rice disease management in the past has not seriously associated seed health with the health of the rice crop. Judging by the number of seed-borne pathogens and their frequency of detection, seed health is essential for crop health for rice farmers on small land-holdings in tropical Asia.

However rice seed harvests contain not only the pathogens but also a large number of non-pathogenic organisms. These micro-organisms associated with rice seed exhibit inhibition/antagonism to the fungal pathogens of rice (Mew & Rosales, 1991). There are more fungal pathogens isolated from seed than the actual numbers of rice diseases observed because of the presence of the many non-pathogens to protect it. How to explore these antagonistic bacteria from rice seed for potential biological control is an area requiring extensive research. Indeed, not only do rice seed harvests contain potential biological control agents, so does the paddy field and the rice plant itself. Mew & Rosales (1986) showed

DISTRIBUTION (infection level)	DETECTION: HIGH	DETECTION: LOW
HIGH	*T. padwickii* *Sarocladium*	*T. barclayana* *A. besseyi*
LOW	*Gerlachia* *Drechslera*	*Pyricularia* *oryzae*

Fig. 3. Detection of rice pathogen and seed infection levels

the presence of a large number of antagonistic bacteria in rice fields in the tropics, which provides a potential alternative to managing rice diseases caused by both specialised and unspecialised pathogens. Distribution of the antagonists is not just limited to the tropics, but it occurs in other rice-growing environments just as well (Mew & Rosales, 1991). The micro-organisms with potential as biological control agents occupy a certain ecological niche, and their functions are very much dependent on specific target rice ecosystems. We have isolated potential biological control agents from various sources, including paddy-water, infected plants from irrigated and upland fields, sclerotia from upland and irrigated soils, healthy plants from irrigated and upland fields, and rhizosphere soils of healthy and infected plants. A collaborative and co-ordinated effort is in progress among rice pathologists to undertake this endeavour to explore the potential of biological control of rice diseases (Mew & Rosales, 1991). The overall research strategy is systematically to identify naturally-occurring antagonistic bacteria from the many local rice-growing ecosystems in Asia and test their effectiveness for disease control. The proper timing and method of application of biological control agents may be the only hope for the control of unspecialised pathogens such as *R. solani*, and other diseases where either host plant resistance has not achieved protection or no major effort is being undertaken to breed resistance against them.

The means and accuracy of the information gathering processes and procedures applied for crop disease assessment have advanced markedly in recent years. Disease management strategies are making increased use of computer-aided technologists for decision making (Teng *et al.*, 1991). As mentioned earlier, because of the change of plant types and cultural practices, the spectrum of rice diseases has also shifted from a scene of a few diseases posing an economic threat in the past to a more complex one in modern rice cultivation. The old but high epidemic potential diseases such as blast have been well controlled, both by using host resistance and chemicals in some ecosystems, while in others, they still take a front seat. The prevalence of many less important diseases in the earlier history of rice cultivation becomes more important in current, and perhaps future, rice cultivation. The multi-disease effect on rice crop production is real and the need to assess the damage caused by this collectively is urgent. The application of categorised information for crop loss analysis in tropical rice is compatible with decision making in pest management (Savary *et al.*, 1991; Dr S. Savary, pers. comm.). This is needed because crop losses in rice, especially in the tropics, require that a wide range of pest constraints and production environments be considered. The method analyses the multi-pest system by conducting surveys in farmers' fields, establishing a database where the levels of pests and production environments are experimentally manipulated.

Recent advances in molecular biology have opened new avenues for understanding the host–parasite interactions, and provided us with new tools for disease management. DNA markers such as restriction fragment length polymorphisms (RFLPs) are being used for analysis of host resistance and of populations of pests and beneficial organisms associated with the rice plant. Increase in rice production in Asia is attributable to the resistance of the modern improved rice cultivars. This will continue to be an important contribution that rice scientists can offer to the farmers. The key issues for the use of resistant cultivars for disease management are then two-fold: selection for durable types of resistance, and the deployment of the host plant resistance. The application of biotechnology is to use the molecular probes to identify and then pinpoint the position of genes on a chromosome. Thus in resistance breeding, the technology will enable scientists to tag or use markers for genes carrying resistance. In rice blast, the effect is to identify and map genes conferring resistance to blast in cultivars which have durable resistance to the disease. Scientists seek the association between RFLP and genes conditioning quantitative and qualitative resistance (Nelson *et al.*, 1991; R. Nelson, pers. comm.). The development will provide plant breeders with a tool based on the molecular marker rather than the conventional selection based on gene expression of the whole plant. Understanding pathogen population dynamics using DNA probes to reveal pathogen diversity, will yield the information which will also assist scientists to design strategies for deployment of resistance so that the durability of the resistance will be improved to target ecosystems with resistant cultivars.

References

De Datta, S.K. (1981) *Principles and Practices of Rice Production.* John Wiley & Sons. New York. 618 pp.

Goto, M., Zeigler, R.S. and John, V.T. (1988) Progress in seed health research on seed-borne bacteria, viruses and nematodes, pp. 131–150. In proc. *Rice Seed Health.* International Rice Research Institute, Los Banos, Philippines, 362 pp.

International Rice Research Institute (1984) *Terminology for Rice Growing Environments.* International Rice Research Institute, Los Banos, Philippines, 35 pp.

International Rice Research Institute (1985) *International Rice Research: 25 Years of Partnership.* International Rice Research Institute, Los Banos, Philippines, 188 pp.

Johnson, R. (1981) Durable resistance: definition of genetic control and attainment in plant breeding. *Phytopathology* 71, 567–568.

Lee, E.J., Zhang, Qi and Mew, T.W. (1989) Durable resistance to rise diseases in irrigated environments. pp. 93–110. In: *Progress in Irrigated Rice Research.* International Rice Research Institute, Los Banos, Philippines, 390 pp.

Mew, T.W. (1990) Disease management in rice. pp. 279–299. (Ed. Pimentel D.) *CRC Handbook of Pest Management in Agriculture* (2nd edn.) Col. III. CRC Press. Boca Raton, 749 pp.

Mew, T. W., Gergon, Evelyn and Merca, S. D. (1990) Impact of seedborne pathogens on germplasm exchange. *Seed Sci. & Technol.* 18, 441–450.

Mew, T.W. and Rosales, A.M. (1986) Bacterization of rice plants for control of sheath blight caused by *Rhizoctonia solani. Phytopathology* 76, 1260–1264.

Mew, T. and Rosales, A.M. (1991). Control of Rhizoctonia sheath blight and other diseases of rice by seed bacterization. In proc. *Biological Control of Plant Diseases: Progress and challenges for the future*. Plenum Press. New York.

Nelson, R., Ardales, E.Y., Vera Cruz, C.M. and Mew, T.W. (1991) Spatial scale of variation of the bacterial blight pathogens of rice. *Phytopathology* (Abstract 456).

Ou, S.H. (1985) *Rice Diseases* (2nd edn.). CAB International Mycological Institute, Kew, 380 pp.

Padmanabhan, S.Y. (1973) The great Bengal famine. *Ann. Rev. Phytopath.* 11, 11–26.

Savary, S., Elazequi, F. and Teng, P.S. (1991) Use of categorised information for crop loss assessment in tropical rice: A perspective. *Phytopathology* (Abstract 159).

Teng, P.S., Klein-Gebbink, H.W. and Pinnschmidt, H. (1991) An analysis of the blast pathosystem to guide modeling and forecasting. pp. 1–31 In: *Rice Blast Modeling and Forecasting*. International Rice Research Institute, Los Banos, Philippines, 99 pp.

Chapter five:
Management of Insect Pests of Rice in 2000

O. Mochida
National Agricultural Research Centre
Ibaraki-Ken, Tukuba-si, Kan'nondai,
305 Japan

Summary

The intensity of pest damage and rice productivity determines the method of insect pest control to be implemented. Rice productivity is affected by climatic conditions, economic and social factors as well as the technical levels of farmers. Acceptance of fragmentary technology associated with pest management is determined by the profitability of rice production resulting from the interactions between inputs for pest management and net returns.

In wet monsoon areas, farmers in Japan produced about five and six tonnes of rough rice/ha in 1965 and 1987, respectively. Insecticides were applied about seven to eight and four to five times in those two respective years, commencing from transplanting through harvest in the south where insect pests were very abundant on rice. In other words, fewer insecticides have been used recently on rice. This trend will probably continue until the year 2000, as long as the pest situation and economic/social conditions do not change very much. Breeding of Japonica rice cultivars resistant to BPH has been tried since 1970 but no commercially viable cultivar has appeared until now. In developing countries of Asia some high-yielding rice cultivars which are resistant mainly to BPH have been cropped widely since IRRI distributed IR26 in 1973. Integrating multiresistant cultivars with low usage of insecticides will become widespread in developing countries producing medium level rice yields (2.5–4 t/ha). In those countries with low level yields (less than 2.5 t/ha), resistant cultivars will be the most widely accepted as control measures. Resistant cultivars will also be cropped in upland and deepwater rice, if such cultivars become available.

In non-monsoon areas, control methods will depend on whether commercial or sustainable agriculture is practised. More chemical pesticides and resistant cultivars will be utilised in commercial rice farming in USA, Australia, etc., where the benefits are higher. Farmers engaged in sustainable agriculture in some Asian and African countries may not use many more chemical pesticides in year 2000. Fewer insect pest problems will occur in countries, e.g. Egypt, which practice rotational cropping systems.

New types of chemicals, for example IGRs (buprofezin (chitin-biosynthesis inhibitor), NC-184 (JHA), etc.) will be developed. Pheromones of some pest species are likely to be used in practice for monitoring but not for controlling pests on rice. Technology associated with pest forecasting and monitoring systems will be developed considerably via computer analysis.

Rice is grown in various countries located from the tropics to some cool zones like USSR and from below the sea level to about 3000 m in the Himalayas. In general, insect pests are more serious on rice in tropical countries than in temperate ones (in higher-yielding than in lower-yielding plots, in the wet season rather than in the dry season crop, and in paddy irrigated rice rather than on the uplands). The methods of insect pest control which are accepted and implemented by rice growers depend on the intensity of pest damage to the rice crop and the crop productivity. The present paper deals with the brief history of rice production, insect pest management and the prospects, mainly in Japan and some other countries.

Rice Production and Insect Pest Management in Japan

Rice Production and Associated Factors

In Japan rice has been cropped in paddy and upland fields. The majority of rice production (97.6–99.5%) was from paddy/lowland fields (Table 1). There is no deepwater, floating or tidal wetland rice.

Rough rice yields have increased from approximately 2 to 6 t/ha for the last century (Fig. 1). Factors and conditions associated with rice production have

Table 1. Area cropped with rice (x1000 ha), rough rice yield (t/ha), and rice production (x1000 t) in Japan (MAFF, Japan)

	Paddy fields			Upland fields			PU/ (PP + PU)%
	Area	Yield	Production	Area	Yield	Production	
1930	3,089	3.3	10,123	133	1.7	230	2.2
1940	3.004	3.7	11,194	148	1.5	220	1.9
1950	2,877	4.1	11,765	134	2.2	298	2.4
1960	3,124	5.0	15,674	184	2.2	400	2.4
1970	2,836	5.5	15,660	87	2.3	201	1.2
1980	2,350	5.2	12,115	27	2.6	73	0.6
1989	2,076	6.2	12,871	22	2.9	62	0.5

PP = Rough rice production in paddy fields.
PU = Rough rice production in upland fields.

changed considerably over the century. Some of the changes of wages with respect to acreage are shown for the last 40 years in Fig. 2. The overproduction of rice has required the Federal Government to revise the policy to allow farmers to convert their lands from rice to other crops since 1970. Crops other than rice were grown on 847,000 ha or 27% of the principal paddy fields in 1990. Working

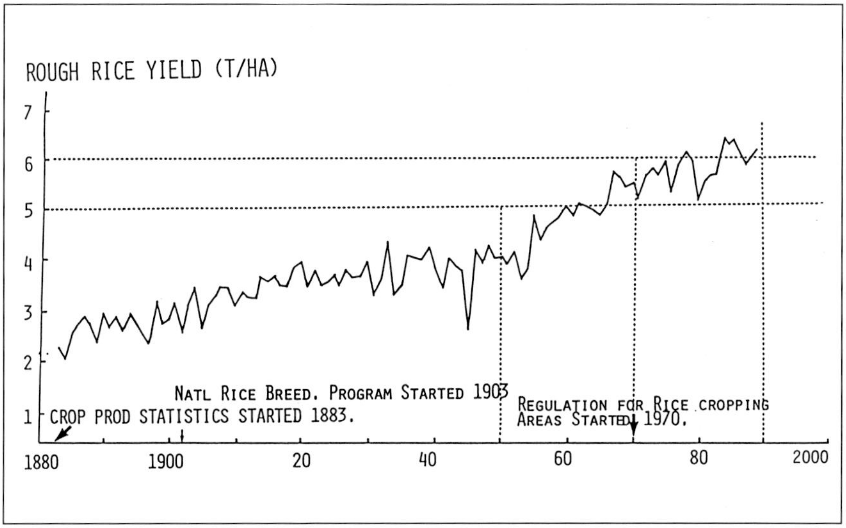

Fig. 1. Development of paddy rice yields in Japan since 1883

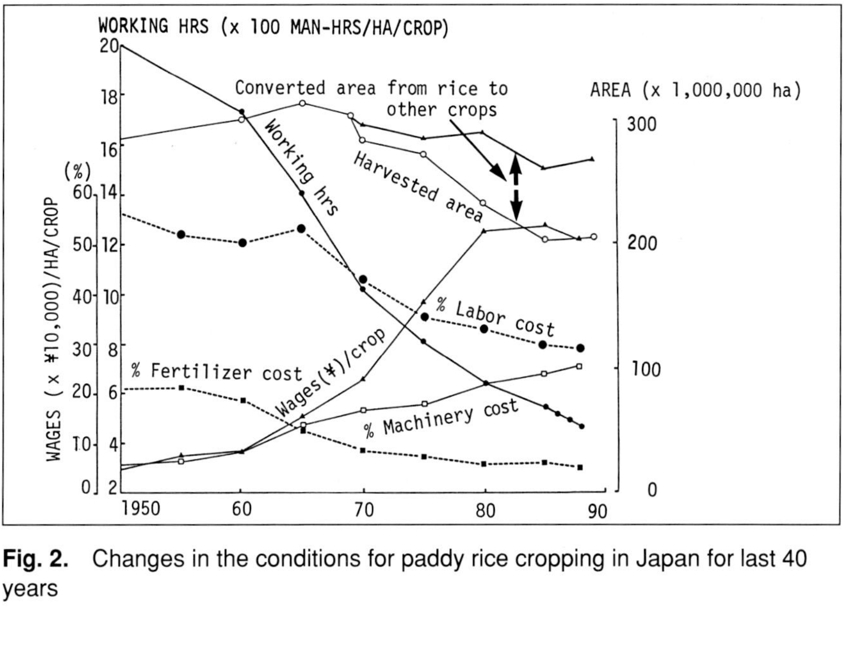

Fig. 2. Changes in the conditions for paddy rice cropping in Japan for last 40 years

Table 2. Pesticide* cost (%) in paddy rice production throughout Japan (MAFF, Japan)

Year	P/TI a	P/TI b	Remarks
1969	3.6	2.9	
70	3.4	2.8	
71	3.7	2.9	
72	3.7	3.0	
73	4.0	3.2	First oil crisis
74	4.6	3.6	
75	5.3	4.0	
76	4.9	3.8	
77	4.8	3.8	
78	4.9	3.8	
79	4.8	3.7	
80	4.9	3.8	Second oil crisis
81	5.1	4.0	
82	5.1	3.9	
83	5.4	4.2	
84	5.3	4.1	
85	5.6	4.3	
86	5.5	4.3	
87	5.6	4.3	
88	5.5	4.3	

* Pesticides includes insecticides, fungicides and herbicides.
P = Pesticide costs.
TI a = Total input excluding capital interest and land rental cost.
TI b = Total input including capital interest and land rental cost.

hours (man-hr/ha/crop) decreased from ca. 2000 in 1950 to less than 500 in 1988. Relative values of labour and fertiliser cost to the total input decreased gradually, whereas those of machinery cost increased gradually and wages sharply. Pesticides including insecticides, fungicides, and herbicides used for paddy rice production increased little in relative values (Table 2). These changes, especially in the sharp increase of wages, decrease of working hours, and shortage of labourers, affected the methods of controlling insect pests on rice.

Occurence of Major Insect Pests on Paddy Rice

One hundred and forty-six species of insects, nematodes and Acarina are known as pests of the rice plant (JSAEZ, 1987). Paddy rice is widely cropped from 24° 15' to 44° N. Thus, key pests differ with locations. Major insect pests of paddy rice are shown in Table 3, together with the abbreviations used below.

Table 3. Important insect pests on paddy rice in Japan

Scientific name	Order & family		English name	Abbreviation*
Chilo suppressalis	Lep.,	Pyralidae	Striped stem borer	SSB
Chlorops oryzae	Dip.,	Chloropidae	Rice stem maggot	RSM
Cnaphalocrocis medinalis	Lep.,	Pyralidae	Rice leaf folder	RLF
Hydrellia griseola	Dip.,	Ephydridae	Smaller rice leaf miner	SRLM
Laodelphax striatellus	Hom.,	Delphacidae	Smaller brown plant hopper	SBPH
Lissorhoptrus oryzophilus	Col.,	Curculionidae	Rice water weevil	RWW
Naranga aenescens	Lep.,	Noctuidae	Rice green caterpillar	RGC
Nephotettix cincticeps	Hom.,	Cicadellidae	Green leaf hopper	GLH
Nezara viridula	Het.,	Pentatomidae	Green stink bug	*Nv*
Nilaparvata lugens	Hom.,	Delphacidae	Brown plant hopper	BPH
Oulema oryzae	Col.,	Chrysomelidae	Rice leaf beetle	RLB
Parnara guttata	Lep.,	Hesperiidae	Rice skipper	RS
Scirpophaga incertulas	Lep.,	Pyralidae	Yellow stem borer	YSB
Scotinophora lurida	Het.,	Pentatomidae	Rice black bug	RBB
Sogatella furcifera	Hom.,	Delphacidae	Whitebacked plant hopper	WBPH

* Abbreviation is used through text.

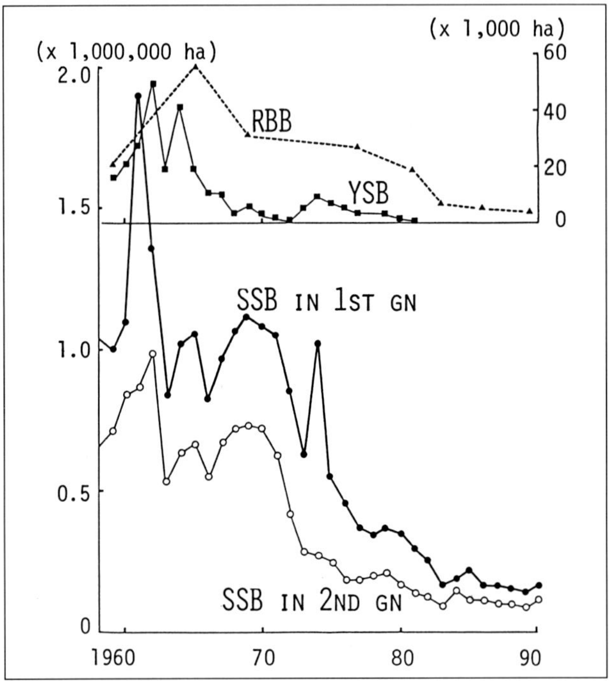

Fig. 3. Occurrence of SSB, YSB, and RBB on paddy rice in Japan for last 30 years

The areas of occurrence of RBB, YSB, and SSB for the last 30 years are shown in Fig. 3. These three species were very serious but are no longer a threat, except for YSB in the Ryukyus. GLH is an insect vector of rice dwarf virus disease. Rice dwarf virus reached a peak occurrence over areas of ca. 77,000 ha in 1974 and decreased to ca. 22,000 ha in 1990. GLH has occurred around 1,000,000 ha recently but is no longer a serious vector of the virus diease (Fig. 4).

WBPH, BPH, RWW, and RLF tend to be widespread in their infestations (Fig. 5). RWW has spread widely year by year since parthenogenic females were found in Tokoname city in May 1976 (Mochida, 1988). These females were suspected to have been imported with hay from California. The seedling box treatment with some effective insecticides has been generally accepted in seriously damaged areas in the north and the central mountainous areas of Honshu main island. RLF feeds on about 20 host plants including rice and overwinters in the larval stage on graminaceous hosts in the Ryukyus but not in the four main islands. Both WBPH and BPH cannot overwinter in the main islands. Thus, three of them are suspected of overwintering in the tropical and/or subtropical zones and invading Japan in the early summer every year on the monsoon fronts

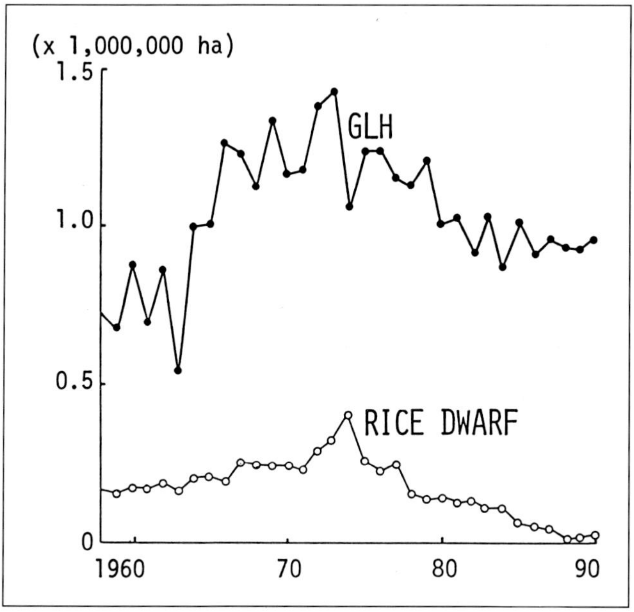

Fig. 4. Occurrence of GLH and rice dwarf virus disease on paddy rice in Japan for last 30 years

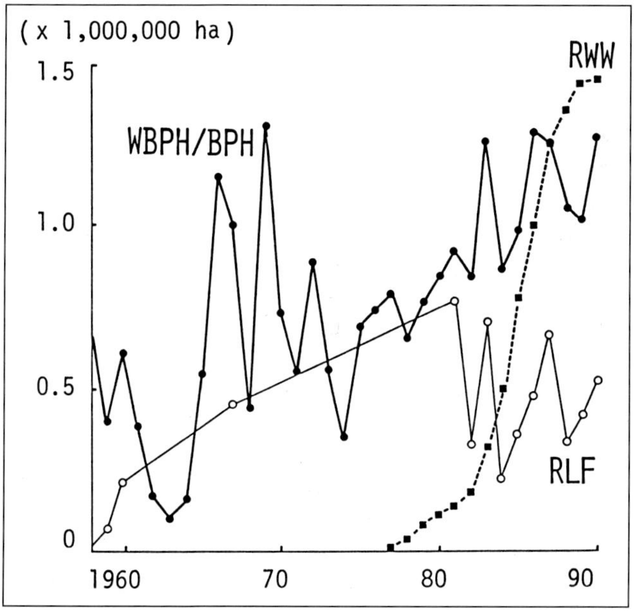

Fig. 5. Occurrence of WBPH/BPH, RWW, and RLF on paddy rice in Japan for last 30 years

from outside Japan. If this supposition is correct, we cannot control the source of infestation.

Forty-four species of heteropteran insects were confirmed to cause pecked rice grains (Hasegawa, 1973). Pecked rice grains are serious for farmers because of the degradation in the standards of rice so caused.

Control Methods

As there is no wild rice in Japan, YSB is monophagous on rice. The delay of seedbed preparation or transplanting was implemented in Kagoshima Prefecture for the first time since 1931. YSB was distributed in 18 prefectures around 1935, in five in 1973, but is in only one of the Ryukyus. When seeding was implemented after 5 June or after the peak of the third generation emergence, YSB populations decreased sharply or no more serious damage appeared. However, the termination of YSB appearance in the main islands around 1980 might be associated with factors other than late transplanting.

Basal application of insecticides against paddy rice insect pests was extended to SSB in most areas of Japan up to the late 1970s. For the first time imported Folidol E-605 was tested to control SSB in Kagawa Prefecture in July 1951. Parathion was popularly used via the "Group Pesticide (mainly insecticide) Application System" until the discontinuance of ethyl and methyl parathion and TEPP at the end of 1969. Pesticides with high human toxicity were required to be applied via the "Group Pesticide Application System" under expert guidance to avoid accidental poisoning and suicide by pesticides. The appearance of such safer insecticides like fenitrothion and caltap does not always require use of the "Group Pesticide Application System".

Shortage of labourers stimulated the development of machines for pesticide applications with higher speed and power. Pipe dusters were developed and accepted popularly by farmers, with the development of dust formulations for ground application. On the other hand, pesticide application by helicopters (heli-application) was practised against rice blast in 1956 and against SSB with blast in 1959. Since then, heli-application has become more popular overall, but was decreasingly used against SSB after 1989 (Fig. 6). This happened because SSB has not been a serious pest in most areas, since around 1980 as mentioned earlier.

Past (1965) and current (1989–90) pesticide application in paddy rice fields is illustrated schematically in Fig. 7. Many farmers applied insecticides eight times a year around 1965, but insecticides twice against hoppers and hoppers/RLF and insecticide–fungicide mixtures twice against stemblight/RLF and neckblast/BPH in 1989–90. Yield loss due to insect pests was estimated to be about half of the potential yield in 1962–73 (Mochida, 1974) but may be much less than 50% around 1989. As no SSB appeared, farmers stopped using insecticides against it. It does not mean that, as farmers stopped using insecticides, no SSB appeared.

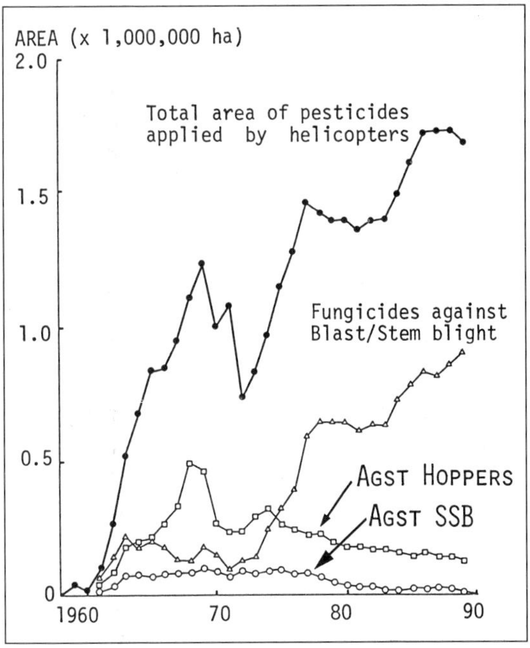

Fig. 6. The trend of helicopter-application of pesticides in paddy rice fields in Japan for last 30 years

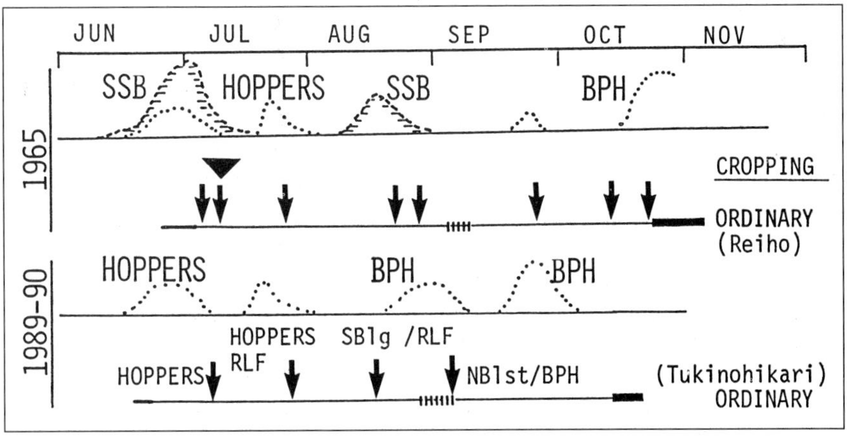

Fig. 7. Past and current pesticide application in paddy rice fields in Hukuoka Prefecture, Japan (schematic).▼: Helicopter application,↓: ground application. NBlst: neckblast, SBlg: sheathblight. Name given in parentheses is the name of a common cultivar

While SSB was the most important insect pest in many regions and grand application of insecticides was popularly implemented via the Group Pesticide Application System before around 1970, the date of the peak of moth appearance needed to be estimated exactly. However, heli-application does not take this into account, because the schedule of heli-application is usually decided by March each year, at least three months ahead of implementation.

Forecasting and Monitoring System

The network system was started in 1941. Regarding insect pests, for example, light trap data on SSB, YSB, WBPH, BPH, GLH, SBPH, have been accumulated. As mentioned in the previous section, precise estimation of the peak period(s) is not always needed since heli-application was introduced in 1958.

Female sex-pheromones are known in six rice insect pests: *Chilo suppressalis, Mythimna separata, Naranga aenescens, Sesamia inferens, Spodoptera exempta,* and *S. frugiperda*. Pheromone traps may be useful for monitoring rice insect pest occurrence but may not be of use in controlling pests.

Factors which May Affect IPM in Rice

Since Carson's *Silent Spring* was published in 1962, there has been much concern over the pollution of foods and the environment by pesticide applications. Some prefer safer foods without pesticide residues. Overproduction of rice in Japan may affect the way IPM develops. Shortage of labourers and high wages require less frequent pesticide applications, labour-saving application methods like "Trebon-surf" (etofenprox (Trebon) spreads on the water surface after being dropped), and simultaneous control of different insect pests with disease(s). For these purposes, mixture formulations of different pesticides (for example, insecticide–insecticide, insecticide–fungicide, and/or insecticide–herbicide) and/or tank mixtures are/will need to be developed in practice.

New chemicals like IGRs (buprofezin (chitin-biosynthesis inhibitor), NC-184 (JHA)), attractants, repellents, which do not harm fish, Crustacea, and other organisms in water, together with natural enemies have been tried and will be developed between now and 2000.

It is said that we are suffering from global warming. If temperatures are 3°C warmer in 2000 than now, some insect pests will spread further north but others will decrease their ranges of distribution or infestation.

Insect Pest Management in Rice in 2000.

The past, current, and future (2000) pest situations are summarised in Table 4. There were more than ten important pests on paddy rice in 1955–65. We anticipate approximately five such pests (WBPH, BPH, RWW, RLF, *Nv*, etc.) in 2000.

Table 4. The situation of paddy rice insect pests in Japan in the past and future. Names given in parentheses indicate insect-borne virus diseases. Items given in square brackets indicate the reason for pest absence

1930	1955	1965	1990	2000
SSB	SSB	SSB	[Insecticides/changes in cropping patterns]	
YSB in 18 p.	YSB in 12 p.	YSB in 5 p.	YSB in one prefecture [Changes in cropping patterns]	
RBB	RBB	[Insecticides/drainage of ill-drained fields]		
WBPH/BPH	WBPH/BPH	WBPH/BPH	WBPH/BPH [buprofezin]	WBPH/BPH [buprofezin, NC184, etc.]
GLH	GLH	GLH	GLH ?	
(Dwarf)	(Dwarf)	(Dwarf)	[Resistant cultivars]	
(Yellow Dwrf)	(Yellow Dwrf)	(Yellow Dwrf)	[Changes in cropping patterns?]	Yellow Dwarf [resistant cultivars?]
SBPH	SBPH	SBPH		
(Stripe)	(Stripe)	(Stripe)	[Resistant cultivars]	
(BSD?)*	(BSD)	(BSD)	[Changes in cropping patterns?]	[Resistant cultivars?]
?	*Nv* in south	*Nv* in south	*Nv* in south	*Nv* in south
?	RSM	RSM	[Resistant cultivars/insecticides]	
SRLM	SRLM	SRLM	[Insecticides/changes in cropping patterns]	
RS	RS	RS partially	RS partially	?
RLB	RLB	RLB	[Insecticides]	
?	?	RLF	RLF	RLF?
			RWW invaded into Japan '76.	[Seedlingbox treatment]

* Black-streaked dwarf. For abbreviated insect names, see Table 3.
p. prefectures

Table 5. Management of paddy rice insect pests in Japan in the past and future

	1930	1955	1965	1990	2000
ROUGH RICE YIELD	3.2 t/ha	4.9	4.8	6.2	[6.0]
YIELD LOSS DUE TO INSECT PESTS*	[40–50%]	[50]	50	[10–20]	[10]
BIOCONTROL	*Trichogramma* agst SSB/ducklings agst RBB				
AGROCONTROL	Handpicking SSB/ YSB eggmasses/late cropping agst YSB	Late cropping agst YSB			
CHEMICAL CONTROL	Pyrethrum-kerosin agst hoppers/kero-sin agst RBB	Synthetic in-secticides (OCH, OP)	OCH, OP	Carbamates/IGR (buprofezin)/ Synpy (etofenprox)	
Formulations		Single	Sing/mixt/granu	Mixt/granu	[Mixt/granu]
Appl. technique	Handsprayer	Handsprayer	Powersprayer/ pipeduster	Powersprayer/pipe duster/granu leappl	[Water-surface/ tablet-water appl.]
Helicopter-appl.			Since 1958	Yes	[Yes]
Appl. frequency/ crop**		4–6	Max. 8–9	4	[Less]
Pesticides/total input (%)		[2.6]	2.5	5.7	[5–6]
RESISTANT CULTIVARS		Agst RSM	Agst virus di-seases/RSM	4 germplasm lines with BPH 1, bph 2, Bph 3, & bph 4 each	[Yes]
FORECASTING/MONI - TORING SYSTEM		Started since 1941	Precise SSB moth peaks	Pheromone-traps ?	[More computerised]

* In Kyusyu. ** From seedbad preparation through harvest in Kyusyu. Estimated situation is given in brackets [ ].

Table 6. Countries with high rough rice yield (t/ha) (IRRI, 1988)

	1987	Maximum record
North Korea (DPR)	7.09	—
South Korea (REP)	6.38	[6.48]
Spain	6.36	[6.44]
Japan	6.22	[6.41]
USA	6.22	[6.33]
Egypt	5.95	[5.95]
Australia	5.72	[7.00]
Italy	5.58	[6.04]
China (mainland)	5.44	—
China (Taiwan)	4.77	[5.05]
Indonesia	4.77	—

The past, current, and future (2000) rice insect pest management situations are summarised in Table 5.

Insect Pest Management outside Japan

Table 6 indicates 11 countries with high rough rice yields: five are located in east Asia in the monsoon area. Indonesia is the only one in southeast Asia with typical monsoon climates. Pest and pest management situations may be different between monsoon and non-monsoon areas, in the economics of their farming systems and probably in grades of yields.

Table 7 summarizes the estimated current and future (2000) IPM situations in monsoon and non-monsoon areas. In some countries with monsoon climates, yields of 6 t/ha in 1990 will probably increase slightly towards 6.5 t/ha in 2000. Yield loss may be about 30% in 1990 and will probably decrease slightly towards 25%. Use of resistant cultivars to insect pests may be less in 1990 and will probably be moderately widespread in 2000. Insecticide use at present is relatively high in 1990 and will remain so in 2000. The ratio of pesticide cost to the total input for paddy rice production will remain around 10% or slightly higher by 2000.

In some countries where yield is less than 2.5 t/ha, low yields in 1990 will not be improved in 2000. Insecticide use, negligible in 1990, will remain so in 2000.

Table 7. Prospect for rice production and insect pest management in the year 2000

	Yield (t/ha) 1990 ➡ 2000	Yield loss due to insects (%)* 1990 ➡ 2000	Use of resistant cultivars 1990 ➡ 2000	Use of insecticides 1990 ➡ 2000	Pesticide costs /total input (%)** 1990 ➡ 2000
IN MONSOON AREA PADDY/IRRIGATED RICE					
Yield: higher than 6 t/ha	6 → 6.5	30 → 25	Less → L/Mod	High → H	10≥ → 10≥
4 – 6	5 → 5.5	25 → 20	High → H	Mod → M/H	7 → 7
2.5 – 4	3.2 → 3.8	20 → 20	High → H	Low → L	5 → 5
less than 2.5	2 → 2	15 → 15	Less → L/M	Neglig → Neg	0 → 0
UPLAND/DEEPWATER RICE	Low → L	5–10 → 5–10	Null → N	Null → Nul	0 → 0
ECONOMIC VIEWPOINT					
Comercial farming	High → H	H/Mod → H/M	L/Mod → L/M	High → H	10≥ → 10≥
Sustainable	Mod/L → M/L	Mod/L → M/L	H/L → H/M	Low → L	Low → L
Subsistence	Low → L	Low → L	Low → L	Neglig → Neg	0 → 0
IN NON-MONSOON AREA					
Rotation in Egypt	6 → 6.5	Low → L	? → ?	Neglig → Neg	Neg → N
In African countries	Low → L	High → H	? → ?	Neglig → Neg	Neg → N
In USA and Australia	High → H	Low → L	Low → L	Mod → M	Mod → M
In Brazil (upland)	1.8 → 1.8	? → ?	? → ?	Low → L	Low → L

* Relative value in untreated plots when yield loss in maximal controlled plots is regarded as null.
** Herbicides will be probably used more in 2000.

Farmers who produced high yields in 1990 will tend to get higher yields in 2000 in commercial farming. They will be prepared to use more resistant cultivars and more sophisticated insecticides provided the returns are commensurate with outlays.

In sustainable farming, farmers used resistant cultivars in 1990 and will accept more resistant ones in 2000.

Rice farming in Egypt produces around 6 t/ha under a four year rotation system. There are no serious insect pests on rice in Egypt, as compared with the pest situations in monsoon Asia (Mochida, 1990). Water is one of the limiting factors controlling yields in this country.

In African countries, sustainable and/or subsistence farming is popular. Commercial farming is carried out with high yields under irrigation in USA and Australia but with low yields under upland conditions in Brazil. If there are resistant cultivars, and the cultivar(s) show high potential for producing higher market prices, or higher net returns to the farmers, is this not good reason for accepting their introduction?

References

Carson, R. (1962) *Silent Spring*. Crest Books, New York. 304 pp.

Hasegawa, H. (1973) [Pecked rice grains caused by Heteropteran insect.] In: Hassei-yosatu-syokuin Tyuuookensyuu Kai text. pp. 1–5. Ministry of Agriculture, Forestry and Fisheries. [in Japanese]

IRRI (1988) *World Rice Statistics 1987*. 257 pp. IRRI, Los Banos.

JSAEZ [Japanese Society of Applied Entomology and Zoology] (1987) *Major Insect and Other Pests of Economic Plants in Japan*, 379 pp. Japan Plant Protection Association. [in Japanese]

Mochida, O. (1974) An estimation of loss of rice yield due to insect pests in Kyushu, Japan. *Rice Entomology Newsletter*. 1,14–17.

Mochida, O. (1988) The rice water weevil (*Lissorhoptrus oryzophilus* Kuschel) and the freshwater snail (*Pomacea canaliculata* (Lamarck) as important pests of crops for plant quarantine in Asia. In: *Movement of Pests and Control Strategies* (Ed. Singh, K.G. *et al.*), pp. 71–80. ASEAN PLANTI.

Mochida, O. (1990) [Agriculture in the world. (14) Africa. 3. Egypt 1.] *Yunyu-Syokuryoo-Kyoogikai Kaihoo* 501,70–78. Tokyo. [in Japanese]

Mochida, O. (1991) [Prospective distributions of insect pests on crops when it will be warmer in 2000.] *Agriculture and Horticulture*, 66,128–136. [in Japanese]

References for the statistical data in Japan are omitted here because all of them are written in Japanese.

Chapter six:
IPM Development in the Indonesian National IPM Program

K.D. Gallagher
Program Nasional Pengendalian Hama Terpadu, Sekretariat Yogyakarta, Bulak Sumur A-16, Yogyakarta, Indonesia

Introduction

The National IPM Development and Training Program in Indonesia is a multi-ministry effort to train farmers, extension workers and government managers for implementation of IPM. The National Development Planning Agency is implementing the program as the co-ordinator of the National Steering Committee which includes officials from the Ministries of Agriculture, Education and Culture, Population and Environment, Home Affairs, Health, Central Bureau of Statistics, and the Co-ordinating Ministry for Finance and Economics for Industry. At the field level, the Program is implemented by staff of the Ministry of Agriculture, and universities. The Food and Agriculture Organisation of the United Nations (FAO) provides technical and management assistance based on its experience through the Regional Inter-Country IPM Program.

The National IPM Program is part of the implementation of the Presidential Instruction No. 3/1986 in which President Suharto banned 57 insecticides for use on rice and called for IPM training efforts. The banning of insecticides in unison with complete removal of pesticide subsidies has led to tremendous reductions nation-wide of pesticide use and production while maintaining yield increases (Fig. 1). *Policy* implementation is one of the cornerstones of IPM development in Indonesia.

Training Program

Another cornerstone of the IPM development Program has been development of IPM training methodologies. These methods are field-based and hands-on. No lectures are involved in the training. The goal of the training is for farmers to become better ecologists with more knowledge about the processes of plant growth and population dynamics. Recognition of pest problems, natural enemies, agronomic-induced problems, and understanding their interactions is learned through field activities. Extension workers learn using the same non-

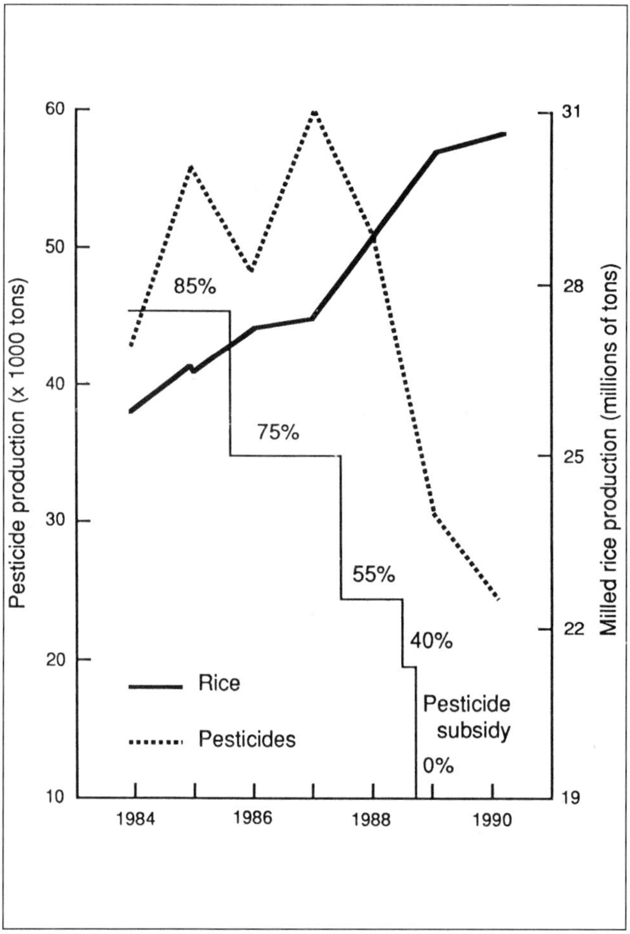

Fig. 1 Pesticide production, rice production and pesticide subsidy — Indonesia 1984–90

formal education methods. "Learning by doing", "discovery learning" and "learning from experimentation" are phrases which have been used to described the National Program.

Training of Trainers

Traditionally, training of trainers has focused on technological aspects of IPM. In the National Program, trainers already have significant field skills related to IPM, so instead we develop the trainer as a *facilitator* of a process. Communication skills, understanding of basic underlying principles of adult education and management skills are the main topics of trainers' training.

IPM Field Skills and Group Dynamics

Pest observers, who are Ministry of Agriculture field staff, study for two complete seasons learning IPM field skills and group dynamic skills. One season is for rice production and the second season is secondary crop production (soybeans, mungbeans, maize, peppers, lowland cabbage, etc.). These pest observers learn IPM and facilitator skills through field exercises and field studies which illustrate basic biological or social processes. The same methods are used by the pest observers with farmers in the field (*teach as you were taught*). After two seasons, these field staff become technical IPM experts and able to facilitate farmer training using field methods.

IPM Farmer Field Schools

An important goal of the National IPM Program is that *Farmers become IPM experts in their own field.* Farmer training is 100% field oriented. Trainers use their facilitator and group dynamics skills to assist farmers to learn recognition of pests and natural enemies, sample fields, make field analyses, and study underlying principles of IPM.

Farmer training lasts for 10 to 12 weeks, with one meeting of 5 hours per week throughout the season. Each field school is comprised of two trainers, and 25 farmers. Graduation from the field school is determined by attendance, and field-based "ballot box" tests.

Farmer to Farmer Training

Graduating farmers often continue to meet and train other farmers in their community. Most often training uses the Farmer Field School Model. This is one unexpected result of field-based training and experiential learning; farmers do not have to give lectures or use expensive materials/tools. Farmers merely need to meet in a field, and use locally available plastic bags, paper and crayons for the majority of activities. Real learning takes place while standing in the field, or discussing problems presented by the field condition itself.

By June 1992, 1000 pest observers will have received one year of basic IPM training. Approximately 3000 extension workers and 150,000 farmers will also have finished a season-long field school. The number of farmers trained by other farmers remains uncounted but is estimated to be at least 300,000. Presently, national, provincial, district, sub-district and village governments are supplying funds for further IPM training by pest observers or farmers. Within five years, it is expected that roughly 2.5 million farmers will have had a season of IPM training.

Old Concepts/New Concepts

The development of national policy, and national training methods are the central part of the National IPM Program. Additionally the Program is also developing the basic concepts of IPM. The development of a new IPM philosophy is essential to maintain conceptual unity with national development goals. Indonesia has stated that national development will include the participation of the people, and that the people will be the subject, not object of development. Thus it is necessary that IPM itself be a means of human development. Table 1 compares new and old concepts related to IPM as they are developing in Indonesia.

Table 1. New and old concepts as they relate to IPM in Indonesia.

Topic	Old concept	New concept
Time and social context	Old concept began after the Second World War, especially with the spread of industrial fertiliser, and pesticides. The 1960s was an era of hunger, communist spread, foreign pesticides corporation dominance, and the Green Revolution in which governments had to increase food production to keep up with population growth. Internationally, pesticide corporations were not being opposed by entomologists and/or environmentalists. Plant breeding was working miracles with high nitrogen-responding varieties. With "Top down" planning, quick-fix plans were put into motion.	New concepts of the 80s and 90s began in the late 70s. Environmental health was seen as important for economic health. People became the focus of development as a way of getting economies and society moving ahead. The Green Revolution was successful but required new thinking about sustainability, health, environmental quality, local responsibility for development and planning. The 40s saw nations becoming independent of foreign colonisers, and the 90s will see individuals gain independence from central planners.
Packet technologies	Essential to get agriculture moving without training mass	Packaged technologies won't work. Farmers develop through

Table 1. *Continued*

	numbers of people. People are *used* to get the economy going...	individual growth and training. People development gets the economy going...
Pesticides	Considered essential to increase yields. Often tied to the use of fertilisers and high yielding varieties. Integral part of package technologies.	Can cause as many problems as they can solve. (e.g. brown plant hopper). Pesticide use reduced to "need only" applications. Need is determined by farmers through analysis of the ecosystem.
Chemical fertilisers	Necessary to increase yields. Must be subsidised to maintain high yields. Integral part of package technologies.	Necessary to increase yields, but should be used on a *need* basis and combined with fertility sustaining methods. Some fertilisers such as P and K do not need to have a continuous high level of application. N fertilisers are important for high yields. Organic fertilisers should be used to heighten the quality of the soil and increase micronutrient levels.
High yelding resistant seeds	Necessary to increase yields without training farmers. The seeds are sufficient in a package with pesticides and fertilisers to give high yields.	High N response and high tillering varieties are useful for compensation against disease and insect damage. New varieties should be able to better compensate for insect damage. "Resistance" is best limited to resistance of diseases and some insects. Most insects remain under existing biological control.

Table 1. *Continued*

Pest and natural enemy recognition	Too difficult. Farmers not capable of complex decision making.	Recognition requires only a little training since farmers have seen them for years in their fields. Moreover, farmers can make an analysis of the agroecosystem to make their own decisions about management of their field.
Definition of IPM and decision making	Count, spray and pray, Largely focuses on pest populations in the field as they relate to a centrally determined economic threshold level (ETL). The ETL is based on conditions found at the central site. Mechanical instructions given to the farmer; count, compare with ETL, spray if necessary. Largely considers pest populations and is based on partial budget calculations.	Profit of the farmer is the central focus. *Growing a healthy crop* that is both resistant to local diseases, and able to compensate for pest attack is the first principle. *Conserving natural enemies* of crop pests so that pest populations will not damage the crop is the second principle. *Weekly observation and analysis* of the field is the third principle. The state of the plant, weeds, rats, variety performance, insects and natural enemies are considered in the analysis. Decision making is based on the integration of plant health/compensation, pest populations, natural enemy populations, potential yield loss, cost of control measures, projected commodity price, farm level economics and previous farmer experience. Thresholds are replaced by "ecosystem analysis methods".

Table 1. *Continued*

Farmer as optimiser	Impossible! Must be done with a highly technical centrally planned package.	Essential. 10 million farmers will be able to optimise their own environments better than one centrally planned package.
Farmers as trainers	Must use T & V system of "top-down" training. Field extension officers bring the knowledge that farmers are supposed to implement. The message is a well planned central message that is given to farmers.	Farmers are very capable of training other farmers because *the process is the message*. Farmers work in groups to pool skills and knowledge. Farmers work in the fields with the assistance of a local IPM expert to extend their field and analytical skills. The local situation defines the topics and direction of training.
Research	Technology is developed in a central research institute then passed to extension planners who then pass the technology down through layers of training and finally to farmers in the T & V system. Farmers and field extension workers are passive (sometimes forced) recipients of packages based on these technologies. On occasion, the research based packages become "menus" from which farmers must select options.	Research is carried out at all levels. Research centres continue to do basic studies, while developing processes which are employed to test research results locally and to initiate local studies. Thus varietal trials, fertiliser trials, sampling methods, natural enemy exlusion, and other research is conducted locally in the field. Field extension workers assist in the process by providing some facilitation and technical support.
Training evaluation	The results of training are evaluated by the extent of adoption of	The result of training should be a modification and integration of

Table 1. *Continued*

	a package. Examples: Have the "messages" been accepted? Are farmers counting 20 hills? Complete adoption just as it was written down at the central planning office of extension or research would be evaluated as successful. The result of evaluations are used to determine if farmers are "modern", "accepting", or "capable".	new ideas and methods being presented in training. "Training" in fact is a farmer evaluation of the new methods and technolologies being suggested from research centers. The result of evaluation should be the extent of integration and results of farmer testing. The results should be returned to researchers for the benefit of the future research agendas. The results of evaluation are used to determine if the research is "useful", "appropriate", or "adaptable".
Future adaptation of new technology	Will need a central message to be passed to lower extension workers to be passed.	Farmers are creating both new technologies and processes locally, and able to evaluate new systems. to farmers in the T & V technologies.

Chapter seven: *Integrated Pest Management Discussion*

Chairman: Professor George Varghese

Dr Peter A.C. Ooi (IIBC, Malaysia) Dr Mochida, you remarked that stem borers have declined as pests of rice in Japan. Would you attribute this to a reduction in pesticide usage?

Dr O. Mochida No. It is due rather to the application of insecticides at the correct time, and also to changes in cultural methods. For example, harvesting is now done by machines, using rice cultivars with small stems to assist in control of the Sesamia Stem Borer. In the case of the Yellow Stem Borer, which is monophagous, late transplanting has also contributed to control.

Mr K.K. Loo (Malaysia Pest Control Sdn Bhd) Pyrethrum is diluted in kerosene and applied using handsprayers. Does the kerosene not damage the plants? Would not a thermal fogger or similar equipment to produce a lighter or indirect spray be more appropriate?

Dr O. Mochida There is a misunderstanding. The pyrethrum–kerosene mixture was applied to the surface of the water, not to the plants by spraying. Thus the plants were not damaged in any way.

Prof D.J. Greenland (CAB International) Could Dr Gallagher comment on how far the higher yields shown after reduction of pesticide use are due to the availability of new highly pest resistant rice varieties, such as IR 64? And could he comment on how far similar results may be expected if pesticide use is reduced on other crops?

Dr K. Gallagher (FAO, Indonesia) Pesticide reduction does not normally lead to increased yields. However the better management and decision making which are causing reduced pesticide use may also be responsible for better crop management leading to higher yields. The use of resistant varieties without pesticides in rice is a very important aspect of high yields. In soybean, better varieties would most certainly lead to better yields, and perhaps lower pesticide usage, although it is not clear that lower pesticides will lead directly to higher yields.

Dr Candida B. Adalla (University of the Philippines) In the Philippines a high percentage of farmers are still low risk takers as indicated by more farmers applying more pesticides (particularly insecticides) after an IPM training as a result of acquiring the skills of pest identification.

Have similar cases been documented in Indonesia?

Dr K. Gallagher The Indonesia National IPM Program was developed with many lessons from the Philippine Program, so this problem was anticipated. Farmer training involves practice fields in which farmers plant and manage rice production for a full season without risk of personal loss. Farmers are willing to take risks in these fields and develop confidence for application in their own fields. Our follow-up studies show most farmers are drastically reducing pesticide applications. Data are currently being collected.

Dr J.D Holloway (CAB International) Dr. Mochida has painted a very intriguing picture of the year 2000 where integrated pest management in Japan is run by technicians, rather than research scientists, with computerised systems for forecasting. Will this system have the flexibility to respond to changes in pest type and pest epidemiology that may arise through the effects of global warming?

Dr O. Mochida There is a possibility that the status of some insect pests will change. In such cases, we need research entomologists but not technicians. However, it is costly to keep any rice entomologists for future problems, the precise nature of which cannot be foreseen.

Prof Varghese Could Dr Gallagher enlighten us as to how much of the rice yield increase in Indonesia is due to: (a) increase in acreage, (b) the growing of higher yielding varieties and (c) the use of plant growth regulators, e.g. Ethepon?

Dr K. Gallagher Yield has increased primarily not due to increased acreage but rather to increased fertiliser use in technical irrigation. The area under rice production has not increased that much. It must be noted that irrigation allows double cropping.

The result of using plant growth regulators is very interesting. In fact a lot of research shows that there is no effect on yields but more recent research in Indonesia shows that probably it is causing greater infection by *Pseudomonas*. So the plant has a low defence mechanism. Thus plant growth regulators, as we see them, may be counter-productive in rice.

Dr Maryati Mohamed (Universiti Kebangsaan Malaysia) It is interesting to note the slight increase of yield when IPM is implemented. Nevertheless, that may not be as important as the quality of the environment which has improved. Is there any quantitative work published regarding the improvement of the environment after the IPM has been implemented, and should not that be highlighted in this Seminar?

Dr T.J. Perfect (NRI, UK) We should be cautious in interpreting the limited data currently available on yield increase, which may not represent a general improvement in crop management. The important point is that there is no severe decline in yield in association with the adoption of IPM. The question in point is well made and a number of people have referred to that this morning. One of the benefits associated with IPM is an improvement in the relationship between the farmer and his farming environment. Another is the improvement of the environment in the chemical sense. Thus the farmer is better able to understand and manage the environment. This kind of relationship I am sure is the main point that Dr Gallagher would like to make. In response to an earlier question about risk exposure, I was very interested to hear that he used compensation techniques, to try to encourage confidence in the idea of using alternative approaches. We all appreciate that this is the fundamental need. We have to develop a willingness on the part of the farmer to take that first step. Again my own experience in the Philippines was with one farmer for five years where we asked him to grow his rice for us while we did our various bits of experimentation on two conditions. One was that he should grow acceptable varieties and the other was that he should never use insecticides, which he thought was a very peculiar set of conditions to put upon him. Within the five years that we worked with him, there was no serious crop protection problem with which he had to contend. I am not in a position to judge the impact of that. That is a very small involvement but it is the kind of thing that is now being adopted much more widely in Indonesia. I think the lesson learnt by farmers is that if they can see it work and if they are cushioned into the process, then they will adopt the techniques.

Prof Varghese We cannot yet be specific about benefits of IPM over other methods. Chemicals will still have to be used in a discriminate way. IPM does not mean necessarily the total exclusion of chemicals. We brought in pests by intensifying agriculture. With scientific research we can greatly improve the position.

PART 3
Pesticides

Chapter eight: **Prospects for Pest Resistance to Pesticides**

W.P. Davies
Royal Agricultural College, Cirencester, UK

Summary

Pesticide treatments are likely to remain the most important component of most crop protection programmes for the foreseeable future. Their continuing use and increasing adoption in integrated pest management will be challenged, however, by the growing problem of resistance in pest, pathogen and weed populations. It has been estimated that 504 species of insects and mites are currently resistant to insecticides; over 100 species of fungi resistant to fungicides; and 81 species of weeds resistant to herbicides.

This problem can be avoided or minimised by more fully exploiting chemicals in combination with biological and physical controls in integrated resistance management programmes. Future needs include improvements in the diagnosis, monitoring and prediction of resistance. Continuing research, training and extension are also required to protect future pesticide inputs.

Introduction

New methods of pest management, exploiting in particular genetic resistance in plants and biological control, will continue to be developed for crop protection. For the foreseeable future, however, pesticide treatments are likely to remain the most important component of pest management programmes in agriculture and horticulture.

Farmers and growers value crop chemicals for a variety of reasons including the ease of application; the rapidity of their effect and convenience for farm management. An over-reliance on pesticide, resulting often in too frequent and heavy usage, will only be discouraged in the future through the greater adoption of integrated pest management (IPM) programmes. More selective use of chemicals applied within a programme of other control measures will increasingly provide a more rational basis for pesticide use.

This scenario is threatened, however, by the continuing development of resistance to key chemicals in pest, pathogen and weed populations. Moberg

(1990) considers pest resistance to chemical control to be one of the most important challenges to sustainable world agriculture.

This account outlines the current status of the pesticide resistance problem as a basis for considering future needs.

Nature of Resistance

The term "resistance" has been carefully defined by Sawicki (1987) as a "genetic change in pest populations in response to toxicants that may impair their control in the field", to which Moberg (1990) suggests adding "when applied at recommended rates". Although this definition is widely accepted, the term "tolerance" in addition to "resistance" continues to feature in the weed literature. Gressel (1987) states that "any decreases in susceptibility to a herbicide, in comparison with a wild type, may be considered as tolerance". Resistance, however, represents complete tolerance to recommended application rates. Use of "tolerance" is generally discouraged in accounts of crop pest and pathogen studies.

The problem of resistance arises through natural selection generated by frequent pesticide use, and is the result of a shift in sensitivity in the population. Survivors of chemical treatment pass on the characteristics of greater resistance to progeny, requiring higher dose rates to achieve the same levels of control.

For resistance to become a practical problem in crop protection, however, it must result in pest control failure in the field. Although small changes in population sensitivity may not initially cause control failure in the field, the further development of resistance will create practical problems.

If resistance to one pesticide also results in resistance to a second chemical it may be described as "cross-resistance" to both materials (Brent, 1987). Such pesticides commonly share a common biochemical basis of toxicity. "Cross-resistance" is sometimes less specifically used to describe resistance of an organism to two or more toxicants (Moberg, 1990), but this may be described by others as simply "multiple resistance".

Status of Resistance

Insecticides

Since the first report in 1908 of resistance to lime sulphur, approximately 504 species of insects and mites are conservatively estimated to have developed insecticide resistance (Georghiou, 1983). Over half of these resistant pests are reported to be agriculturally important and 3% are beneficial predators, parasites and pollinators.

Most of these species exhibit resistance to more than one group of chemicals. Georghiou (1990) states that fully susceptible species of insects to insecticides have become rare, and are now the exception.

Development of resistance has become more prevalent during the last 40 years, but the number of new resistant species recorded recently is less than in previous decades (Table 1). This is due to many recent reports adding to the resistance spectrum in species already reported as resistant to other compounds.

Table 1. Chronological increase in number of insects and mites with documented cases of resistance to pesticides (after Georghiou, 1990)

	Number of species	
Period	New in decade	total
1908	1	1
1909–1918	2	3
1919-1928	2	5
1929-1938	2	7
1939-1948	7	14
1949-1957	62	76
1958-1967	148	224
1968-1978	190	414
1979-1988	90	504

Resistance has developed particularly to cyclodiene insecticides, DDT and organophosphates (Table 2). Substantial numbers of resistant species have developed within the Lepidoptera (67 species), Coleoptera (66 species), Acarina (58 species), Homoptera (46 species) and Heteroptera (20 species).

Table 2. The occurrence of resistance in insects and mites according to pesticide chemical group (after Georghiou, 1990)

Chemical group	Number of species	Per cent of total
Cyclodiene	291	57.7
DDT	263	52.2
Organophosphate	260	51.0
Carbamate	85	16.9
Pyrethroid	48	9.5
Fumigant	12	2.4
Other	40	7.9

Critical situations have now arisen (e.g. Fig. 1) in which many previously applied insecticides, the use of which could be economically afforded, are no longer effective. Most of these problems have arisen with damaging pest problems in intensive cropping, where insecticide usage and selection pressures

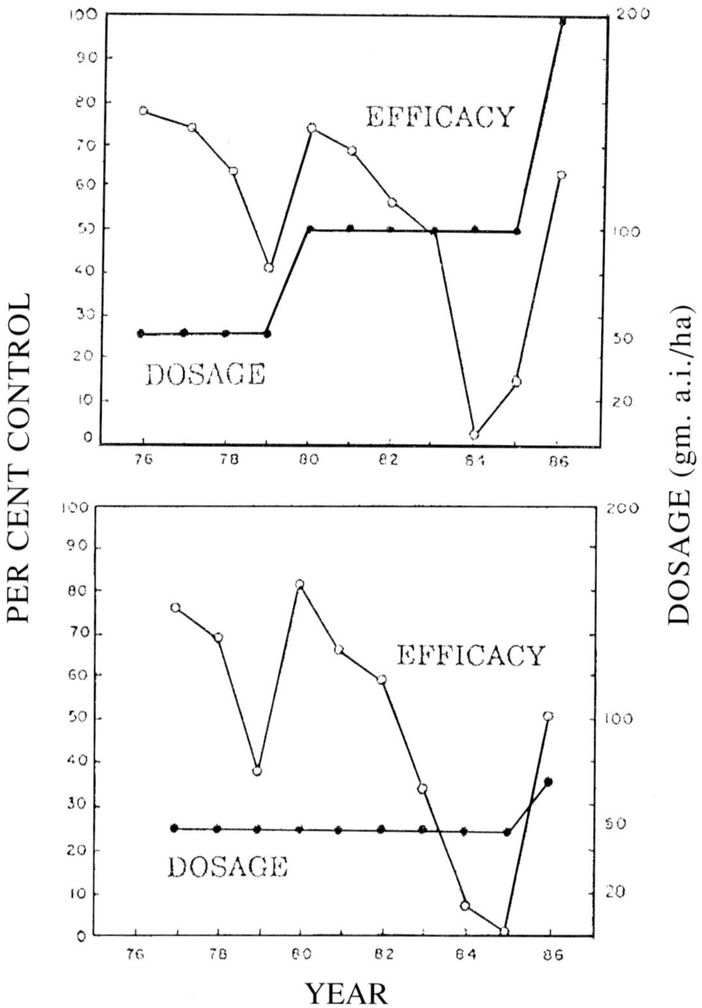

Fig. 1. Development of insecticide resistance (after Georghiou, 1990). Decline in fenvalerate (top) and cypermethrin (below) efficiency against *Heliothis armigera* on cotton, and dosages of insecticide applied, 1976-86. (Compiled from data provided by Wangboonkong S., Thailand Department of Agriculture, Bangkok, 1987)

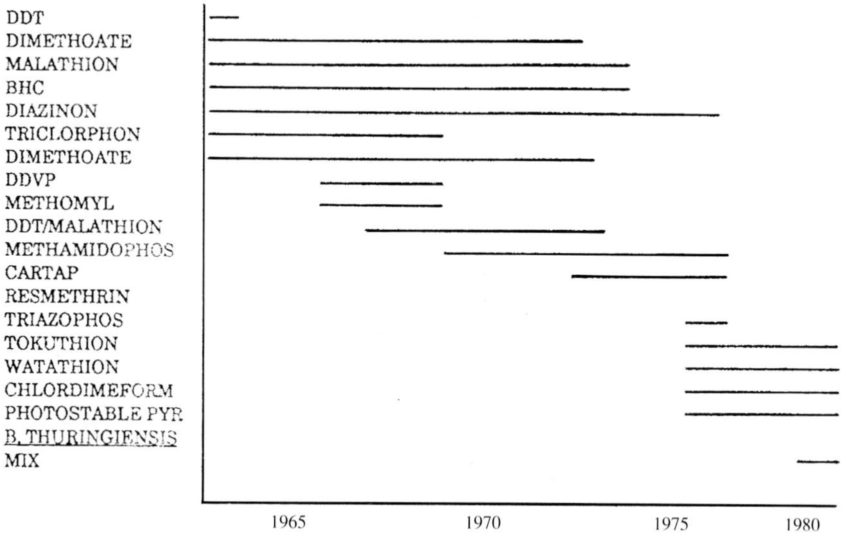

Fig. 2. Use and disuse of insecticides against diamondback moth *(Plutella xylostella)* on brassicae in Malaysia (Cameron Highlands) (after Ruscoe, 1987)

have been substantial. These include the diamondback moth (*Plutella zylostella*); whitefly (*Romicia tabaci*); peach-potato aphid (*Myzus persicae*); leaf miner (*Liriomyza trifolii*); budworms (*Heliothis virescens* and *H. armigera*); Colorado beetle (*Leptirotarsa decemlineata*); two-spotted mite (*Tetranychus urticae*) and the fruit tree spider mite (*Panoncychus ulmi*).

One of the greatest tendencies to develop resistance to different insecticides has been shown by the diamondback moth (Fig. 2), a major problem on cruciferous crops in the Far East (Ruscoe, 1987). The problem is particularly acute in the Cameron Highlands in Malaysia where heavy usage of chemicals has resulted in the rapid development of multiple resistance now also including resistance to new formulations of *Bacillus thuringiensis* and to the relatively new insect growth regulators chlorfluazuron and teflubenzuron (Syed, 1990). Continued resistance monitoring is now required in support of integrated control strategies.

Fungicides

Resistance of plant pathogens to fungicides has developed during the last 25 years in particular, with the increasing availability of systemic chemicals for disease control (Brent, 1987). Previous occurrences of fungicide resistance were rare, local in distribution and took many years to develop (Table 3). These included resistance of *Penicillium* spp. on citrus fruit to biphenyl and sodium orthophenylphenate; resistance of wheat bunt (*Tilletia foetida*) to hexachlorbenzene, and organomercury resistance in *Pyrenophora graminea*, the cause of leaf stripe of oats. These problems resulted in disease control failure, but were only of local importance. In contrast, resistance developed rapidly and extensively in the 1970s within two years of the commercial release of benzimidazole and 2-amino-pyrimidine fungicides.

Table 3. Occurrence of fungicide resistance problems in crop (after Brent, 1987)

Date (approx)	Fungicide or fungicide class	Years of commercial use prior to resistance (approx)	Main crops, diseases and pathogens affected
1960	Aromatic hydrocarbons	20	Citrus storage rots: *Penicillium* spp.
1964	Organomercurials	40	Barley leaf-stripe *Pyrenophora graminea*
1969	Dodine	10	Apple scab: *Venturia inequalis*
1970	Benzimidazoles	2	Many target crop diseases and pathogens
1971	2-Amino-pyrimidines	2	Cucumber powdery mildew: *Sphaerotheca fuliginea*
1972	Kasugamycin	5	Rice blast: *Pyricularia oryzae*
1976	Phosphorothiolates	9	Rice blast: *Pyricularia oryzae*
1980	Acylalanines	2	Potato blight: *Phytophthora infestans*
1982	Dicarboximides	5	Grapevine: *Botrytis cinerea*
1982	Sterol demethylation inhibitors,	7	Cucurbit and barley powdery mildews: *Sphaerotheca fuliginea* and *Erisyphe graminis*

The continuing effectiveness of many major surface protectant fungicides, such as copper compounds; sulphur; dithiocarbonates; phthalimides; quinones and chlorothalonil has been partly attributed to their multi-site toxicity (Dekker, 1981). This is unlikely, however, to be the only reason for resistance failing to develop to these particular chemicals (Georgopoulos, 1986).

The greater vulnerability of systemic fungicides to the development of resistance is associated with their highly specific, single site toxicity rather than systemicity (Georgopoulos, 1986). Systemic morpholine and Fosetyl Al fungicides have not yet developed resistance problems.

The evolution of fungicide resistance has been described as "single-step" or "multi-step" (Fig. 3) depending on the rate and magnitude of change and control by different numbers of genes (Georgopoulos, 1986; Brent *et al.*, 1990).

The rapid development of resistance to benzimidazole and phenylamide fungicides by several pathogens was associated with sudden disease control failure in a "single-step" manner. This contrasts with pathogen populations where genetic control is polygenic and change to fungicide selection is more gradual, resulting in slower progressive loss of disease control. Resistance of pathogens to 2-amino-pyrimidine fungicides and sterol demethylation inhibitors (DMI) has been of this more gradual "multi-step" type.

Both these types of resistance are of economic importance to agriculture. The "one step" form, which is stable and can develop so rapidly, seems to be the more difficult form to control in practice.

In all cases, the development of fungicide resistance is determined by the nature of the pathogen and type of chemical and application environment (Dekker, 1981). It tends to develop more rapidly in pathogen populations which reproduce frequently, producing many spores, such as powdery and downy mildews. Pathogens which reproduce more slowly, such as loose smut of barley (*Ustilago nuda*), also tend to develop fungicide resistance more slowly.

Over 100 species of fungi resistant to particular fungicides have now developed in all of the major groups of fungal plant pathogens (Georghiou, 1983). Most of these resistance problems have occurred in western Europe, where there is heavy fungicide use. Particular problems have also developed, however, in plantation crops in the tropics, such as the resistance of banana leaf spot (caused by *Mycosphaerella fijiensis*) and coffee berry disease (caused by *Colletotrichum coffeanum*) to benzimidazole fungicides.

Herbicides

The discovery of herbicides took place later than the development of many organic insecticides and certain fungicides, and for some time there did not appear to be any threat of a herbicide-resistance problem. The rapid evolution of resistant weed species to phenoxy-herbicides, predicted by Harper (1956), did not materialise. Triazine resistance in *Senecio vulgaris* on an ornamental

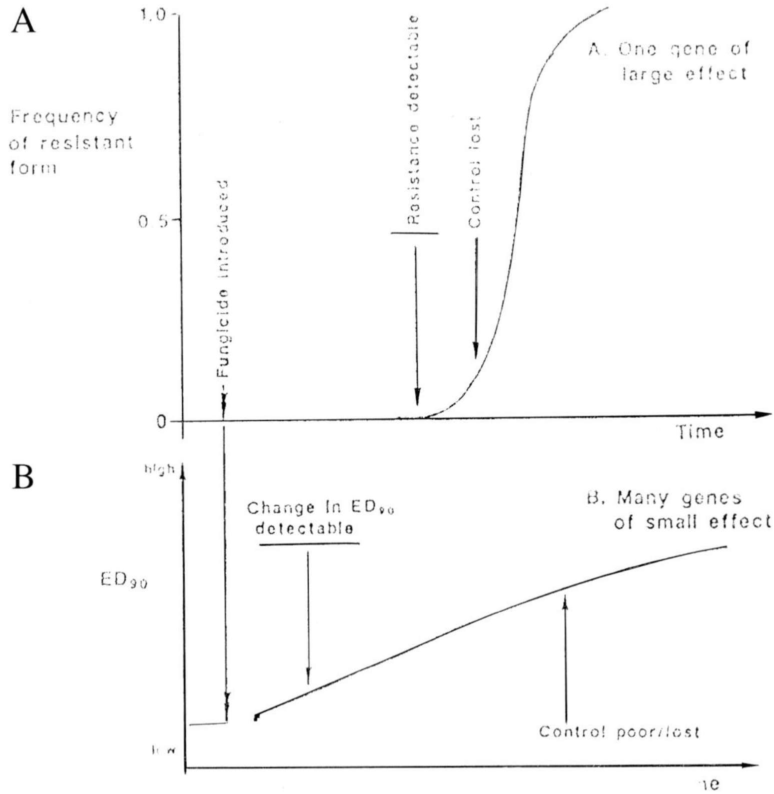

A. Exponential increase where only a single gene is involved.

B. Gradual increase in resistance where many genes are involved.

Fig. 3. Evolution of fungicide resistance (after Brent *et al.,* 1990)

nursery in North America in the 1960s was the first reported problem (LeBaron & Gressel, 1982). Resistance to herbicides has since become widespread and more economically damaging (LeBaron & McFarland, 1990; Holt & LeBaron, 1990; Rubin, 1991).

Estimating the actual numbers of herbicide-resistant weeds is difficult. The number of cases relates to the intensity of the search, and it has been suggested that many new cases are probably not now being reported due to a lack of novelty (Gressel, 1987). However, LeBaron (1991) reports 107 resistant biotypes. This includes 57 species (40 broad-leaved weeds and 17 grasses) with biotypes resistant to triazines, and 50 species (33 broad-leaved weeds and 17 grasses) with resistant types to 14 other classes of herbicides. Because some weeds are resistant to more than one type of herbicide, LeBaron (1991) estimates a total of 81 resistant species worldwide.

Although the problem is widely distributed in almost all continents, resistance has developed particularly in North America and Europe to simazine, atrazine and related triazine herbicides. Other major herbicides, which include soil- and foliar-applied materials, to which resistant biotypes have evolved include chlortoluron, dichlofop methyl, chlorsulfuron, mecoprop, paraquat and trifluralin (Table 4).

No clear relationships have been reported between plant families, genera or taxa in the tendency of weeds to evolve resistance. Triazine resistance has developed particularly, however, in biotypes of *Senecio vulgaris, Chenopodium album. Poa annua, Solanum nigrum, Amaranthus retroflexus, Kichia scoparia* and *Amaranthus hybridus* (LeBaron & McFarland, 1990). Resistant populations of these particular weeds have sometimes been found to be inferior in competitiveness and fitness, but this has not been generally reported in weeds resistant to non-triazines.

Most triazine-resistant weed populations have evolved in maize fields, orchard situations, on railway tracks and roadside verges where atrazine and simazine are commonly used. The popularity of triazine herbicides is attributed to their low cost, persistence, broad spectrum activity and potential for good weed control (Gressel, 1987) resulting often in frequent, large dose, monoherbicide applications and strong selection pressures. In Hungary large areas of maize growing land are infested by triazine-resistant *Amaranthus* species and serious problems have also developed in other countries such as Switzerland, which limit effective triazine use. This problem has been confined within acceptable limits, however, in parts of North America and it has not been considered necessary or desirable to discontinue the use of triazine herbicides for weed control in maize.

A more recent problem is the development of chlorotoluron-resistant blackgrass (*Alopecurus myosuroides*) populations in cereals in England, first reported in 1982 (Moss & Cussons, 1991). The degree of resistance has been found to vary considerably between populations, and to be most severe in fields where intensive winter cereal cropping and non-ploughing cultivation techniques have been practised, and herbicides to control blackgrass have been used frequently (Table 5). Most chlorotoluron-resistant blackgrass populations show cross-resistance to many other herbicides (Moss, 1987) complicating alternative control strategies.

In addition, the use of straw burning and minimum tillage systems can result in the development of an adsorptive surface which can reduce the activity of many soil-acting herbicides (Moss, 1987), encouraging the greater use of herbicides even in the absence of a resistance problem.

More than half of the pesticides currently used in agriculture are herbicides. As a result of their greater use it has been predicted that herbicide resistance will become a much more serious economic problem within the next five to ten years (LeBaron & McFarland, 1990).

Integrated Resistance Management

The problem of pesticide resistance can be avoided or reduced by exploiting a wide range of crop protection measures, rather than over-relying on the chemical input. Brent (1987) coined the term "Integrated Resistance Management" by analogy with, and as a part of IPM, to describe the integration of physical and chemical control measures (Table 6). The importance of

Table 4. Herbicides with high risk for weed resistance (after LeBaron & McFarland, 1990)

Herbicide classification	Example
Bioyridiliums	diquat paraquat
Dinitroanilines	benefin oryzalinte pendimethalin trifluralin
Diphenyl ethers	acifluorfen diclofop methyl lactofen oxyfluorfen
Imidazolinones	imazapyr imazaquin
Sulfonylureas	chlorsulfuron chlorimuron sulfometuron-methyl triasulfron
Triazines	atrazine cyanazine metribuzin prometon prometryn simazine
Uracils	bromacil terbacil
Ureas	diuron fluometuron linuron tebuthiuron

Table 5. Summary of cultivation, cropping and herbicide history for fields in England with herbicide-resistant *Alopecurus myosuroides* populations (after Moss & Cussons, 1991)

			Number of herbicide applications[1]		
No of harvest years	No of years ploughed	Years in winter cereals	Chlorotoluron or Isoproturon	Diclofop methyl	Others[2]
211	16	192	194	51	97

[1] Only herbicides used for *A. myosuroides* control.

[2] Terbutryn and metoxuron, triallate, chlorsulfuron/metsulfuron, pendimethalin, flamprop-isopropyl, propyzamide, fluazifop-*p*-butyl.

appropriate crop management as a main determinant of crop protection is stressed by Schwinn & Morton (1990) (Table 7).

This diversity is the key to most anti-resistance strategies, and is increasingly being exploited by farmers and growers.

Future Needs

Prediction Systems

It has been realised for some time that an ability to predict accurately the build-up of resistance to pesticides would be valuable. This information could be used, in particular, to formulate anti-resistance strategies for the commercial

Table 6. Components of Integrated Resistance Management (after Brent, 1987)

1	Predict risks for chemical
2	Obtain 'baseline data' on response of field samples
3	Monitor performance
4	Monitor sensitivity
5	Introduce avoidance strategies (e.g. minimising use, applying mixed formulations or programmes)
6	Use complementary non-chemical methods
7	Establish co-operation between personnel and organisations in the public and private sectors.

Table 7. Elements of anti-resistance strategies utilising integrated crop production practices (after Schwinn & Morton, 1990)

Crop management			
*	Resistant cultivars	→	Lower disease levels
*	Sites with low disease occurrence	→	Reduce selection pressure
*	Lower N fertilisation	→	Lower disease levels
*	Sound sanitation	→	Less inoculum
Fungicide use			
*	Based on forecasting	→	Reduce selection pressure
*	Protective use	→	Hit small pathogen populations
*	Mixtures	→	Reduce risk of resistance selection
*	Alternating use	→	Reduce risk of resistance selection

application of new materials. Attempts, though, to develop prediction systems applicable to field situations have not generally been successful.

Most mathematical models for predicting an increase in resistant forms of a pest are difficult to validate, and remain largely theoretical. Denholm *et al.* (1990) have emphasised the need for good experimental work to allow models to be adequately tested during development, and they advocate an integrated approach. With increasing availability of data, and the aid of computers, it should be possible in future to produce more realistic resistance development models.

Predicting the risk of resistance to a particular pesticide under specified conditions is becoming easier, in the light of experience and increasing research. "High", "medium" and "low" risk categories have been identified, based on knowledge of mode of action of the chemical; the degree of pesticide use; the fecundity and ease of dispersal of the target organism, and ease of resistance development in controlled environment studies (Brent, 1987). It is possible on this basis to understand the non-development of resistance to certain fungicides, e.g. dithiocarbamates, and to understand better high risk situations such as the rapid resistance development of *Phytophthora infestans*, the potato late blight fungus, to metalaxyl (Brent, 1987). A good understanding of pest–pesticide–

crop interactions will still be required, however, to predict new risk potential and the likelihood of control failure.

It cannot be assumed that the build-up of resistance will automatically result in failure of the chemical control measures. Increasing selection on low levels of such resistance would probably, however, lead to a future problem.

Diagnosis and Monitoring Techniques

Simple techniques for determining the frequency of resistance and the subsequent monitoring of any resistance build-up in pest populations would be particularly useful as a basis for more rational resistance management. Knowing background resistance in past populations before chemical selection is applied would be particularly helpful, together with information on any regional differences.

For these reasons a great deal of research is being carried out on diagnostic techniques, to develop in particular, biochemical, immunological and bioassay tests. Brent *et al.* (1990) point out that such systems will respond to known resistance factors, but new resistances may not be so readily detectable, and that this possibility needs to be considered. The rapid development of "single-step" resistance (Brent *et al.*, 1990) to certain fungicides may also not be easily detected at an early enough stage to avoid control failure.

Increased availability, and the more widespread adoption, of inexpensive, field-based diagnostic techniques will undoubtedly contribute in the future to better pesticide use.

Improving Communications

Sharing information on pesticide resistance will continue to support the adoption of improved control strategies. It is very important, therefore, to continue to encourage co-operation between users, the agrochemical industry, government and researchers to ensure freedom of information.

There is a continuing need for this interchange at local, national and international levels to foster information exchange, with pesticide resistance problems being no respecter of boundaries. Unless conscious efforts are made to support continuing information flow, commercial interests of competing companies, restricted release of research findings and legislation might limit the future development of common anti-resistance strategies.

Industry groups (Ruscoe, 1987) such as the Fungicide Resistance Action Committee (FRAC); the Insecticide Resistance Action Committee (IRAC); the Herbicide Resistance Action Committee (HRAC), and the Pyrethroids Efficiency Group (PEG) currently foster co-operation between otherwise competing agrochemical manufacturers. A further body, the International Organisation for Resistant Management (IORPM) has also recently been formed with wider representation including government, academics and the agrochemical industry.

Training Initiatives

To continue to foster rational pesticide use, and more careful integration of chemicals into IPM programmes, there will be a continuing need for training farmers, advisers and pesticide suppliers. To gain a better understanding of IRM (as part of IPM) broader training in agricultural systems would beneficial, in addition to that of pesticide use. It will be particularly important to continue to provide opportunities for specific training on anti-resistance strategies.

Research Requirements

As well as continuing research on such topics as the mode of pesticide action, pest biology, fitness of pesticide resistance strains, the nature of resistance and agroecology of cropping systems, there is also an increasing need for broader issues to be studied. These could include: farmers' perceptions of crop damage, attitudes to pesticide use, and even perhaps the influences of secondary selection pressures (such as government legislation and policies) on pesticides.

Discussion

With fewer "new chemistry" pesticides becoming available on the market for pest control, there will be increasing pressures on existing materials for crop protection. Where high risk situations are identified, appropriate integrated resistance management strategies must be promoted through advice and training to farmers and growers. The economics of production of particular crops can sometimes encourage over-reliance on cheap pesticides, or even the increased adoption of reduced dose rate applications. With increasing knowledge and experience of the development of pesticide resistance, it is becoming more possible to reduce control problems.

Although it seems likely that pesticide resistance situations will continue to arise, the increasing adoption of IPM, and more rational plant protection, should ensure, however, that the impact will be less than in former years.

References

Brent, K.J. (1987) Fungicide resistance in crops - its practical significance and management. In: *Rational Pesticide Use* (Eds. Brent, K.J. and Atkin, R.J.), pp. 137–151. Cambridge University Press, Cambridge.

Brent, K.J., Hollomon, D.W. Shaw, M.W. (1990) Predicting the evolution of fungicide resistance. In: *Managing Resistance to Agrochemicals: From fundamental research to practical strategies* (Eds. Green, M.B., LeBaron, H.M. and Moberg, W.K.), pp. 303–319. American Chemical Society *Series 421,* Washington, DC.

Dekker, J. (1981) Impact of fungicide resistance on disease control. *Proceedings of the 1981 British Crop Protection Conference – Pest and Diseases*, 3, pp. 850–860.

Denholm, I., Rowland, M., Farnham, A.W. and Sawicki, R.M. (1990) Laboratory evaluation and empirical monitoring of resistance-counting strategies. In: *Managing Resistance to Agrochemicals: From fundamental research to practical strategies* (Eds. Green, M.B., LeBaron, H.M. and Moberg, W.K.), pp. 92–104. ACS *Series 421,* Washington, DC.

Georghiou, G.P. (1983) In: *Pest Resistance to Pesticides* (Eds. Georghion, G.P. and Saito, T), Plenum, New York, 14 pp.

Georghiou, G.P. (1990) Overview of insecticide resistance. In: *Managing Resistance to Agrochemicals: From fundamental research to practical strategies* (Eds. Green, M.B., LeBaron, H.M. and Moberg, W.M.), pp. 18–41. ACS Symposium *Series 421*, Washington DC.

Georgopoulos, S.G. (1986) Plant pathogens. In: *Pesticides Resistance Strategies and Tactics for Management. Proceedings of a Conference on Management of Pesticide. Resistant Pest Populations* (Chairman: Glass, E.H.), pp.100–110. National Academy Press. Washington.

Gressel, J. (1987) Strategies for prevention of herbicide resistance in weeds. In: *Rational Pesticide Use* (Eds. Brent, K.J. and Atkin, R.K.), pp.183–196. Cambridge University Press, Cambridge.

Harper, J.L. (1956) The evolution of weeds in relation to resistance to herbicides. *Proceedings of the 1956 British Weed Control Conference,* 1, pp. 179–188.

Holt, J.S. and LeBaron, H.M. (1990). Significance and distribution of herbicide resistance. Weed Technology, 4, 141–149.

LeBaron, H.M. (1991) Distribution and seriousness of herbicide-resistant weed infestations worldwide. In: *Herbicide Resistance in Weeds and Crops* (Eds. Caseley, J.C., Cussons, G.W. and Atkin, R.K.), pp. 27–45. Butterworth-Heinemann, Oxford.

LeBaron, H.M. and Gressel, J. (Eds.) (1982) *Herbicides Resistance in Plants.* Wiley-Interscience, New York.

LeBaron, H.M. and Mcfarland, J. (1990) Overview and prognosis of herbicide resistance in weeds and crops. In: M*anaging Resistance to Agrochemicals: From fundamental research to practical strategies* (Eds. Green, M.B., LeBaron, H.M. and Moberg, W.M.), pp. 336–352. ACS Symposium *Series 421*, Washington DC.

Moberg, W.K. (1990) Understanding and combating agrochemical resistance. In: *Managing Resistance to Agrochemicals: From fundamental research to practical strategies* (Eds. Green. M.B., LeBaron, H.M. and Moberg, W.M.), pp. 1–15. ACS Symposium *Series 421*. Washington DC.

Moss, S.R. (1987) Herbicide resistance in blackgrass. *(Alopecurus mycsuroides). Proceedings of the 1987 British Crop Protection Conference– Weeds,* 3, pp. 879–886.

Moss, S.R. and Cussons, G.W. (1991) The development of herbicide-resistant populations of *Alopecurus mycsuroides* (Blackgrass) in England. In: *Herbicide Resistance in Weeds and Crops* (Eds. Caseley, J.C., Cussons, G.W. and Atkin, R.K.), pp. 45–56. Butterworth-Heinemann, Oxford.

Rubin, B. (1991) Herbicide resistance in weeds and crops: progress and prospects. In: *Herbicide Resistance in Weeds and Crops* (Eds. Caseley, J.C., Cussons, G.W. and Atkin, R.K.), pp. 387–416. Butterworth-Heinemann, Oxford.

Ruscoe, C.N.E. (1987) Pesticide resistance: strategies and co-operation in the agrochemical industry. In: *Rational Pesticide Use* (Eds. Brent, K.J. and Atkin, R.K.), pp.197–210. Cambridge University Press, Cambridge.

Sawicki, R.M. (1987) In: *Combating Resistance to Xenobiotics: Biological and chemical approaches* (Eds. Ford, M.G., Hollowman, D.W., Whambay, B.P.S. and Sawicki, R.M.), pp. 105–117. Ellis Harwood, Chichester.

Schwinn, F.J. and Morton, H.V. (1990). Anti-resistance strategies: design and implementation in practice. In: *Managing Resistance to Agrochemicals from Fundamental Research to Practical Strategies* (Eds. Green, M.B., LeBaron, H.M. and Moberg, W.M.), pp. 170–183. ACS Symposium *Series 421*, Washington DC.

Syed, A.R. (1990) Insecticide resistance in the diamondback moth, *Plutella xylostella* (L) in Malaysia. Proceedings of the 1990 International Workshop on Diamondback moth Management.

Chapter nine: Semiochemicals and Related Compounds in Insect Pest Management—Malaysian Experiences

W.H Loke[1], K.H. Tan[2] and S. Vijaysegaran[1],
[1] *MARDI, Serdang, Selangor*
[2] *School of Biological Sciences*
Universiti Sains Malaysia, Minden, 11800 Penang

Abstract

Awareness of novel approaches offered by the use of semiochemicals and related compounds is present among pest management scientists in Malaysia. Starting with the sex pheromones, there are two notable examples that can be mentioned. The first is *Plutella xylostella* L. Studies showed that the use of a pheromone as a component of the current Integrated Pest Management (IPM) programme for this pest is considered to hold promise in terms of its use for biomonitoring and population suppression. The second case of sex pheromone use in Malaysia concerns the cocoa pod borer, *Conopomorpha cramerella* Snellen. With the identification and synthesis of the sex pheromone of the cocoa pod borer, prospects of controlling it by mass trapping and mating disruption seem promising.

Fruit fly attractants (methyl eugenol and Cue-lure), have been found to be very useful for surveying and monitoring six *Dacus* species in various local habitats. However, large-scale use of these lures (parapheromones) has not taken off as farmers still prefer to practise conventional spraying and wrapping of fruits. Recently, a locally formulated improved protein bait spray, PROMAR, has been found to be effective, cheap and easy to use with almost no side-effects.

The last example of studies on semiochemicals in Malaysia involves studies on the behavioural manipulation of natural enemies using kairomones. Results obtained showed that elucidating the host-finding and behavioural sequence for oviposition as well as identification of key steps within the "find and attack" cycle of *Apanteles plutellae* Kurdj., an important parasitoid of *P. xylostella*, is of great importance. The ability to manipulate behaviour of parasitoids at selected points in the host-finding sequence must be achieved if more effective utilisation of parasitoids in pest management is to be realised.

It is generally felt that semiochemicals offer good potential as tools in insect pest management programmes as they are environmentally friendly and ecologically sound. One main drawback is that these chemicals are too specific and commercial exploitation by multinational companies is limited by the issues of economics.

Introduction

Traditionally, Malaysia is an agricultural country where some seven million hectares representing about 22% of available land are being cultivated. Agriculture in Malaysia is confronted by several constraints, and pests and diseases are considered of prime importance. It is estimated that pests and diseases can cause losses between 15 and 30% of total production per annum. Currently, pest and disease control has become such an important part of Malaysian agriculture that it is a major requisite for the economic and successful production of agricultural commodities. However, some of the methods used during the past four decades have produced many undesirable effects.

Since the 1940s, the main method of control of insect pests practised by farmers or growers has been the use of synthetic pesticides. The demand for chemical treatment has been enormous and seems endless, generally accounting for about 30% of the production costs. Due to overdependence on and misuse of chemicals, several insecticide-related problems such as resistance development, hazards to non-target organisms, environmental pollution, poisoning and residues in crops have surfaced and become serious. There is a general concurrence among crop protection workers that there is an urgent need to develop more effective and efficient methods for insect pest management based on sound ecological principles. Current methods may vary greatly, but they have been developed to reflect compromises involving technological capability, economic feasibility and socio-environmental acceptability. Future developments in insect pest control can be expected to continue to centre around Integrated Pest Management (IPM) systems that emphasise the use of biological, cultural and target-specific control strategies such as insect-active chemicals (chemicals which have the ability to evoke a specific action in insects).

In the search for insect-active chemicals with relatively low toxicity to vertebrates and non-target species, two highly specific groups of compounds have been identified, *viz.* insect growth regulators (IGRs) and semiochemicals. This paper aims to address the topic of semiochemicals and related compounds in insect pest management in Malaysia.

Semiochemicals

Semiochemicals are chemicals that are able to modify the behaviour of a perceiving organism at sub-micro/nanogram levels (Tan & Loke, 1985). Broadly, semiochemicals can be categorised based on the type of behaviour each chemical induces, its intra- or inter-species activity and whether it is naturally secreted or not (Fig. 1). Such behaviour-modifying chemicals (BMCs) include pheromones, kairomones, allomones, attractants, deterrents and other para-pheromones. Sometimes, a single chemical secreted may have both intra- and inter-species activities. Chemical communication has been known to play a vital role in several aspects of insect life. Many of the chemicals involved in insect communication may be exploited as potential pest detection and control agents. Thus, semiochemicals are becoming increasingly valuable in insect pest management and offer enormous potential for controlling many devastating insect pests through a range of environmentally and ecologically sound pest management techniques.

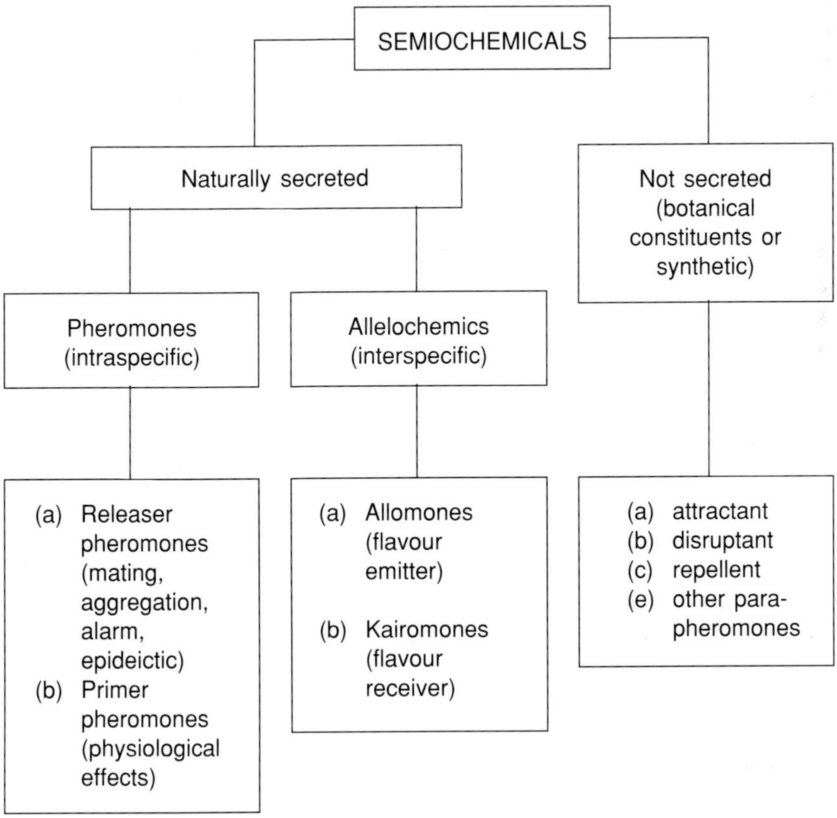

Fig. 1. Classification of semiochemicals

Considerable efforts have been expended towards developing BMCs for pest surveillance and monitoring (population study) and to time conventional insecticide spraying. Additionally, efforts are also directed at using BMCs to annihilate or suppress pest populations through approaches such as mass trapping, mating disruption and various combinations of attractant baits, repellents and deterrents. A rather innovative approach of exploiting BMCs which has also received a fair amount of attention is the use of kairomones for modifying behaviour and stimulating searching by natural enemies. Instead of attempts to elicit an incorrect response or inhibit a correct one, the goal of entomophagous insect manipulation is to redirect an incorrect response or to stimulate the correct one. The behavioural manipulation of certain arthropod natural enemies of insect pests through the use of kairomones has provided new opportunities for the manipulations of these arthropods for the benefit of mankind. Such a pest control strategy is compatible with pest management philosophies and fits well into an IPM scheme.

Biomonitoring and Control Using Sex Pheromones

For many years the control of the diamondback moth (DBM) relied solely on the use of chemicals. Any attempt to disrupt the cycle of pesticide dependency would require that the new approach (IPM) incorporate alternative control components that can be shown to be just as effective as pesticides at farm level. The initial IPM approach consisted essentially of need-based chemical treatment when tentative economic thresholds of 5 larvae per 10 plants were exceeded (Lim *et al.*, 1988). Over the years as additional information became available, the original IPM approach was continually improved through modifications. In 1987, an improved package for DBM management was developed by MARDI (Loke *et al.*, 1990). The development of such a package is by no means an end by itself. In order to stay ahead of the pest, the package is constantly under scrutiny and subject to more refinement. The continuous improvement of the DBM IPM package is two-pronged, involving further exploitation of available control options and exploring new or novel control tactics.

The sex pheromones of *P. xylostella* have been well studied by several researchers (Chow *et al.*, 1974, 1977; Tamaki *et al.*, 1977; Chisholm *et al.*, 1979; Koshihara & Yamada, 1981). Two major components and a minor component have been identified and synthesised. There are:

cis-11-hexadecenal (Z11-16-Ald)

cis-11-hexadecenyl acetate (Z11-16-Ac)

cis-11-hexadecenol (Z11-16-OH)

Most synthetic formulations use a binary mixture of the two major components either in a 5:5 or 6:4 ratio. However, it has been observed that with addition of a small amount (1%) of the minor component (Z11-16-OH), attractancy was enhanced.

Studies conducted so far involve evaluating various commercial formulations. Tables 1 and 2 show some of the recent blends that have been tested. Earlier blends such as those from Takeda and Montedison were also found to be effective. More blends are being obtained for testing. The most effective blend will be selected for use as a biomonitoring tool as well as for mass trapping. Locally, the pheromone has been synthesised by a university team. A recent innovation, the pheromone impregnated hollow fibre formulation, will be tried for mating disruption.

Table 1. Sex pheromone trapping of *Plutella xylostella* in two farm locations

Type of trap	Mean no. of plutella caught	
	Sg. Ikan	Brinchang
IAEA-DJ	472.77a	311.90a
IAEA-SJ	468.00a	273.50a
Reagron	300.87b	134.63b
Px 010D	266.33b	74.50b

Means followed by different letters are significantly different (P ≤ 0.0.5; Duncan's Multiple Range Test).

Table 2. Comparison of different traps baited with Reagron *Plutella* sex pheromone in MARDI, Tanah Rata

Trap type	No. caught (1 week)	
	Plutella	Diadegma
BCS Cross-line wing trap (white)	6116	21
BCS Delta trap (white)	6774	16
IPS Delta trap (yellow)	6554	338
Hoechst Delta trap (green)	4366	13
Reagron wing trap (white)	7927	80
Reagron wing trap (yellow)	8009	494

BCS = Biological Control Systems
IPS = International Pheromone Systems

The second case of sex pheromone studies in Malaysia concerns the cocoa pod borer (CPB), *Conopomorpha cramerella* Snellen. The cocoa pod borer is to date the most serious pest of cocoa in the oriental region. Early studies on CPB adults in Sabah have indicated that the female moth produced a sex pheromone to attract male moths (Mumford, 1986). Beevor *et al.* (1986a) succeeded in identifying and synthesising a sex pheromone blend that is very attractive to male CPB moths in Sabah:

E, Z, Z AND E, E, Z, isomers of:
4, 6, 10-hexadecatrienol acetate
4, 6, 10-hexadecatrienol plus
hexadecyl alcohol
(40:60:4:6:10 ratio)

Traps baited with the synthetic pheromone (1 mg loading) caught more that 10 times as many male CPB as traps baited with virgin CPB (Beevor *et al.*, 1986b). Much field work was undertaken on trap design, colour, height and loading of dispensers. Trials were also carried out on mass trapping using grid, systems of 4, 8 and 16 traps per hectare. Results to date indicate that trapping alone can provide partial control. Better control is achieved where trapping is integrated with other measures such as chemical spraying and release of the egg parasitoid, *Trichogrammatoidea bactrae fumata* Nagaraja (Ho *et al.*, 1987).

The availability of such an attractive synthetic pheromone source opened up several possibilities of using it to monitor and control CPB. These include using the pheromone to trap out the male moths (mass trapping) or to saturate the area with the pheromone to confuse the male moths (mating disruption). Both these approaches are aimed at affecting a sufficient proportion of the adult males to reduce mating, hopefully leading to a reduction in CPB population.

The other option of using the CPB pheromone is as a monitoring tool to decide when and where to initiate control measures as well as to evaluate or assess the status of the pest population and success of control programmes. The use of the pheromone as a monitoring device in quarantine situations has been quite useful although in some instances the Sabah pheromone blend was found to be either partially effective or not effective in Peninsular Malaysia. Further studies by Beevor (unpublished) resulted in the development of another blend with a different component ratio. Currently, two effective blends are available.

A new area of research is currently in hand whereby the pheromone is impregnated in hollow fibres and hung in the field on cocoa trees as a mating disruption system. Details of trials have not been documented.

Fruit Fly Attractants and Protein Bait

Tephritid fruit flies are known to be highly injurious pests of fruits and vegetables in Malaysia. Light infestations may leave some edible crop, but heavier ones can ruin it entirely. Losses affect produce for both domestic consumption and export. In the latter case, even very light infestation can cause a country to lose its export markets because of stringent quarantine restrictions imposed by importing countries.

Fruit fly lures have been found to be very useful for surveying and monitoring purposes (Tan, 1987). Male flies of the D*acus dorsalis* complex are very strongly attracted to methyl eugenol while male melon flies, *D. cucurbitae,* are attracted to cue-lure. Locally, cue-lure and methyl eugenol have been extensively studied and found to be effective attractants of males of six *Dacus* species (Table 3). These chemicals have been shown, recently, to be sequestered and converted to other compounds, which act as allomones, probably, for the protection of fruit fly species against predation (Nishida *et al.,* 1988; Nishida *et al.,* 1990).

Table 3. *Dacus* species attracted to different lures in Penang, Malaysia

Species	C	CL	D	ME	T
D. caudatus	–	+	+	–	–
D. cucurbitae	–	+	–	–	–
D. dorsalis	–	–	+	+	–
D. frauenfeldi	–	+	+	–	–
D. mcgregori	–	–	+	+	–
D. nigrotibialis	–	+	+	–	–
D. occipitalis	–	+	–	–	–
D. tau	–	+	–	–	–
D. umbrosus	–	–	+	+	–

C - capilure; CL - cue-lure; D - dorsalure;
ME - methyl eugenol; + = attraction;
– = no attraction

Fruit growers, particularly carambola growers are highly impressed by the large numbers of males attracted to traps baited with methyl eugenol. It is not uncommon to find individual carambola growers using an assortment of methyl eugenol baited traps (tin cans, used plastic containers, plastic bags, etc.) distributed all over their orchards in an attempt to control fruit flies.

However, when used in this manner for the protection of individual orchards, especially in areas where the flies are endemic, male lures are of little value. Experiments conducted locally using different trap densities and trapping continuously over a period of one year did not provide any significant control of fruit flies infesting carambola (Vijaysegaran, 1985). The traps provided impressive catches of male flies but gravid females were still present to oviposit in host fruits. The male lure, methyl eugenol, however, has been very useful for research purposes and has been used successfully for the estimation of native populations of *D. dorsalis* and *D. umbrosus* in different ecosystems in Malaysia (Tan, 1985; Tan & Jaal, 1986) and relative abundance of fruit fly species in different ecosystems (Tan & Lee, 1982; Vijaysegaran, 1985) (Tables 4 and 5).

Table 4. Mean number of male *Dacus* spp. caught for six weeks in methyl eugenol or cue-lure baited traps at six different locations in Selangor

Dacus spp.	Star-fruit orchard	Village dusun	Housing estate	City street area	Forest reserve	Rice field farm
Species attracted to methyl eugenol						
D. dorsalis						
mean/week	1561.0	237.7	855.0	12.7	45.0	8.3
± s.e.	91.6	27.5	92.7	2.1	5.7	2.6
D. umbrosus						
mean/week	0	246.7	64.2	0.7	0.3	0
± s.e.	22.7	5.1	0.2	0.2		
Species attracted to cue-lure						
D. caudatus						
mean/week	trap damaged,	5.2	0	0	0	0.3
± s.e.	no data available	7.9				0.2
D. cucurbitae						
mean/week	"	4.0	0	0	0.3	0.5
± s.e.		2.5			0.3	0.2
D. frauenfeldi						
mean/week		24.0	4.3	0	0	0
± s.e	"	14.0	0.8			
D. occipitalis						
mean/week	"	144.5	0	0	23.7	0
± s.e.		110.0	0	0	4.0	0

Large-scale use of the above male attractants (parapheromones) has, therefore, not taken off and most farmers still prefer to use conventional tactics such as chemical sprays and wrapping of fruits. However, bagging is labour intensive and difficult to practise for large areas while cover spraying has many undesirable consequences such as being harmful to beneficial insects (e.g. honeybees) besides posing toxic residue problems. The successful growth and expansion of the fruit and vegetable industry in Malaysia, particularly for the export sector, is thus largely dependent on a suitable method of fruit fly control.

To fill this need, the Malaysian Agricultural Research and Development Institute (MARDI) has formulated an improved bait spray technique which is

Table 5. Mean number of male *Dacus* spp. caught for ten weeks in methyl eugenol or cue-lure baited traps placed at various elevations in primary forest along the Tapah–Cameron Highlands road

Dacus spp	Elevation above sea level (metres)						
	305	579	914	1158	1185	1829	2030
Species attracted to methyl engenol							
D. dorsalis							
mean/two week period	81.4	80.6	222.4	33.0	5.4	1.0	1.2
± s.e.	20.3	15.7	21.9	5.9	1.6	0.5	0.5
species attracted to cue-lure							
D. caudatus							
mean/two week period	1.2	0.2	4.8	5.4	0	0	0
± s.e.	1.0	0.2	2.2	5.4			
D. occipitalis							
mean/two week period	48.6	22.6	3.2	27.8	0.8	0.8	0.6
± s.e.	15.4	6.0	0.7	11.6	0.2	0.4	0.4
D. tau							
mean/two week period	1.8	5.0	47.0	0.8	0.2	0	0
± s.e.	1.1	1.9	9.8	0.4	0.2		

both cheap and easy to use with almost no side-effects. Currently, bait spraying, particularly the use of hydrolysed protein or autolysed yeast, which has been proven to be highly successful in other countries is not widely practised by fruit growers in Malaysia. The major constraint to the widespread use of bait sprays was the high cost of the imported protein hydrolysates, *viz*. Staley PIB-7 and Nu-lure (Vijaysegaran, 1991). With the successful formulation of PROMAR by MARDI, hopefully, this constraint will be overcome.

The bait spray formulated by MARDI is made up of the following constituents:

1. By-product of the brewing process 1500 ml.
2. Synthetic thickening agent 36 ml.
3. Buffer 14 g.

The above mixture is diluted with 3 litres of water to give 4.5 litres of bait spray mixture. An insecticide, malathion, is added at a concentration of 0.1% active ingredient. Four and a half litres (one gallon) of bait spray are then applied to one acre (108 trees) of carambola at weekly intervals. This works out to 40 ml of spray per plant. This small volume of spray covers only a small portion of the foliage and the fruits and flowers are generally not affected. Using such spot applications at weekly intervals fruit fly damage in carambola orchards may be reduced to less than 5% in one season (8 weeks).

Bait spraying should be introduced and promoted for the cultivation of carambola for the following reasons:

- Bait spraying is safe to honeybees and other pollinators which are very important in carambola cultivation.
- As the main constituent of the bait spray, i.e. the soluble protein in the form of autolysed yeast is available locally as an industrial waste-product, the cost of the formulated bait is extremely low.
- By effectively controlling field populations of adult flies, bait spraying enables carambola production without manual bagging or enables extension of the period during which bagging needs to be completed. This would have an indirect effect of enabling carambola to be cultivated over much larger areas than is currently feasible.

Tables 6 and 7 show the comparative costs of cover insecticide spraying versus PROMAR for carambola and mango. It can be seen that PROMAR is definitely superior.

Table 6. Comparative cost ($) of insecticide spraying and PROMAR spraying for 1 hectare of carambola

	Material cost/ spray	Annual material cost	Labour per spray	Annual labour	Total cost
Insecticide	90.00	2250.00 (5 sprays x 5 rounds)	5.0 h	391.00	2641.00
PROMAR	18.35	917.50 (10 sprays x 5 rounds)	0.7 h	104.00	1021.50

Table 7. Comparative cost ($) of insecticide spraying and PROMAR spraying for 1 hectare of mango

	Material cost/ spray	Annual material	Labour per cost	Annual labour spray	Total cost
Insecticide	90.00 (8 sprays)	720.00	5.0 h	125.00	845.00
PROMAR	18.35	146.80 (8 sprays)	0.7 h	17.00	164.00

Behavioural Manipulation of Natural Enemies Using Kairomones

Apanteles plutellae Kurdj. is an important endolarval parasitoid of *P. xylostella*. *Apanteles plutellae* is now well established in cabbage ecosystems but populations are rather low. Efforts are being undertaken to enhance its early season population build-up and efficiency. One way to do this is via mass release. Another potential method is to use semiochemicals such as kairomones to manipulate its host-finding behaviour. The first step taken was to elucidate its host-finding behaviour and ovipositional process.

The host-finding behaviour of *A. plutellae* was studied in the laboratory (Loke, 1986). Results obtained showed that the host-finding and behavioural

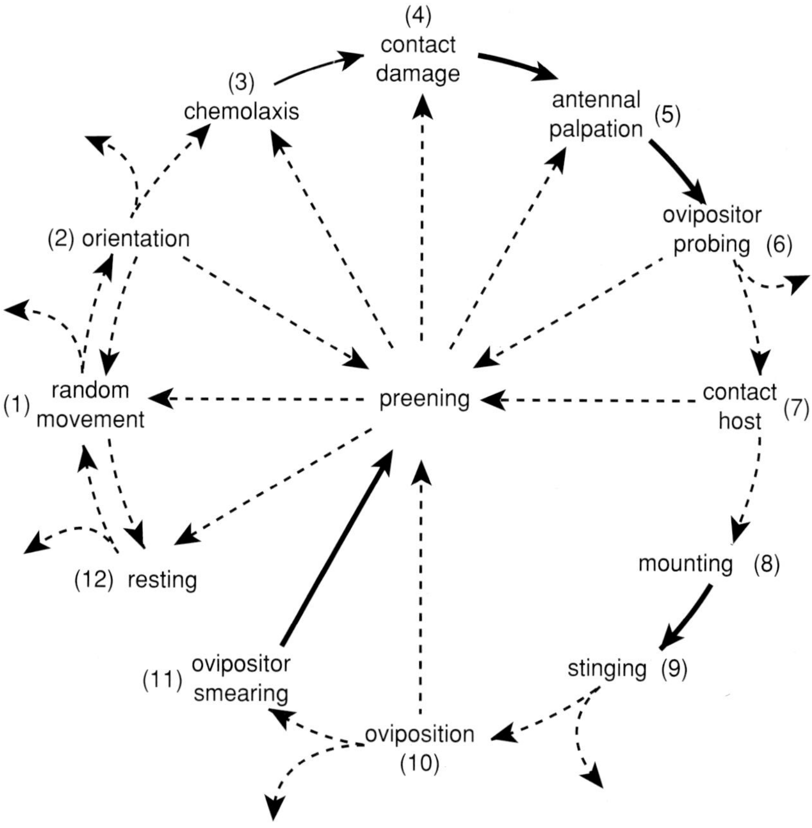

Fig. 2. Behavioural ethogram of the host-finding and ovipositional sequence of *Apanteles plutellae* on DBM-damaged *sawi* plants
Solid arrows – invariable pathways
Dashed arrows – variable/alternate pathways

sequence for oviposition of *A. plutellae* on DBM-damaged *sawi* plants consisted of 12 or more steps as shown in Fig. 2. The entire sequence can be divided into four phases, *viz.* random search, non-random search, oviposition and non-search. The non-random search (steps 4, 5, 6 and 7) and oviposition phase (steps 8, 9 10 and 11) constitute the "find and attack" cycle. Analysis of the cycle showed that there are key points in which fixed-action patterns, mediated by releasing stimuli, are evoked (marked with an asterisk on the ethogram, Fig. 2). Antennal palpation and ovipositor probing are of particular interest because these actions are released primarily by contact chemical cues (kairomones) or chemical cues in conjunction with physical ones, e.g. shot-holes and crevices. The identification of these key points within the "find and attack" cycle is of great importance because the ability to manipulate behaviour at selected points in the host-finding sequence must be achieved if more effective utilisation of parasitoids in pest management programmes is to be realised (Loke *et al.,* 1983).

To understand further the host-finding behaviour of *A. plutellae,* research to study the attraction and retention of *A. plutellae* on *sawi* plants was conducted. Results obtained are shown in Figs. 3 and 4. It can be seen that the highest proportion of released female parasitoids was attracted to and retained on DBM-damaged plants. This was also the case for the proportion of parasitoids actively searching for hosts. The results indicated that a combination of the host (DBM) and plant (*sawi*) factors is important in attracting, retaining and activating searching behaviour of *A. plutellae.*

The behavioural modifications exhibited by *A. plutellae* under different sets of situations, *viz.* DBM-damaged *sawi* plant, artificially-damaged *sawi* plant and no-damaged *sawi* plant were next studied. Results obtained (Table 8) showed that the time spent for various activities varied according to the type of *sawi* plant the parasitoid was placed on. Generally, parasitoids put onto DBM-damaged plants spent most of the time actively searching for hosts to

Table 8. Behavioural modification of *Apanteles plutellae* on three types of *sawi* plant with regard to time spent (s) for various activities

	Activity			
Type of *sawi* plant	Searching & walking	Preening	Resting	Take-off
No damage	40.16	45.96	98.73	86.52
Artificially damaged	71.88	32.55	76.63	71.80
DBM-damage	127.36	47.96	31.47	15.20

DBM = Diamondback moth

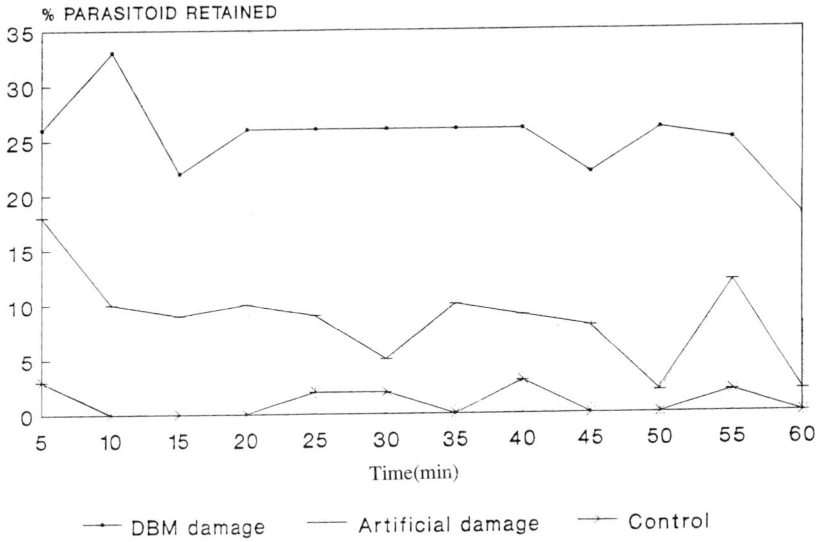

Fig. 3. Attraction of *Apanteles plutellae* to different types of *sawi* plant damage

parasitoids. This concurs with the hypothesis that parasitoids when presented with the right hierarchy of cues will actively go into a "find and attack" cycle.

Further research to study the influence of kairomones on parasitisation of DBM by *A. plutellae* was conducted. This consisted of determining the levels of parasitisation of DBM larvae placed on the three types of *sawi* plants arranged in triangular patterns (choice test situations with rotation to reduce positional bias). Results obtained showed that overall parasitisation was highest in DBM larvae placed on DBM-damaged plants (Table 9). This further

Table 9. Influence of type of *sawi* plant damage on parasitisation of DBM by *A. plutellae*

Replicated	None OP(SP)	Type of damage: Artificial OP(SP)	Type of damage: DBM OP(SP)
I	31.7 (0)	31.7 (1.7)	31.7 (3.3)
II	56.7 (11.7)	60.0 (3.3)	63.3 (10.0)
IV	38.3 (1.7)	35.0 (1.7)	45.0 (3.3)
V	13.3 (0)	11.7 (0)	18.3 (0)
Mean	34.3 (2.7)	37.0 (1.7)	43.3 (3.3)

OP = Overall parasitisation
SP = Superparasitisation
DBM = Diamondback moth

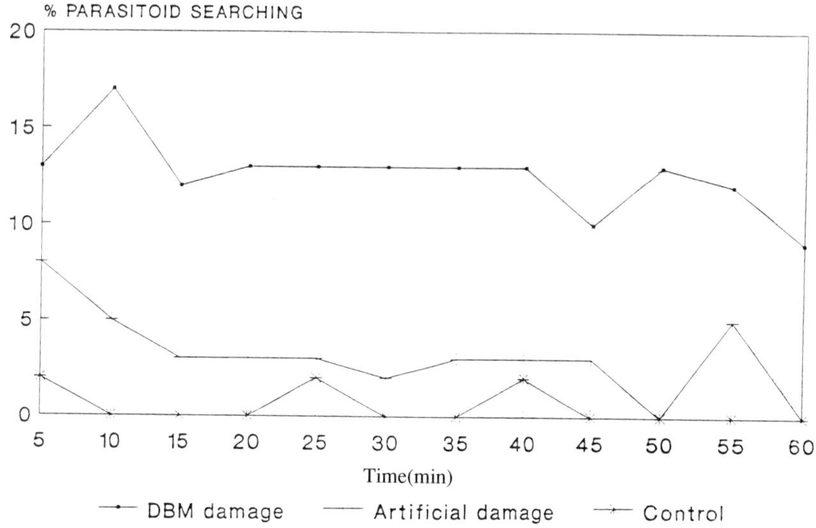

Fig. 4. Searching by *Apanteles plutellae* in relation to different types of *sawi* damage

confirmed that parasitoids are more effective in environments where kairomones are present to provide the "find and attack" stimulation.

Concluding Remarks

Though Malaysian studies and exploitations in the field of semiochemicals have been rather limited, it is generally felt that such novel chemicals offer good potential as tools in insect pest management programmes as they are environmentally friendly and ecologically sound. One big drawback is that these chemicals are constrained by the issue of economics. Sex pheromones, in particular, are so highly specific that commercial production and exploitation may not be seriously pursued by established companies unless the local pest is of widespread importance both nationally and transnationally. BMCs for fruit flies appear to offer greater potential as these are technologically simpler to produce (methyl eugenol is naturally found in several plants in Malaysia), and can be used in a wider range of situations for more than one species of the pests.

It is further felt that local research/academic institutions, especially the universities, should be strongly encouraged to pursue research in this area of insect control. The relevant expertise and facilities are already present in many university laboratories, and concerted work by postgraduate level students can generate much useful information and technology for further development and exploitation in this field.

Acknowledgements

The authors are grateful to the Director-General of MARDI and the Dean of the School of Biological Sciences, Universiti Sains Malaysia for permission to present this paper.

References

Beevor, P.S., Day, R.K. and Mumford, J.D. (1986a) Female sex pheromone of cocoa pod borer moth, *Acrocercrops cramerella:* Identification and field evaluation. In: *Cocoa and Coconuts: Progress and outlook* (Eds. Pushparajah, E. and Chew, P.S.). Incorporated Society of Planters, Kuala Lumpur, pp. 287–292.

Beevor, P.S., Cork, A., Hall, D.R., Nesbitt, B.F. and Mumford, J.D. (1986b) Components of female sex pheromone of cocoa pod borer moth, *Conopomorpha cramerella. J. Chem. Ecol.* 12,1–23.

Chisholm, M.D., Underhill, E.W. and Steek, W.F. (1979) Field trapping of the diamondback moth *Plutella xylostella* using synthetic sex attractants. *Environ. Entomol.* 8, 516–518.

Chow, Y.S., Chin, S.C. and Chien, C.C. (1974) Demonstration of a sex pheromone of the diamondback moth (Lepidoptera: Plutellidae). *Ann. Entomol. Soc. Am.* 67, 510–512.

Chow, Y.S., Tin, Y.M. and Hsu, C.T. (1977) Sex pheromone of the diamondback moth (Lepidoptera: Plutellidae). *Bul. Inst. Zool. Academia Sinica* 16, 99–105.

Ho, S.H., Beevor, P.S. and Mumford, J.D. (1987). A practical approach to the control of the cocoa pod borer moth using synthetic sex pheromone in an integrated system. In: *Management of the Cocoa Pod Borer* (Eds. Ooi, P. A. C. *et al.*). Malaysian Plant Protection Society, Kuala Lumpur, pp. 53–70.

Koshihara, T. and Yamada, H. (1981) Female sex pheromone of the diamondback moth, *Plutella xylostella. JARQ* 15 (1), 22–28

Lim, G.S., Loke, W.H., Chan, H.H. and Syed A. Rahman (1988) Pest Management of Vegetables in Malaysia. Paper presented at the Informal Expert. Consultation on Integrated Pest Management in Major Vegetable Crops in Asia. 14–16 November 1988. FAO, RAPA, Bangkok. 79 pp.

Loke, W.H. (1986) Host-finding behaviour of *Apanteles plutellae* L. *2nd Int. Conf. Plant Protection in the Tropics.* 17–20 March, 1986. Genting Highlands, Pahang, Malaysia.

Loke, W. H., Ashley, T.R. and Sailer, R.I. (1983) Influence of fall armyworm, *Spodoptera frugiperda* (Lepidoptera : Noctuidae), larvae and corn plant damage on host-finding in *Apanteles marginiventris* (Hymenoptera: Braconidae). *Environ. Entomol.* 12, 911–915.

Loke, W.H., Lim, G.S., Syed, A. R., Abdul Aziz, A.M., Rani, M.Y., Md. Jusoh, M. and Cheah, U.B. (1990) IPM on the Diamondback Moth in Malaysia: Development, Implementation and Impact. 2nd Int. Workshop on the Management of Diamondback Moth and other Crucifer Pests. 10–14 December 1990, AVRDC, Shanhua, Taiwan.

Mumford, J.D. (1986) Control of the cocoa pod borer (*Acrocerops cramerella*): A critical review. In: *Cocoa and Coconuts: Progress and outlook* (Eds. Pushparajah, E. and Chew, P.S.). Incorporated Society of Planters, Kuala Lumpur, pp. 277–286.

Nishida, R., Tan, K.H., Serit, M., Lajis, N.H., Sukaro, A . M., Takahashi S. and Fukami,

H. (1988) Accumulation of phenylpropanoids in the rectal glands of males of the Oriental fruit fly, *Dacus dorsalis*. *Experientia* 44, 534–536

Nishida, R., Tan, K.H., Serit, M., Lajis, N.H., Sukaro, A . M., Takahashi S. and Fukami, H. (1990)Volatile components of male rectal glands of the melon fly *Dacus cucurbitae* Coquillett (Diptera : Tephritidae). *Appl. Ent. Zool.* 25 (1), 105–112.

Tamaki, Y., Kawasaki, K., Yamada, H., Koshihara, T, Osaki, N., Ando, T., Yashida, S. and Kakinohama, H. (1977) Z–11–hexadecenal and Z–11–hexadecenyl acetate: sex pheromone components of the diamondback moth (Lepidoptera: Plutellidae). *Appl. Ent. Zool.* 12, 208–10.

Tan, K.H. (1985) Estimation of native populations of male *Dacus* spp. by Jolly's stochastic method using a new designed attractant trap in a village ecosystem. *J. Plant Protection in the Tropics* 2, 87–95.

Tan, K.H. (1987) Surveying and monitoring techniques for fruit flies. Paper presented in ASEAN PLANTI Meeting on Suitability of Different Plant Quarantine Treatments on Mangoes. 13–15 May 1987, Penang.

Tan, K.H. and Jaal, Z. (1986) Comparisons of male adult population densities of the Oriental and Artocarpus fruit flies, *Dacus* spp. (Diptera: Tephritidae) in two nearby villages in Penang, Malaysia. *Res. Popul. Ecol.* 28, 85–89.

Tan, K.H. and Lee, S.L. (1982) Species diversity and abundance of *Dacus* (Diptera: Tephritidae) in five ecosystems of Penang, West Malaysia. *Bull. Ent. Res.* 72, 709–716.

Tan, K.H. and Loke, W.H. (1985) Insect growth regulators and semiochemicals in integrated pest management. In: *Integrated Pest Management* (Eds. Lee, B. S., Loke, W.H. and Heong, K.L.). Malaysian Plant Protection Society, Kuala Lumpur, pp. 105–131.

Vijaysegaran, S. (1985) Management of fruit flies. In: *Integrated Pest Management in Malaysia* (Eds. Lee, B.S., Loke, W.H. and Heong, K.L.). Malaysian Plant Protection Society, pp. 231–254.

Vijaysegaran, S. (1991) The current situation on fruit flies in Peninsular Malaysia. In: *Fruit Flies in the Tropics* (Eds. Vijaysegaran, S. and Ibrahim, A.G.). Malaysian Agricultural Research and Development Institute and Malaysian Plant Protection Society, Kuala Lumpur, pp. 125–139.

Chapter ten:
Pesticides for the Year 2000: Mycochemicals and Botanicals

M.S.J. Simmonds[1], H.C. Evans[2] and W.M. Blaney[3]

[1] *Jodrell Laboratory, Royal Botanic Gardens, Kew, Richmond, Surrey, TW9 3DS, U.K.*

[2] *CAB International Institute of Biological Control, Silwood Park, Ascot, Berkshire SL5 7TA, U.K.*

[3] *Department of Biology, Birkbeck College, University of London, Malet Street, London, WC1E 7HX, U.K.*

Summary

The next generation of pesticides for use in the year 2000 will most likely be based on secondary metabolites isolated from fungi and higher plants. We review the use of fungal metabolites as pesticides, here termed mycochemicals. Examples are given of mycochemicals used as fungicides, bactericides, insecticides, acaricides and herbicides. Pesticides derived from higher plants are often referred to as "botanicals". We have surveyed 59 higher plant families and identified 188 genera with potent anti-insect activity. Some of these genera, such as *Chrysanthemum* and *Azadirachta*, contain compounds with known anti-insect activity, whereas others have been shown to affect insects, but the compounds responsible for that activity have not yet been identified. The potential uses of mycochemicals and botanicals are discussed.

Introduction

The scientific and economic challenges that face pest management programmes in the year 2000 are numerous, especially in the developing countries. At the moment about 80% of the world's population is in the developing countries where there is an urgent need to increase agricultural productivity in order to feed the fast-growing population. However, many of the methods currently used to control pests are health hazards and because of the indiscriminate use of broad-spectrum pesticides the environment has been and is being damaged. Therefore, it is highly desirable that alternative pest control methods are developed to reduce pest populations without harming people, their livestock or the environment.

Many farming institutes and agrochemical companies are looking again at natural products for novel leads to provide farmers in both the Western World and Third World with alternatives to the present broad-spectrum synthetic pesticides. Within this review we will focus on the identification and potential use of compounds from higher plants (botanicals) that can be used as antifeedant or insect growth regulators (IGRs) and of compounds from fungi (mycochemicals) that might have a wider spectrum of activity.

Mycochemical Pesticides (Mycochemicals)

The true fungi or Eumycota have been actively exploited as sources of medically important metabolites since the discovery and subsequent manufacture of penicillin more than 50 years ago. Since this time, most of the microbial metabolites with commercial potential have been found in the Actinomycetales, a group of micro-organisms producing coenocytic mycelium and having a similar in vitro habit to fungi, but which are often classified within the bacteria (Hawksworth *et al.*, 1983). Because of their affinities with fungi, and the fact that many of their derived products have mycological connotations (myco–; –mycin), they are included within the present review.

Metabolites of microbial origin are usually referred to as antibiotics and were developed specifically for medical purposes. The first attempts to utilise these microbial products in agriculture were directed against plant pathogens (Dekker, 1963), and involved the use of streptomycin, tetracycline and chloramphenicol to control plant diseases of bacterial origin in a manner analogous to the treatment of human bacterial diseases. In theory, they possess all the attributes necessary for the eradication of systemic pathogens, being highly selective and easily absorbed by plant tissue (including seeds), even into actively growing meristems. The extensive literature on this subject has been reviewed by Dekker (1963) and more recently by Misato & Yamaguchi (1984). The latter authors have concluded that, despite the screening of many fungal products for plant disease control, the practical problems, in particular their high phytotoxicity and low stability under field conditions, have proven to be almost insurmountable. Moreover, they were competing with synthetic pesticides with more consistent or reliable activity and lower cost. A further disadvantage was the concern that the widespread use of antibiotics could lead to the development of resistance in bacteria of medical importance. Despite these problems, research continued in Japan, where the major expertise in the use of microbial products in agriculture is currently to be found. This research now seems to be paying dividends as the pressures increase for easily degradable pesticides with greater selectivity and less harmful effects on non-target organisms. Many conventional chemical pesticides, even though their acute toxicities are low, are highly persistent in nature since they are "un-natural" compounds and few micro–organisms have evolved enzyme systems to degrade them. This results in a build-up in food chains and an eventual hazard to mammals. Paradoxically, the biological instability of microbial products, which was one of the main reasons why they were rejected for agricultural use, may now become a positive attribute. We conclude, therefore, that mycochemical products have the potential to contribute to a significant part of the pesticide market in the next century, and we review here their present and future prospects.

Before defining the source and function of microbial metabolites, the problems of terminology need to be addressed. Misato & Yamaguchi (1984) rejected the term "agricultural antibiotic" and proposed the rather unwieldy "microbio-origin pesticides". Properly defined, antibiotics are chemicals of microbial origin which inhibit the growth of other micro-organisms: hence the term cannot be applied to the field of general pest control (plant pathogens, arthropods, weeds). If the term biopesticide, or more specifically mycopesticide, is used more confusion is generated since these terms are based on the presence of a living or renewable micro-organism within the product (see J. Waage, this volume), rather than a metabolite derived from a micro-organism, either naturally or synthetically. Since the processes governing their manufacture, use and mode of action are so different, it is considered highly desirable to distinguish between these two types of natural pesticides. The term "mycochemical pesticide" is proposed here as a convenient means of distinguishing the pesticides that are based on micro-organism-derived metabolites from either the biopesticides or the synthetic chemical pesticides, which all fall within the general description of agricultural pesticides. Further subdivisions are necessary depending on the origin of the metabolite and the target pest involved. In the present context these are listed as: mycochemical fungicides or bactericides (active against plant pathogens); mycochemical insecticides or acaricides (active against arthropods); and mycochemical herbicides (active against weeds).

1. Mycochemical Fungicides or Bactericides

As previously mentioned, Dekker (1963) evaluated the agricultural potential of a number of these products which had originally been developed for use against human rather than plant diseases. Most had been isolated from various species of the genus *Streptomyces* (Actinomycetales), as for example streptomycin, which proved to be active against a wide range of bacterial plant diseases, particularly fireblight (*Erwinia amylovora*) of rosaceous tree crops and walnut blight (*Xanthomonas juglandis*). Streptomycin sulphate sprays have been routinely employed in the USA in fireblight and walnut blight control programmes, being marketed under the names Agrimycin and Phytomycin. Other *Streptomyces*-derived metabolites, with activity against bacterial plant pathogens, include the tetracyclines, especially oxytetracycline (Terramycin) which has been widely used in conjunction with streptomycin to avoid the possibility of resistant pathogen strains developing. Antifungal activity has been shown by cycloheximide (Actidione), another metabolite of *Streptomyces* origin, which has been successfully employed against a wide range of fungal diseases, including both powdery and downy mildews, once problems with phytotoxicity and poor translocation had been resolved. Griseofulvin, derived from *Penicillium griseofulvum*, has strong fungistatic properties and inhibits the growth of a number of plant pathogens, notably *Botrytis* diseases. Brian (1960) concluded, however, that this mycochemical offered no better control than conventional chemical fungicides and was more expensive to produce. Around this period, Japanese workers began to investigate the potential of *Streptomyces* as sources of new mycochemicals for plant disease control specifically against their major rice pathogens. This resulted in the development of a number of products active against blast disease (*Pyricularia oryzae*) and sheath blight (*Rhizoctonia solani*). The most successful has been blasticidin S which can eradicate the blast pathogen from infected tissues. However, despite its favourable environmental characteristics of being easily degraded by soil organisms and sunlight and absence of phytotoxicity, the initial products proved to have some mammalian toxicity, causing conjunctivitis. Later formulations have virtually eliminated this problem. Other mycochemicals active against rice diseases have been marketed, as for example kasugamycin, the polyoxins and validamycins. These all affect the growth of fungi by disrupting the synthesis of cell wall chitin with little or no effect on mammalian tissues.

2. Mycochemical Insecticides or Acaricides

Entomopathogenic fungi, in their role as obligate pathogens of arthropods, produce powerful insecticides which paralyse or kill their hosts, and bactericides or fungicides which preserve the cadaver from saprophytic competitors. However, as Roberts (1981) concluded "virtually no searches have been conducted for toxins in fungus-infected insects". The few toxins that have been investigated are from the ubiquitous entomopathogens, *Beauveria bassiana* and *Metarrhizium anisopliae*, and have been identified as depsipeptides, such as destruxin and beauverin, which appear to disrupt the sites of energy production within the mitochondria and endoplasmic reticulum of the host. The possibility of mammalian toxicity, therefore, has limited their development as mycochemical pesticides. In fact, the most fruitful use of mycochemicals produced by this group of fungi has been in medicine rather than in pest control, where their use as antibiotics has superseded the development of penicillin. There is abundant evidence to show that a number of ancient civilisations were indirectly exploiting metabolites from entomopathogenic fungi for medicinal purposes (Milner, 1987; Samson *et al.*, 1988), whilet cordycepin (extracted from *Cordyceps militaris*) and cyclosporin (extracted from *Tolypocladium inflatum*) are now being employed in transplant surgery to suppress invading micro-organisms and the body's immune response. However, within the present context of pest control other *Streptomyces*-derived compounds, the avermectins and milbemycins, have stimulated great interest in mycochemicals for both insect and mite control. The avermectins consist of eight closely related compounds of which Avermectin B1a is the major component, and is considered to have more potential as a pest control agent than the milbemycins. Avermectins function by disrupting the pest's nervous system by acting on the GABA-receptor-chloride ion channels which results in paralysis, as signal transmission is inhibited at the neuromuscular junction. Because the avermectins are unable to cross the blood–brain barrier in mammals they do not affect the GABA-ergic neurones which in mammals are restricted to the central nervous system. Thus, they are selective to arthropods, and non-toxic to mammals, and because they act on a novel site they can be used to control arthropod pests which have become resistant to conventional insecticides. As an added bonus, the avermectins have a low solubility in water and are rapidly degraded in the soil, presenting negligible environmental risks. One such avermectin product, Affiom, has been specifically formulated for use against fire ants and acts by inhibiting reproduction in the queen ant. Analogues are currently being developed to expand its spectrum of activity to include other agricultural pests and success has been achieved against citrus and pear rust mites, spider mites, leaf miners, diamondback moth (*Plutella xylostella*) and the Colorado potato beetle (*Leptinotarsa decemlineata*).

The evidence clearly indicates that the continuing screening of specialist groups of fungi will yield further novel mycochemicals with insecticidal and acaricidal activity, as well as compounds with pharmaceutical potential, the antibiotics *sensu strictu*. Other genera, such as *Aspergillus* and *Penicillium*, are known to produce lethal insecticidal toxins (the so-called mycotoxins) but "The use of mycotoxins for insect and mite control appears to be impractical considering their effects in higher animals. The potential for other insecticidal materials from micro-organisms is considerable" (Wright *et al.*, 1982).

3. Mycochemical Herbicides

During the course of screening metabolites from *Streptomyces,* a number of compounds were found to possess significant phytotoxic activity a negative attribute of course for mycochemical fungicides, bactericides, insecticides and acaricides but highly positive for weed control. Among these mycochemicals with herbicidal activity are cycloheximide, anisomycin, herbimycin and bialaphos which have been or are being marketed as herbicides in Japan (Misato & Yamaguchi, 1984). An increasing number of phytotoxic compounds with herbicidal potential are being reported in plant pathogenic fungi, for example in the genera *Cercospora* (Blaney C.L. *et al.*, 1988); *Bipolaris* (Sugawara *et al.*, 1985) and *Curvularia* (Ellison & Evans, 1990). Although many of these mycochemicals are non-specific, the possibility exists that they can be used in integrated systems for weed control, especially since they pose no threat to the environment, the major constraint to the future of chemical pesticides.

Botanical Pesticides (Botanicals)

Pesticide activity in plants has chiefly been sought for in the form of antifeedants or repellents to control insect pests. Very little attention has been directed towards a search for plant-derived compounds with fungicidal or herbicidal activity. The remainder of this chapter reflects this trend and deals almost exclusively with botanical insecticides.

Botanical Insecticides from Higher Plants: First to Fourth Generation Insecticides

The "first generation" insecticides include plant extracts and plant-derived compounds, often referred to as "botanicals", such as Derris, pyrethrum, rotenone, nicotine, sabadilla, physostigmine and quassin, as well as compounds like lead arsenate and calcium arsenate. Some of these compounds formed the basis for the synthetic "second generation" insecticides. For example, pyrethrum from chrysanthemum flowers gave rise to the synthetic pyrethroids, and physostigmine from *Physostigma venenosa* gave rise to the structurally similar synthetic carbamates.

Since the development of DDT, more that 50 years ago, over 200 "second generation" insecticides have been marketed. Most of them are nerve poisons and have the disadvantage of killing not only target insects but also non-target insects which are neutral, or even beneficial to man. Many of these insecticides have enabled farmers to increase crop yields and have helped bodies such as the World Health Organisation to control vectors of diseases which otherwise would be very detrimental to man and his livestock. Although there is no doubt that many of them have helped to control pests in the past, and are still useful in that context, their extensive and sometimes indiscriminate use has led to environmental repercussions.

For example, in many areas the situation has arisen in which farmers, in order to maintain crops free from pests, increase the frequency with which the crops are sprayed. This often results in the target insects developing resistance, whereupon more powerful insecticides are used, which often cause the complete destruction of the insect's natural enemies as well as the development of "super-resistant" pest insects. Such insects show resistance to a broad spectrum of insecticides. Clearly pest management programmes in the year 2000 will have to re–evaluate the use of these insecticides. However, until some realistic alternatives are found, many of the currently used insecticides will still play an important role in pest management programmes.

In the last decade there has been a move away from the "second generation" synthetic insecticides to the "third generation" insecticides, which include insect growth regulators such as the precocenes and phytoecdysones. Although these compounds are generally fairly selective and environmentally safe, they act slowly and do not prevent insects from destroying a large amount of a crop before dying. The next or "fourth generation" of insecticides include the so-called behaviour-modifying compounds, such as the antifeedants and deterrents.

Within each generation of insecticides there are examples of products that are derived from natural plant compounds and at the moment there is a trend to return to the "botanicals", not only as sources of novel compounds but as products in their own right. Many of the plant chemicals that could have potential as botanical pesticides are the secondary plant metabolites or allelochemicals, which serve no apparent function in the fundamental physiological or biochemical processes of the plant. These chemicals include the terpenoids, flavonoids, phenolics, amines and alkaloids. Advances in techniques for isolation and analysis have led to the identification of many new compounds. However, our knowledge about how these compounds affect insects and how to develop them as pest control agents is only beginning to advance.

Botanical pesticides, although natural, should not be thought of as "safe". However, unlike many synthetic pesticides, natural compounds are usually biodegradable and are potentially environmentally safer than many of the synthetic products on the market today. In order to evaluate fully the potential of a botanical pesticide, it is desirable that the plant material should be readily available or easily cultivated, the extraction and isolation of the active material should not be costly or, alternatively, it should be possible to establish synthetic procedures to produce similar but safe and biodegradable compounds (Jacobson, 1989).

Many agrichemical companies have taken up the search for molecules with novel modes of action, especially as there has been a decline recently in the number of novel synthetic insecticides that have been successfully developed for marketing. In the period from 1961 to 1970, 16 new synthetic insecticides were placed on the market (Berenbaum, 1989) but in the last 10 years only four novel insecticides have been marketed. This decrease in the manufacture of new products has been paralleled by an increase in the number of insect species that have developed resistance to those insecticides available to control them (Engel, *et al.*, 1990). Thus the agrochemical industries have returned to the plant kingdom to provide them with new leads and ideas. We predict that pest management programmes in the year 2000 will benefit from, on the one hand an increase in the use of new technology to identify compounds from plants, especially from those plants that have been used for centuries by farmers in Africa, Asia and the Americas to control pests, and on the other hand an increase in the use of crude plant products such as oils, extracts and cakes.

There are a variety of ways in which botanical insecticides could be used; as crude formulations or semipurified or purified extracts that contain compounds active as antifeedants, ovicides, repellents or growth disrupters.

Evaluation of Plant Material for Insecticidal Activity

Since 1979 our research efforts, and those of other laboratories, have been directed towards the evaluation of the insecticidal potential of extracts and compounds isolated from higher plants. We have employed different approaches to select the plants to be investigated. For example, we have investigated the active components in "botanicals" used to control pests in various parts of the world or we have tested speculatively material collected on expeditions. Once activity has been found, we undertake a chemotaxonomical study to find species within the same genus or related genera that might have higher concentrations of an active compound or contain related compounds that enable us to undertake structure–function studies (Blaney *et al.*, 1987, 1988; Simmonds *et al.*, 1985, 1989, 1990 a,b).

Acanthaceae

Justicia spp. contain compounds that have insecticidal activity against *Spodoptera litura* as well as antifeedant activity against a range of stored product pests (see Grainge & Ahmed, 1988).

Aceraceae

Acer saccharium contains compounds that deter *Leptinotarsa decemlineata* and *Manduca sexta* from feeding (see Grainge & Ahmed, 1988).

Amaranthaceae

Gomphrena globosa and species of *Amaranthus* contain compounds with antifeedant activity against a range of insects including *Leptinotarsa decemlineata* (Jermy, 1966). *Achyranthes* and *Cyathula* contain ecdysteroids that could act as IGRs (Camps, 1991).

Survey of Plant Families that Contain Compounds with Insecticidal Activity (Table 1)

Table 1. Plant genera with potent anti-insect activity

Family	Genus	Insect activity	Reference
Acanthaceae	*Justicia*	antifeedant	Grainge & Ahmed, 1988
Aceraceae	*Acer*	antifeedant: IGR	Grainge & Ahmed, 1988
Amaranthaceae	*Achyranthes*	antifeedant: IGR	Camps, 1991
	Amaranthus	antifeedant	Jermy, 1966
	Cyathula	antifeedant	Camps, 1991
	Gomphrena globosa	antifeedant	Jermy, 1966
Amaryllidaceae	*Allium*	antifeedant	Harmatha *et al.*, 1987
	Crinum	antifeedant	Simmonds & Blaney, (unpublished)
	Hymenocallis	antifeedant	Simmonds & Blaney, (unpublished)
Annonaceae	*Annona*	IGR: antifeedant	Jacobson, 1989.
Apocynaceae	*Acokanthera*	antifeedant	Grainge & Ahmed, 1988
	Apocynum	IGR: antifeedant	Grainge & Ahmed, 1988
	Haplophyton	insecticidal	Bell *et al.*, 1990
	Holarrhena	IGR: antifeedant	Thappa *et al.*, 1989
	Mandevilla	antifeedant	Grainge & Ahmed, 1988
	Nerium oleander	antifeedant	Grainge & Ahmed, 1988
Araceae	*Acorus*	antifeedant: IGR	Koul *et al.*, 1990
Aristolochiaceae	*Aristolochia*	antifeedant	Grainge & Ahmed, 1988
Asclepiadaceae	*Asclepias*	antifeedant: insecticidal	Harborne, 1991
	Calotropis	IGR: antifeedant	Grainge & Ahmed, 1988
Asteraceae	*Ageratum*	insecticidal: IGR	Bowers, 1977
	Arctium	antifeedant	Grainge & Ahmed, 1988
	Artemisia	antifeedant	Yano, 1987
	Baccharis	antifeedant	Grainge & Ahmed, 1988
	Bellis perennis	antifeedant	Grainge & Ahmed, 1988
	Carlina acaulis	IGR: antifeedant	Grainge & Ahmed, 1988
	Centaurea	antifeedant	Grainge & Ahmed, 1988
	Chrysan-themum	insecticidal	Elliott, 1977

Table 1. *Continued*

	Cichorium intybus	antifeedant	Rees & Harborne, 1985
	Cirsium	antifeedant	Grainge & Ahmed, 1988
	Dahlia	antifeedant	Grainge & Ahmed, 1988
	Encelia	antifeedant	Isman, 1989
	Erigeron	antifeedant	Grainge & Ahmed, 1988
	Eupatorium	antifeedant	Grainge & Ahmed, 1988
	Helianthus annuus	antifeedant	Gershenzon *et al.*, 1985
	Momordica	antifeedant	Grainge & Ahmed, 1988
	Onopordum acanthium	antifeedant	Grainge & Ahmed, 1988
	Schkuria pinnata	antifeedant	Grainge & Ahmed, 1988
	Senecio	antifeedant	Grainge & Ahmed, 1988
	Solidago	antifeedant	Grainge & Ahmed, 1988
	Tagetes	insecticidal	Grainge & Ahmed, 1988
	Taraxacum officinale	antifeedant	Grainge & Ahmed, 1988
	Vernonia	antifeedant	Rodriguez, 1985
	Xanthium	IGR	Kawazu *et al.*, 1979
Balsaminaceae	*Impatiens*	antifeedant	Grainge & Ahmed, 1988
Bignoniaceae	*Catalpa*	antifeedant	van Beek & de Groot, 1986
Boraginaceae	*Heliotropium*	antifeedant	Grainge & Ahmed, 1988
Caesalpiniaceae	*Cassia*	insecticidal	Grainge & Ahmed, 1988
Canellaceae	*Canella*	insecticidal	Grainge & Ahmed, 1988
	Warburgia	antifeedant	Blaney *et al.*, 1987
Cannaceae	*Canna*	antifeedant	Grainge & Ahmed, 1988
Celastraceae	*Celastrus*	antifeedant	Chiu, 1989
	Maytenus	antifeedant	Delle Monache *et al.*, 1984
	Tripterygium	antifeedant	Swingle *et al.*, 1941
Chenopodiaceae	*Anabasis*	insecticidal	Bell *et al.*, 1990
	Beta	antifeedant	Grainge & Ahmed, 1988
	Chenopodium	antifeedant	Grainge & Ahmed, 1988
	Spinacia	IGR	Simmonds & Blanet (unpublished)
Clusiaceae	*Mammea americana*	antifeedant	Grainge & Ahmed, 1988
Combretaceae	*Combretum*	antifeedant	Grainge & Ahmed, 1988

Table 1. *Continued*

Compositae	*Ageratina adenophora*	antifeedant	Proksch *et al.*, 1990
	Cichorium intybus	antifeedant	Rees & Harborne, 1985
	Eupatorium	antifeedant	Grainge & Ahmed, 1988
	Helenium amarum	antifeedant	Arnason *et al.*,1987
	Helianthus annuus	antifeedant	Ellinger *et al.*, 1976
	Melampodium divaricatum	antifeedant	Harborne, 1991
	Polymnia udevali	antifeedant	van Beek & de Groot, 1986
	Tithonia diversifolia	antifeedant	van Beek & de Groot, 1986
	Vernonia	antifeedant	Burnett *et al.*, 1978
	Wedelia	antifeedant	Simmonds *et al.* (unpublished)
Convolvulaceae	*Calystegia*	antifeedant	Grainge & Ahmed, 1988
	Convolvulus arvensis	antifeedant	Grainge & Ahmed, 1988
	Ipomoea purpurea	antifeedant	Grainge & Ahmed, 1988
Crucifereae	*Brassica*	antifeedant	Simmonds *et al.* (unpublished)
	Cheranthus cheiri	antifeedant	Grainge & Ahmed, 1988
	Diethyrea wislizenii	antifeedant	Powell *et al.*, 1991
	Erysimum	antifeedant	Grainge & Ahmed, 1988
	Iberia	antifeedant	Grainge & Ahmed, 1988
	Lepidium	antifeedant	Grainge & Ahmed, 1988
Cucurbitaceae	*Momordica charantia*	antifeedant	Chandravadana, 1987
Cycadaceae	*Cycas revoluta*	antifeedant	Grainge & Ahmed, 1988
Dioscoreaceae	*Dioscorea*	antifeedant	Grainge & Ahmed, 1988
Euphorbiaceae	*Alchornea triplinervia*	antifeedant	Miles *et al.*, 1985
	Aleurites fordii	antifeedant	Grainge & Ahmed, 1988
	Croton aromaticus	insecticide	Bandara *et al.*, 1987
	Euphorbia	antifeedant	Grainge & Ahmed, 1988

Table 1. *Continued*

Fagaceae	*Fagus grandifolia*	antifeedant	Grainge & Ahmed, 1988
Geraniaceae	*Geranium*	IGR	Grainge & Ahmed, 1988
	Pelargonium	antifeedant	Grainge & Ahmed, 1988
Ginkgoaceae	*Ginkgo biloba*	antifeedant	Grainge & Ahmed, 1988
Iridaceae	*Iris*	antifeedant	Grainge & Ahmed, 1988
Juglandaceae	*Juglans*	antifeedant	Norris, 1986
Labiatae	*Ajuga*	antifeedant: IGR	Blaney *et al.*, 1988
	Lamium maculatum	antifeedant	Grainge & Ahmed, 1988
	Lavandula angustifolia	antifeedant	Grainge & Ahmed, 1988
	Mentha	antifeedant	Simmonds *et al* . (unpublished)
	Nepeta	pheromone	Dawson *et al.*, 1990
	Ocimum	IGR	Grainge & Ahmed, 1988
	Plectranthus	insecticidal	Bell *et al.*, 1990
	Salvia	antifeedant	Grainge & Ahmed, 1988
	Scutellaria	antifeedant	Cole *et al.*, 1990
	Teucrium	antifeedant	Simmonds *et al.*, 1989
Leguminosae	*Acacia*	antifeedant: IGR	Bell, 1977
	Aglaia	antifeedant	Mikolajczak *et al.*, 1989
	Albizia	antifeedant	Grainge & Ahmed, 1988
	Amorpha	antifeedant	Grainge & Ahmed, 1988
	Astragalus	antifeedant	Bell, 1977
	Calliandra	antifeedant	Romeo & Simmonds, 1989
	Copaifera	antifeedant	Langeheim *et al.*, 1986
	Derris	insecticidal	Grainge & Ahmed, 1988
	Dysoxylum	antifeedant	Mikolajczak *et al.*, 1988
	Glycine max	antifeedant	Grainge & Ahmed, 1988
	Hymenaea	antifeedant	Langenheim *et al.*, 1986
	Medicago	antifeedant	Sutherland, 1980
	Mundulea suberosa	antifeedant	Grainge & Ahmed, 1988
	Lonchocarpus	antifeedant: insecticidal	Simmonds *et al.*, 1990
	Lotus	antifeedant	Sutherland *et al.*,1980
	Lupinus	antifeedant	Sutherland *et al.*,1980
	Melilotus	antifeedant	Grainge & Ahmed, 1988
	Robinia	antifeedant	Grainge & Ahmed, 1988
	Tephrosia	antifeedant	Simmonds *et al.*, 1990b
	Willardia mexicana	antifeedant	Grainge & Ahmed, 1988

Table 1. *Continued*

Liliaceae	*Lilium longiflorum*	antifeedant	Grainge & Ahmed, 1988
	Schoenocaulon	antifeedant	Grainge & Ahmed, 1988
	Urginea maritima	antifeedant	Grainge & Ahmed, 1988
Magnoliaceae	*Liciodendron tulipifera*	antifeedant	Grainge & Ahmed, 1988
Malvaceae	*Gossypium hirsutum*	antifeedant	Jacobson, 1989
	Hibiscus	antifeedant	Grainge & Ahmed, 1988
	Sphaeralcea angustifolia	antifeedant	Grainge & Ahmed, 1988
Meliaceae	*Aglaia cordata*	antifeedant	Mikolajczak *et al.*, 1989
	Azadirachta indica	IGR: antifeedant	Simmonds *et al.*, 1990c; Blaney *et al.*, 1990
	Carapa	antifeedant	Grainge & Ahmed, 1988
	Cedrela odorata	antifeedant	Jacobson, 1989
	Chickrassia tabularis	antifeedant	Grainge & Ahmed, 1988
	Dysoxylum	antifeedant	Grainge & Ahmed, 1988
	Lansium domesticum	antifeedant	Mikolajczak *et al.*, 1989
	Melia	antifeedant	Chiu, 1989
	Trichilia	antifeedant	Nakatani *et al.*, 1981
	Xlyocarpus	antifeedant	Grainge & Ahmed, 1988
Melianthaceae	*Bersama pauliniodes*	insecticidal; antifeedant	Bell *et al.*, 1990
Myrtaceae	*Eucalyptus cinerea*	antifeedant	Grainge & Ahmed, 1988
Onagraceae	*Epilobium*	antifeedant	Grainge & Ahmed, 1988
Papaveraceae	*Chelidonium majus*	antifeedant	Grainge & Ahmed, 1988
Piperaceae	*Piper*	antifeedant: insecticide	Miyakado *et al.*, 1989
Plantaginaceae	*Plantago*	antifeedant	Grainge & Ahmed, 1988
Plumbaginaceae	*Plumbago zeylanica*	antifeedant	Hassanali & Lwande, 1989
Poaceae	*Bothrichloa ischaemum*	antifeedant	Grainge & Ahmed, 1988

Table 1. *Continued*

	Echinochloa crus-galli	antifeedant	Grainge & Ahmed, 1988
	Lolium	antifeedant	Grainge & Ahmed, 1988
Podocarpaceae	*Podocarpus gracilior*	antifeedant	van Beek & de Groot, 1986
Polygonaceae	*Polygonum*	antifeedant	Blaney *et al.*, 1987
	Rheum	antifeedant	Grainge & Ahmed, 1988
	Rumex	antifeedant	Grainge & Ahmed, 1988
Ranunculaceae	*Clematis*	antifeedant	Grainge & Ahmed, 1988
Rosaceae	*Fragaria ananassa*	antifeedant	Grainge & Ahmed, 1988
	Purshia tridentata	antifeedant	Grainge & Ahmed, 1988
Rutaceae	*Atalantia racemosa*	antifeedant	Grainge & Ahmed, 1988
	Clausena aniseta	antifeedant	Grainge & Ahmed, 1988
	Fagara chalbea	antifeedant	Grainge & Ahmed, 1988
	Zanthoxylum holstii	antifeedant	Grainge & Ahmed, 1988
Saxifragaceae	*Heuchera sanguinea*	antifeedant	Grainge & Ahmed, 1988
Simaroubaceae	*Quassia*	antifeedant	Bell *et al.*, 1990
Solanaceae	*Lycopersicon solanum*	antifeedant	Grainge & Ahmed, 1988
	Physalis	antifeedant	Grainge & Ahmed, 1988
	Nicotiana	insecticidal	Grainge & Ahmed, 1988
	Atropa	insecticidal	Murdock *et al.*, 1985
	Datura metel	insecticidal	Grainge & Ahmed, 1988
	Hyoscyamus datura	insecticidal	Grainge & Ahmed, 1988
	Petunia	antifeedant	Grainge & Ahmed, 1988
	Solanum	deterrent	Grainge & Ahmed, 1988
	Nicandra physalodes	antifeedant	Grainge & Ahmed, 1988
Umbelliferae	*Angelica silvestris*	antifeedant	van Beek & de Groot, 1986
Verbenaceae	*Stachytarpheta mutabilis*	antifeedant	van Beek & de Groot, 1986

Table 1. *Continued*

	Callicarpa	antifeedant	Grainge & Ahmed, 1988
	Clerodendron	antifeedant	van Beek & de Groot, 1986
	Caryopteris divaricata	antifeedant	Grainge & Ahmed, 1988
	Lantana camara	antifeedant	Grainge & Ahmed, 1988
	Vitex	IGR	Kubo *et al.*, 1990
Vitaceae	*Parthenocissus*	antifeedant	Grainge & Ahmed, 1988
Winteraceae	*Drimys lanceolata*	antifeedant	Grainge & Ahmed, 1988

Amaryllidaceae

Many species of *Allium* have been shown to have insecticidal and antifeedant activity against a wide range of insects. Harmatha *et al.* (1987) have shown that the high levels (0.2–0.4% dry weight) of a saponin in the flowers of *Allium porrum* deter the leek moth from feeding. Species of *Crinum* and *Hymenocallis littoralis* contain alkaloids that have antifeedant activity against *Spodoptera exempta* at 100 p.p.m. (Simmonds & Blaney, unpublished results) and *Schistocerca gregaria* (see Grainge & Ahmed, 1988).

Annonaceae

Ethanolic extracts of the fruits of species of *Annona* (*A. reticulata, A. glaba* and *A. purpurea*) contain several kaurene diterpenes which may be responsible for the IGR effect of these plants against *Diabrotica* (Jacobson, 1989).

Apocynaceae

Hollarrhena antidysenterica contains a steroidal alkaloid, conessine which is a potent growth disrupter, sterilant and antifeedant (Thappa *et al.*, 1989). This compound is active at low concentrations (0.5–10 p.p.m.). Its activity has been compared favourably with that of azadirachtin (see below) and would justify further evaluation. Other genera within this family (*Acokanthera, Apocynum, Mandevilla* and *Nerium*) have potent antifeedant compounds (see Grainge & Ahmed, 1988).

Araceae

Acorus calamus contains asarones which inhibit insect growth (60–300 p.p.m.), gonadal development and are antifeedant against the variegated cutworm *Peridroma saucia* and *Spodoptera litura* (Koul *et al.*, 1990). The growth inhibition activity of the oil from this plant is attributed to the phenylpropene *cis*-asarone which can occur at concentrations amounting to 65% of the weight of the resin, whereas the *trans*-asarones act as feeding deterrent. Ecdysteroids from *Arisaema ambiguum* have potential IGR activity (Chou & Lu, 1980).

Aristolochiaceae

Aristolochia spp. contain compounds that have potent insecticidal activity (see Grainge & Ahmed, 1988).

Asclepiadaceae

Asclepias contains cardenolides and alkaloids which are utilised by the Monarch butterfly *Danaus plexippus* for its protection from bird predation. These compounds are toxic to other species of insects: however they are also toxic to mammals (Harborne, 1991). *Calotropis* contains compounds that have potent IGR and antifeedant activity (see Grainge & Ahmed, 1988).

Asteraceae

This family contains many genera with compounds that affect insect growth, such as the ecdysteroids (Camps, 1991), pyrethrins from *Chrysanthemum cinerariaefolium* (Elliot, 1977) and photo-activated thiophenes, that could have potential as insecticides. When alpha-terthienyl (alpha-T) is subjected to near-UV light it becomes very toxic to many herbivorous insects, although some insects have been able to circumvent the toxic activity. However, there appears to be no cross-resistance with other insecticides (Arnason *et al.*, 1989).

The desert sunflower *Encelia* contain activity against a range of noctuid caterpillars at concentrations as low as 3–18 μmol/g fresh weight (Isman, 1989). It is doubtful whether these compounds could be used commercially; they are not as active as azadirachtin. Other chromenes, such as the precocenes isolated from the ornamental plant *Ageratum houstonianum*, differ in their activity, and have been shown to induce precocious moulting by inhibiting the synthesis of Juvenile Hormone in the corpora allata (Bowers, 1977). Unfortunately these compounds are also toxic to rats (Isman, 1989). Whether the chromenes in the desert sunflower would have similar toxicity to mammals is unknown. Chromenes are susceptible to degradation by sunlight which may decrease their potential use as commercial insecticides. Species of *Encelia* contain sesquiterpene lactones. Encelin isolated from *E. actoni* and *E. asperifolia* was toxic (LD50 = 60 μg/larvae) when injected into *Spodoptera littoralis* and also showed antifeedant activity (1 μmol g^{-1} fresh weight) (Srivastava *et al.*, 1990). *Helenium amarum* contains tenulin, a sesquiterpene lactone, which acts as an antifeedant against *Ostrinia nubilalis* and also reduces growth and fecundity (Arnason *et al.*, 1987). The sesquiterpene lactones, lactupicrin and 8-deoxylactucin in *Cichorium intybus* deter *Schistocerca gregaria* from feeding at 0.2% (dry weight) (Rees & Harborne, 1985). *Momordica charantia* contains triterpenoids (momordicine II active at 32 mg/ml, the level found in leaves) that deter the red pumpkin beetle *Aulacophora foveicollis* from feeding. Other genera with potent anti-insect activity are listed in Table 1.

Balsaminaceae

Species of *Impatiens* have been shown to contain antifeedants against oligophagous insects such *Pieris brassicae* and *Plutella xylostella* (see Grainge & Ahmed, 1988). To evaluate the commercial potential of these extracts, they need to be tested against more polyphagous species.

Bignoniaceae

The nectar of *Catalpa speciosa* contains the iridoid glycosides catalpol and catalposide (0.4%) that deter the common skipper (*Poanes hobomok*) from feeding on nectar (see van Beek & de Groot, 1986). Catalpol at 100 p.p.m. deters *Spodoptera exempta* from feeding (Simmonds & Blaney, unpublished results). *Catalpa* also contains the iridoid, specionin, which has antifeedant properties (50–100 p.p.m.) against eastern spruce budworm *Choristoneura fumiferana* (Chan *et al.*, 1990).

Boraginaceae

Species of *Symphytum* contain phytoecdysteroids that have not been fully evaluated but could have potential as IGRs (Takemoto *et al.*, 1967). Extracts from *Heliotropium* spp. have been shown to have insecticidal activity and *H. subulatum* extracts deter *Locusta migratoria* from feeding (see Grainge & Ahmed, 1988).

Caesalpiniaceae

Species of *Cassia* have compounds that have IGR and antifeedant activity (see Grainge & Ahmed, 1988).

Canellaceae

Species of *Warburgia* contain sesquiterpenoid dialdehydes (warburganal, muzigadial, cinnamolide and polygodial) which deter many lepidopterous larvae from feeding at concentrations as low as 0.1 p.p.m. (see van Beek & de Groot, 1986; Blaney *et al.*, 1987). Some of the active compounds have been synthesised and shown to have many other biological activities and have potential as fungicides, molluscicides and plant growth inhibitors. However, it is doubtful whether warburganal would be useful as a crop protection agent for a number of reasons. First, it is not readily available, occurring at low concentrations in the plant (0.01%). Second, the synthesis of the compound is expensive and complex. Third, although it deters insects from feeding, if it is consumed by insects it is not toxic to them but it does show some cytotoxicity against humans. Species of *Warburgia* contain other drimanes that might combine potent antifeedant activity with no cytotoxicity (Kioy *et al.*, 1990).

Cannaceae

Canna spp. contain compounds that deter a range of insects from feeding (see Grainge & Ahmed, 1988).

Celastraceae

Chinese bittersweets *Celastrus angulatus* and *C. glaucophyllus* have been used by farmers for years to control pests (Chiu, 1989). Active extracts contain sesquiterpenoid alkaloids with a dihydro-beta-agrofuran nucleus, such as celangalin, maytansine, triptolide, maytoline and evonine. Insects treated with these compounds show narcosis then die. The compounds are thought to act on the insect nervous system but not, as many other neurotoxic compounds, on cholinesterase activity (Chiu, 1989).

The genus *Maytenus* contains compounds with potent antifeedant activity, including the alkaloid wilforine (Delle Monache *et al.*, 1984) which is also thought to be the active component in the root powders of another member of the family, *Tripterygium wilfordii*, which is used as an insecticide by the Chinese (Swingle *et al.*, 1941). Its effectiveness varies greatly between species, although it is as potent an antifeedant as azadirachtin against some oligophagous insects such as *Locusta migratoria* and *Pieris rapae*, but it is less effective against certain polyphagous species (*Schistocerca gregaria, Heliothis zea*) (Delle Monache *et al.*, 1984). Although effective, wilforine would be difficult to synthesise or extract in bulk. Therefore, it would be worth undertaking a chemotaxonomic study on other genera within the Celastraceae to see if other species contain analogues of wilforine which might occur in higher concentrations in the plant or be simpler to synthesise.

Chenopodiaceae

Spinacia spp. contain phytoecdysteroids (Bathori *et al.*, 1982) that could have potential IGR activity (Simmonds & Blaney, unpublished results). Extracts from species of *Beta* and *Chenopodium* have shown potent antifeedant activity against *Leptinotarsa decemlineata* and *Pieris brassicae* (see Grainge & Ahmed, 1988).

Clusiaceae

Extracts from the Mammee apple tree *Mammea americana* are insecticidal to a range of pests, including *Culex* spp., *Diabrotica, Myzus persicae, Plutella xylostella* and *Spodoptera* spp. (see Grainge & Ahmed, 1988).

Combretaceae

Combretum spp. contain compounds that deter *Locusta migratoria* from feeding (see Grainge & Ahmed, 1988).

Compositae

The genus *Wedelia* contains terpenoids that show potent antifeedant activity against species of *Spodoptera* and the Thai plant *Wedelia biflora* contains compounds that have IGR activity against *Heliothis virescens* (Simmonds & Blaney, unpublished results). *Eupatorium adenophorum* contains the sesquiterpene cadinene which is a potent antifeedant (see Grainge & Ahmed, 1988). Leaves of the East African plants *Vernonia glauca* and *V. gigantea* are bitter and contain sesquiterpenes, such as glaucolide A, that are IGRs and deter lepidopterous larvae from feeding but the compounds show cytotoxic activity (Burnett *et al.*, 1978). Another species of *Vernonia* contains sesquiterpene lactones which deter *Spodoptera exempta* from feeding (Simmonds & Blaney, unpublished).

So far only a few of the sesquiterpenes isolated from the Compositae have been tested for antifeedant activity. Many of them occur in high concentrations in the plant and are thought to have a defence role. Those showing activity include tagitinin from *Tithonia diversifolia* (see van Beek & de Groot, 1986); lactupicrin, 8-deoxy lactucin as well as the hydroxy coumarin cichorin from *Cichorium intybus* (Rees & Harborne, 1985); diketone from *Ageratina adenophora* (Proksch *et al.*, 1990); tenulin from *Helenium amarum* (Arnason *et al.*, 1987).

The sunflower *Helianthus annuus* contains the diterpenoids kaurenoic and trachylobanoic acid which are IGRs (Elliger *et al.*, 1976).

Convolvulaceae

Extracts of *Ipomoea calonyction* contain phytoecdysteroids (Canonica *et al.*, 1973) that might be responsible for the observed IGR activity against *Spodoptera exempta* (Simmonds & Blaney, unpublished results) and extracts of *I. purpurea* and *Calstergia sepium* deter *Locusta migratoria* from feeding (see Grainge & Ahmed, 1988). *Convolvulus arvensis* contains compounds that show potent antifeedant activity against *Leptinotarsa decemlineata* and *Pieris brassicae* (see Grainge & Ahmed, 1988).

Cruciferae

The Cruciferae contain a range of sulphur-containing secondary metabolites that are important in host selection behaviour (Chew, 1988). A novel sulphur-containing indole alkaloid, dithyreanitrile, isolated from the seeds of *Dithyrea wislizenii*, has potent antifeedant activity against *Spodoptera frugiperda* and *Ostrinia nubilalis* (Powell *et al.*, 1991). It is postulated that its mode of action is hydrolysis, similar to that of the icynanogenid glycosides. Unlike many potent antifeedants it is a simple compound and could be synthesised easily.

Brassica (Matsuoka *et al.*, 1969) contain phytoecdysteroids that could have IGR activity.

Cucurbitaceae

Species of *Momordica* show insecticidal activity (see Grainge & Ahmed, 1988) and *M. charantia* contains compounds with antifeedant activity (Chandravadana, 1987).

Cycadaceae

Cycas revoluta contains compounds that show potent antifeedant activity against *Manduca sexta* (see Grainge & Ahmed, 1988).

Dioscoreaceae

Dioscorea spp. contain compounds, as yet unidentified, that have insecticidal and antifeedant activity (see Grainge & Ahmed, 1988)

Euphorbiaceae

Euphorbia, Alchornea and *Aleurites* contain compounds with antifeedant activity (see Grainge & Ahmed, 1988). The diterpenoid (–) -hardwickiic acid, isolated from *Croton aromaticus* has potent insecticidal activity (Bandara *et al.*, 1987).

Fagaceae

Fagus grandifolia contains compounds that deter *Manduca sexta* from feeding (see Grainge & Ahmed, 1988) and *Castanea* spp. contain phytoecdysteroids (Yokota *et al.*, 1982) that could have IGR activity.

Geraniaceae

Geranium spp. contain compounds with potent IGR activity and species of *Pelargonium* contain compounds like limonene, which have antifeedant activity (Simmonds *et al.* unpublished results), and which could be obtained in bulk from plant tissue cultures (Charlwood & Charlwood, 1991).

Ginkgoaceae

Ginkgo biloba contains compounds with potent insecticidal and antifeedant activity (see Grainge & Ahmed, 1988).

Iridaceae

Iris spp. contain phytoecdysteroids (Imai *et al.*, 1969) that could be involved in the reported IGR activity (see Grainge & Ahmed, 1988).

Juglandaceae

Juglans spp. contain quinones that have been shown to have potential antifeedant activity (Norris, 1986).

Labiatae

Many clerodanes with antifeedant and IGR activity have been isolated from species of *Ajuga, Teucrium, Scutellaria* and *Ocimum*. Some of these compounds have been synthesised (Blaney *et al.*, 1988) and others are been produced in cell cultures (Charlwood & Charlwood, 1991).

Ajuga chamaepitys contains a range of clerodanes, one of the most active being ajugapitin which deters *Spodoptera littoralis* from feeding at 0.3 p.p.m. and 14, 15-dihydro-ajugapitin, from *A. pseudiva* has similar activity. These compounds are potent enough to justify field trials as activity at 0.3 p.p.m. equates to 0.3 mg/l. The activity of the clerodane diterpenes in the Labiatae have been reviewed by van Beek & de Groot (1986) and Simmonds & Blaney (in press).

Monoterpenes from *Mentha* have antifeedant and insecticidal activity against *Spodoptera frugiperda* and *Delia floralis* (Simmonds & Blaney, unpublished results) and the variegated cutworm *Peridroma saucia* (Harwood *et al.*, 1990). The monoterpene pulegone shows potent antifeedant activity (0.05% and 0.1%) and toxicity when topically applied (1077 μg/g) to larvae, menthone shows antifeedant activity (0.2%) and contact toxicity (2478 μg/g). whereas menthol only showed antifeedant activity (0.1% and 0.2%) (Harwood *et al.*, 1990). Very little work has been done on the anti-insect activity of genera such as *Salvia, Sideritis* and *Ocimum* that contain many novel monoterpenoids and diterpenoids (Simmonds *et al.*, unpublished results), which if shown to be active could be obtained in bulk from tissue cultures (Charlwood & Charlwood, 1991).

Some genera produce terpenoids that attract insects and can be used either to increase the number of beneficial insects to an area or to trap pest insects. For example, *Melissa officinalis* produces terpenoids that are similar to those in the Nasonov pheromone, and these terpenoids have been used to attract bees to empty hives (Pickett, 1991). Nepetalactones obtained from species of *Nepeta* have been shown to be in the pheromones for a number of aphid species (Dawson *et al.*, 1990).

Leguminosae

Many species within the family Leguminosae contain allelochemicals that have potential use as insecticides. Species of *Lonchocarpus* and *Tephrosia* contain a range of anti-insect compounds including rotenones (Hassanali & Lwande, 1989), flavonoids (Simmonds *et al.*, 1990b) non-protein amino acids (Bell, 1977) and polhydroxyalkaloids (Fellows *et al.*, 1989).

The genera *Medicago, Lotus* and *Lupinus* contain phytoalexins which have been shown to be active against fungi and insects (Sutherland *et al.*, 1980. For example, some of the isoflavonoids isolated from *Medicago sativa, Lotus pedunculatus* and *Lupinus angustifolius* are antifeedants (Sutherland *et al.*, 1980) and some of these compounds, such as vestitol and phaseollin, are active against fungi as well as insects.

There are many reports that extracts from species of *Tephrosia* have been used as insecticides in Africa, Central and South America and Asia. These extracts are usually rich in flavonoids, including rotenoids (Hassanali & Lwande, 1989; Gomes *et al.*, 1981). Chemotaxonomic studies have shown that 20 of the 300 species of *Tephrosia* contain flavonoids (Gomes *et al.*, 1981). However, although many of these compounds have been shown to be active in the laboratory against a range of insects, only rotenone has been used commercially. *T. elata*, which is found in Kenya, contains isopongaflavone which deters *Eldana saccharina* and *Maruca testulalis* from feeding and tephrosin, which deters the African armyworm *Spodoptera exempta* from feeding (Bentley *et al.*, 1987). Acetone extracts of *T. vogelii* have potent insecticidal and antifeedant activity against *Pieris rapae*. As an ovicide the extract was as effective (0.5–0.1% = 0.5 µl/egg) as Dimilin (Chiu, 1989).

The isoflavonoid glyceollin from soybeans (*Glycine max*) deters insects from feeding and could represent a common defence against phytophagous insects which could be utilised in plant breeding programmes (Fisher *et al.*, 1990).

Species of *Calliandra* contain non-protein amino acids and sulphur amino acids that exhibit moderate insecticidal activity against *Spodoptera littoralis, Heliothis virescens* and *Heliothis zea*. When the compounds were combined to mimic the composition found in the plant they were more effective at deterring aphids from feeding than were the individual compounds when tested alone. Additive and synergistic interactions may commonly occur in nature and it is possible that many active compounds have been discarded as being inactive because they have been tested in isolation and not in combination with other compounds that occur in plants (Romeo & Simmonds, 1989).

The non-protein amino acids, such as mimosine, canavanine and L-dopa, that occur in many legume genera have been shown to have potent IGR and antifeedant activity (Bell, 1977). However, although they show potent anti-insect activity many of them are toxic to mammals. The polyhydroxy alkaloids are potent glycoside inhibitors and antifeedants; they were initially isolated from *Streptomyces* in 1966 and since 1977 they have been found to occur in five families of the higher plants (Fellows *et al.*, 1989). A chemotaxonomic study of the Leguminosae sub-family Papilionoideae led to the discovery of several polyhydroxy alkaloids which inhibit digestive disaccharidases of *Spodoptera littoralis* (Fellows *et al.*, 1989) and deter *S. littoralis, S. frugiperda, Heliothis virescens* and *Helicoverpa armigera* from feeding (Simmonds *et al.*, 1990a).

Sesquiterpenes can be potent antifeedants and the trees *Copaifera langsdorfii* and *Hymenaea stigonocarpa* contain sesquiterpenes such as caryophyllene that deter insects from feeding (Langenheim *et al.*, 1986).

Extracts of a range of different legumes have been tested against *Spodoptera frugiperda*. Those that have shown some promise include: *Aglaia cordata*, which is effective at 16 p.p.m., *Sandoricum koetjape* (2000 p.p.m.), *Swietenia mahogani* (2000 p.p.m.), *Chickrassia tabularis* (2000 p.p.m.), *Dysoxylum malabaricum* (2000 p.p.m.), *D. spectabile* (2000 p.p.m.), *Lansium domesticum* (2000 p.p.m.) (Mikolajczak *et al.*, 1989). Extracts from these plants, which require little equipment to prepare, could be made locally and used in simple formulations to control insects.

Overall, species within this family have already provided many novel compounds that have been used as pesticides and they are also providing leads for use in biotechnological crop protection, for example the trypsin inhibitors isolated from *Vigna* which are being used in transgenic plants (Hilder *et al.*, 1990).

Liliaceae

Lilium longiflorum contains compounds that deter *Plutella xylostella* from feeding, whereas other genera such as *Schoenocaulon* contain compounds that are insecticidal and antifeedant against a range of insects (see Grainge & Ahmed, 1988). The antifeedant activity associated with extracts of *Urginea maritima* could be due to the alkaloid caffeine (see Grainge & Ahmed, 1988).

Magnoliaceae

Extracts from the tulip tree *Liriodendron tulipifera* deter *Lymantria dispar* from feeding (see Grainge & Ahmed, 1988).

Malvaceae

The levels of gossypol in *Gossypium hirsutum* are associated with resistance to *Spodoptera littoralis*. Gossypol, alone and in combination with other terpenoids, condensed tannins and flavonoids deters many species from feeding (Jacobson, 1989). The genera *Hibiscus* and *Sphaeralcea* contain compounds that have insecticidal and antifeedant activity (see Grainge & Ahmed, 1988).

Meliaceae

The seeds and leaves of the neem tree *Azadirachta indica* contain terpenoids with potent anti-insect activity. Neem products have been shown to have a multitude of anti-insect activities including repellent and antifeedant activity, growth inhibition, suppression of reproduction, mating disruption and ovicidal activity. One of the most active terpenoids in neem seeds is azadirachtin which acts as an antifeedant and growth disrupter against a range of insects at microgram levels. The active terpenoids in neem leaves include nimbin, deacetylnimbin and thionemone.

Home-made formulations of neem have been used by farmers in the Third World for centuries and recent studies have shown that neem extracts and neem-derived compounds can be active against 198 different species of insects (Saxena, 1988). Although neem products have been shown to be active against such a large number of insect species they apparently do not kill many of the beneficial insects, such as the predators and parasitoids. The apprehension that large-scale use of neem-based insecticides may lead to resistance among pests has not, so far, been substantiated. Studies with two genetically different strains of *Plutella xylostella* exposed to neem products in feeding bioassays for 35 generations showed no signs of resistance. Whether insects will show resistance to neem will depend partly on how frequently it is used (Saxena, 1988).

The interest in neem has been increasing and since 1980 there have been eight national and international conferences dedicated to review the potential of neem as a pest control agent. Neem trees are at present being cultivated in parts of Central America and South America as well as in India. Neem products from these plantations include neem seed oils, neem seed cakes as well as "suspensions" and alcoholic extracts from the kernels and leaves. These products have been used alone or in combination with synthetic insecticides and synergistic compounds. Margason O and Azatin are examples of neem products being marketed for use on non-food crops.

Overall, field trials with neem products have shown, not only a decrease in damage by pests but an increase in crop yield when compared with the use of recommended synthetic insecticides. In Togo a methanol suspension (2–4%) of neem leaves was used in a field trial against *Plutella xylostella* and it was as active as either the synthetic insecticides mevinphos (0.05%) or deltamethrin (0.02%) (Dreyer, 1987). Field trials in Thailand showed that piperonyl butoxide increased the efficacy of neem and the combination was as active as cypermethrin (0.025%) against *P. xylostella* and *Spodoptera litura* (Sombatsiri & Tigvattanont, 1987). In the Philippines neem was compared with BAY SIR 14591, an IGR marketed by Bayer AG, and the synthetic compound was only marginally better than neem (Kirsh, 1987). In Mauritius field trials showed that neem was as active as the deltamethrin, Decis (Fagoone, 1987). In the USA neem extracts were shown to be effective against *Tichoplusia ni* (Reed & Reed, 1986) and azadirachtin, when systemically applied to corn, was effective against *Spodoptera frugiperda* at 0.51 b/ha (Klocke *et al.*, 1989). Neem has been shown to be effective against the citrus red mite (*Panonychus citri*) in the USA (Jacobson *et al.*, 1978) and in China (Chiu, 1984). Various neem formulations were found to be effective against *Pieris brassicae* in Germany (Lange, 1985).

In Tanzania neem extracts were compared against the synthetic insecticide Lindane in the control of pests on beans, and a seed extract of neem was as effective as Lindane against thrips (50% control) and gave up to 40% control of *Heliothis armigera* and the pod sucking bug, *Acanthomia horrida* (Hongo & Karel, 1986). *Heliothis armigera* has also been effectively controlled by neem extracts in India and the extracts were as active as many conventional insecticides, such as pyrethroids, cypermethrins, malathion and endosulfan (Parmar & Srivastava, 1987).

Powdered neem seed kernels or leaves have been used in seed stores and warehouses and can reduce insect infestation and damage to grain during a three to twelve month storage period (Makanjuola, 1989). Neem products have also been shown to decrease the spread of insect-transmitted viral diseases (Ascher & Meisner, 1989).

The potency of neem extracts, especially those from seeds, is often positively associated with the presense or absence of azadirachtin, which occurs in the plant at very low concentrations and is difficult and expensive to extract. Azadirachtin is also a very complex molecule to synthesise. In order to synthesise a simpler molecule with the same potency as azadirachtin, structure–function studies are being undertaken on derivatives of azadirachtin to characterise the important functional groups (Ley *et al.*, 1988, 1989; Blaney *et al.*, 1990; Simmonds *et al.*, 1990c). These derivatives differ in their potency. Some, such as dihydroazadirachtin, are as potent as azadirachtin against *Spodoptera littoralis, S. frugiperda, Heliothis virescens* and *Helicoverpa armigera* (Blaney *et al.*, 1990), other derivatives vary in their activity against *Leptinotarsa decemlineata* (Klocke *et al.*, 1989).

Other genera within this family contain compounds with potent antifeedant activity. Species of *Trichilia* contain limonoids called trichilins A–F, which are antifeedant against species of *Spodoptera* (Nakatani *et al.*, 1981). *Melia azedarach* and *M. toosendan* contain the antifeedant toosendanin. The concentration of this compound in the plants varies from area to area. The bark of *M. azedarach* growing in southwestern and northwestern China contain a greater amount of toosendanin (5.2–4.7 mg/g) than is obtained from the bark of trees growing in south China (1.3 mg/g). However, the toxicity and antifeedant activity associated with extracts from these plants do not correlate with the levels of toosendanin in the extracts. This suggests that either toosendanin is not the most active compound in the plant or it is interacting with other compound(s) in the extracts. For example, *M. azedarach* contains other limonoids called meliatoxins which deter *Spodoptera litura* from feeding and are structurally similar to the trichilins in *Trichilia* (Nakatani *et al.*, 1981). Species of *Carapa* contain the tetranortriterpenoid carapolides A, B, C, D, E, and F (Jacobson, 1989), which deter *Spodoptera exempta* from feeding (Simmonds & Blaney, unpublished).

As the concentrations and proportions of these terpenoids, within a species, vary between geographical areas, it is important, when searching for sources of botanical insecticides, to be aware of phenotype and genotype variability. Environmental factors, such as climate and soil type, could also influence the production of active secondary metabolites. Before we can optimise the harvest of compounds, like azadirachtin, from plants, we need to know far more about the phenology and genetics of plants that contain these potential insecticides.

Other genera within this family have species which have potent anti-insect activity. For example, ethanol extracts of *Cedrela odorata*, the Spanish cedar, deter the striped cucumber beetle, *Acalymma vittatum* (Jacobson, 1989) and *Spodoptera littoralis* (Simmonds & Blaney, unpublished) from feeding. Extracts of *Aglaia cordata* have been shown to be as active as those from *Azadirachta indica* (Mikolajczak *et al.*, 1989).

Overall, it would appear from antifeedant studies on limonoids from the Meliaceae and Rutaceae that an epoxide and furan group are essential structural requirements, for potent antifeedant activity (Bentley *et al.*, 1988; Blaney *et al.*, 1990). The IGR activity associated with many of these compounds is thought to be associated with the keto moiety (Macleod *et al.*, 1990; Simmonds *et al.*, 1990c).

Melianthaceae

Bersama spp. contain compounds that have antifeedant activity (see Grainge & Ahmed, 1988).

Myrtaceae

Eucalyptus spp. contain a range of alkaloids but it is not known if these compounds are responsible for observed antifeedant activity (see Grainge & Ahmed, 1988).

Onagraceae

Epilobium spp. contains compounds that deter *Locusta migratoria* from feeding (see Grainge & Ahmed, 1988).

Papaveraceae

Chelidonium majus deters many insects from feeding; this could be due to the presence of alkaloids, such as proptopine, sparteine and stylopine, which deter *Spodoptera exempta* from feeding (Simmonds & Blaney, unpublished results).

Piperaceae

Over 50 unsaturated amides have been isolated from this plant family but few have been evaluated for insecticidal activity. Of those that have been examined, guineensine (LD50 = 0.11 μg/insect) was more active against the bruchid beetle *Callosobruchus chinesis* than pipercide (LD50 = 0.56 μg/insect), but it was nearly as active as pyrethrin (LD50 = 0.1 μg/insect) (Miyakado *et al.*, 1989). Pipercide exhibited potent knockdown activity against a range of insects and early reports suggested that it was more active than pyrethrum against house flies: however, this was not confirmed by later studies (Miyakado *et al.*, 1989). It would appear that the amides from Piperaceae do not act on the same site within the nervous system as the pyrethrins and it is highly likely that the Piperaceae amides will provide leads for the agrochemical industry.

Plantaginaceae

Plantago spp. contain compounds that deter a wide range of insects from feeding (see Grainge & Ahmed, 1989).

Plumbaginaceae

Plumbago zeylanica contains the naphthoquinone plumbagin which is a potent antifeedant against *Spodoptera exempta* and IGR against a number of lepidopterous larvae (Hassanali & Lwande, 1989).

Poaceae

Bothriochloa spp. and *Echinochloa crus-galli* contain compounds that deter *Leptinotarsa decemlineata* from feeding (see Grainge & Ahmed, 1988). *Lolium* spp. contain the alkaloids perlolidine and peroline that at 100 p.p.m. deter *Spodoptera exempta* from feeding (Simmonds & Blaney, unpublished results).

Podocarpaceae

Methanolic extracts of *Podocarpus gracilior* have been shown to have antifeedant activity against *Heliothis zea* and *Spodoptera frugiperda*. The extract contained the diterpines nagilactone C and D which deterred larvae from feeding at concentrations of 15 to 38 μg/disc (see van Beek & de Groot, 1986).

Polygonaceae

The marsh pepper (*Polygonum hydropiper*) contains the drimane-type sesquiterpene polygodial which shows antifeedant activity against a range of insects (Blaney *et al.*, 1987).

Ranunculaceae

Clematis spp. contain compounds that are insecticidal and antifeedant against a range of insects (see Grainge & Ahmed, 1988) and *Helleborus* contain ecdysteroids that could have IGR activity (Camps, 1991).

Rosaceae

Only a few genera within this family have been tested for anti-insect activity. *Fragaria* spp. and *Purshia tridentata* contain componds that deter *Leptinotarsa decemlineata* from feeding (see Grainge & Ahmed, 1988).

Rutaceae

Atalantia racemosa contains the coumarins xanthotoxin and racemosin which deter *Spodoptera litura* from feeding. Structure–function studies undertaken on plant-derived coumarins have shown that the active compounds are those with a linear fused furan ring and alkoxy group at positions C-9 and C-4 (Luthria *et al.*, 1989). The limonoid, limonin, which has been isolated from several species of Rutaceae, is not as active an antifeedant, against *Spodoptera littoralis*, as other limonoids such as azadirachtin. However, it is very active against *Leptinotarsa decemlineata* (Jacobson, 1989).

Alkaloids with antifeedant activity have been isolated from *Zanthophyllum monophyllum*. For example, zanthophylline deterred *Melanoplus sanguinipes* and *Hypera postica* from feeding; n-methyflindersine from *Fagara chalybea* deterred the Mexican bean beetle (*Epilachna varivestis*) and the beet armyworm from feeding; the isobutylamide herculin, isolated from *Zanthoxylum clava-herculis* is as toxic as pyrethrins but it is unstable.

Saxifragaceae

Heuchera sanguinea contains compounds that deter *Plutella xylostella* from feeding (see Grainge & Ahmed, 1988).

Simaroubaceae

Genera within this family contain quassinoids which show potent antifeedant activity against *Spodoptera frugiperda, S. littoralis* and *Heliothis virescens* (Simmonds & Blaney, unpublished results). Quassin is an effective antifeedant against aphids (Polonsky *et al.*, 1989).

Solanaceae

Some genera within this family can be characterised by the type of alkaloids they contain. For example, *Lycopersicon solanum* and *Physalis* contain steroidal glycoalkaloids (tomatine and solanine); *Nicotiana* the pyridine alkaloids (nicotine); *Atropa* and *Hyoscyamus* the tropane alkaloid (atropine); *Solanum* the glycoalkaloids (tomatine, solanine, chaconine, demissine and leptine). All these compounds deter *Spodoptera exempta* from feeding and in the electrophysiological studies undertaken on the sensilla used by this species to taste its food, they all caused injury to the receptor cells (Simmonds & Blaney, unpublished results).

A number of steroidal compounds, such as the withanolides, have been isolated from various solanaceous plants belonging to the genera *Withania, Nicandra, Datura, Jaborosa, Petunia* and *Physalis* (Ascher *et al.*, 1987; Elliger & Waiss, 1989). The steroids isolated from these plants showed antifeedant activity against a range of pests. Elliger & Waiss (1989) suggested that this type of compound could be utilised by plant breeders when selecting crop plants resistant to insect attack. Plant breeders have been attempting either to increase the content of active substances within a particular crop or to introduce resistant factors found in closely related plant varieties. Modern cell biology and molecular genetics have provided the techniques that enable the transfer of active substances not only between species within the same genus but between genera in the same family and between families. If it is possible to transfer an array of chemicals to crop plants it would require multiple changes to occur in the insects before they become adapted to or are able to tolerate these transformed crop plants.

Umbelliferae

Angelica silvestris contains the sesquiterpenoid bisabolangelone which acts as an antifeedant at 1000 p.p.m. against *Leptinotarsa decemlineata* and *Mythimna unipuncta* (see van Beek & de Groot, 1986).

Verbenaceae

Stachytarpheta mutabilis contains the bitter iridoid glycoside ipolamiide which deters *Locusta migratoria, Schistocerca gregaria* and *Spodoptera littoralis* from feeding at concentrations of 0.2%, 2% and 0.1%, respectively (Bernays & De Luca, 1981). This family, along with Labiatae, contains a range of iridoid glycosides which deter *Spodoptera littoralis* from feeding (Simmonds & Blaney, unpublished results).

The phytoecdysteroid, vitexirone, 20-hydroxyecydsone, ajugasterone c and turkestrerone isolated from *Vitex fisherii* all have IGR activity (Kubo *et al.*, 1990).

Species of *Clerodendron* and *Caryopteris* contain clerodane diterpenes which have been shown to have potent antifeedant and IGR activity. Clerodin isolated from *Clerodendron infortunatum* showed potent antifeedant activity against *Spodoptera litura* at 80 p.p.m. and clerodin and dihydro-clerodin from *Caryopteris divaricata* showed similar activity at 50 p.p.m. (see van Beek & de Groot, 1986).

Vitaceae

Parthenocissus contains compounds with potent activity against a range of insects (see Grainge & Ahmed, 1988) including *Spodoptera exempta* (Simmonds & Blaney, unpublished results).

Winteraceae

Drimys lanceolata contains the active antifeedant drimane diterpenoid polygodial (see above).

Conclusions

In the literature there are many examples of micro-organisms and higher plants that contain compounds with pesticide activity. Within this review, we do not claim to have presented an exhaustive account of this literature. Instead we have concentrated on those genera of higher plants that we feel justify further study and have outlined the present use of mycochemicals as pesticides.

The continual use of broad-spectrum persistent pesticides over the last 50 years has resulted in many problems. Today there is an urgent need to develop pesticides for the next century that are effective against specified target organisms, are benign to non-target organisms and are biodegradable. Natural products, either from higher plants or from fungi, could fulfil these criteria and are thus of enormous potential, either as pesticides in their own right or as models for the synthetic chemist.

References

Arnason, J.L., Isman, M.B., Philogene, B.J.R. and Waddell, T.G. (1987) Mode of action of the sesquiterpene lactone teulin, from *Helenium amarum* against herbivorous insects. *Journal of Natural Products* 50, 690–695.

Arnason, J.L., Philogene, B.J.R., Morland, P., Imrie, K., Iyengar, S., Duval, F., Soucy-Breau, C., Sciano, J.C., Werstiuk, N.H., Hasspieler, B. and Down, A.E.R. (1989) Naturally occurring and synthetic thiopenes as photoactivated insecticides. In: *Insecticides of Plant Origin* (Eds. Arnason, J.T., Philogene, B.J.R. and Morland, P.), pp. 164–172.

Ascher, K.R.S. and Meisner, J. (1989) The effects of neem on insects affecting man and animals. In: *Focus on Phytochemical Pesticides, Vol. 1, The Neem Tree* (Ed. Jacobson, M.), pp. 113–131. CRC Press, Florida.

Ascher, K.R.S., Eliyahu, M., Glotter, E., Goldman, A., Kirson, I., Abraham, A., Jacobson, M. and Schmutterer, H. (1987) The antifeedant effect of some new withanolides on three insect species, *Spodoptera littoralis, Epilachna varivestis* and *Tribolium castaneum. Phytoparasitica* 15, 15–29.

Bandara, B.M.R., Wimalasiri, W.R. and Bandara, K.A.N.P. (1987) Isolation and insecticidal activity of (-) -hardwickiic acid from *Croton aromaticus. Planta medica* 53, 575.

Bathori, M., Toth, I., Szendrei, K. and Reisch, J. (1982) Ecdysteroids in *Spinacia oleracea* and *Chenopodium bonus-henricus. Phytochemistry* 21, 236–238.

van Beek, T.A. and de Groot, Ae. (1986) Terpenoid antifeedants, part 1. An overview of terpenoid antifeedants of natural origin. *Recueil des Travaux Chimiques des Pays-Bas* 105, 513–527.

Bell, E.A. (1977) The possible significance of uncommon amino acids in plant–vertebrate, plant–insect and plant–plant relationships. In: *Natural Products and the Protection of Plants* (Ed. Marini-Bettolo, G.B.), pp. 571–595. Pontificia Academia Scientiarvm, Citta del Vaticano.

Bell, E.A., Fellows, L.E. and Simmonds, M.S.J. (1990) Natural products from plants for the control of insect pests. In: *Safer Insecticides: Development and Use* (Eds. Hodgson, E. and Ruhr, R.J.), pp. 337–350. Marcel Dekker Inc, New York.

Bentley, M.D., Hassanali, A., Lwande, W., Njoroge, P.E.W., Ole Sitayo, E.N. and Yatagai, M. (1987) Insect antifeedants from *Tephrosia elata deflers. Insect Science its Application* 8, 85–88.

Bentley, M.D., Rajab, M.S., Alford, A.R., Mendel M.J. and Hassanali (1988) Structure activity studies of modified citrus limonoids as antifeedants for Colorado potato beetle, *Leptinotarsa decemlineata. Entomologia experimentalis et applicata* 49, 189–193.

Berenbaum, M.R. (1989) North American ethnobotanicals as sources of novel plant-based insecticides. In: *Insecticides of Plant Origin* (Eds. Arnason, J.T., Philogene, B.J.R. and Morand, P.), pp. 11–24. ACS 387, Washington.

Bernays, E. A. and De Luca, C. (1981) Insect antifeedant properties of an iridoid glycoside: ipolamiide. *Experientia* 37, 1289–1290.

Blaney, C.L., Van Dyke, C.G. and Grand, L.F. (1988) *Cercospora caricis* from *Cyperus esculentus* (yellow nutsedge): morphology and cercosporin production. *Mycologia* 80, 418–421.

Blaney, W.M., Simmonds, M.S.J., Ley, S.V. and Katz, R.B. (1987) An electrophysiological and behavioural study of insect antifeedant properties of natural and synthetic drimane-related compounds. *Physiological Entomology* 12, 281–291.

Blaney, W.M., Simmonds M.S.J., Ley S.V. and Jones, P.S. (1988) Insect antifeedants: a behavioural and electrophysiological investigation of natural and synthetically derived clerodane diterpenoids. *Entomologia experimentalis et applicata* 46, 267–274.

Blaney, W.M., Simmonds, M.S.J., Ley, S.V., Anderson, J.C. and Toogood, P.L. (1990) Antifeedant effects of azadirachtin and structurally related compounds on lepidopterous larvae. *Entomologia experimentalis et applicata* 55, 149–160.

Bowers, W.S. (1977) Anti-juvenile hormones from plants: chemistry and biological activity. In: *Natural Products and the Protection of Plants* (Ed. Marini-Bettolo, G.B.), pp. 129–142. Pontificia Academia Scientiarvm, Citta del Vaticano.

Brian, P.W (1960) Presidential address: Griseofulvin. *Transactions of the British Mycological Society* 43, 1–13.

Burnett, W.C., Jones, S.B. and Mabry, T.J. (1978) The role of sesquiterpene lactones in plant-animal coevolution. In: *Chemical Aspects of Plant and Animal Coevolution* (Ed. Harborne, J.B), pp. 233–257. Academic Press, London.

Camps, F. (1991) Plant ecdysteroids and their interaction with insects. In: *Ecological Chemistry and Biochemistry of Plant Terpenoids* (Eds. Harborne, J.B. and Tomas-Barberan, F.A.), pp. 331–376. Clarendon Press, Oxford.

Canonica, L., Danieli, B., Hainanova, M.A. and Krepinsky, D. (1973) The structure of a new phytoecdysone, kaladasterone: an application of 13[c] magnetic resonance to structural problems. *Experientia* 29, 1062–1063.

Chan, T.H., Guertin, K.R., Prasad, C.V.C., Thomas, A.W., Strunz, G.M. and Salonius, A. (1990) Spruce budworm (*Choristoneura fumiferana*) antifeedants 2. Structure–activity relationship among functional decalin compounds. *Canadian Journal of Chemistry* 68, 1170–1177.

Chandravadana, M.V. (1987) Identification of triterpenoid feeding deterrent of red pumpkin beetles (*Aulacophora foveicollis*) from *Momordica charantia*. *Journal of Chemical Ecology* 13, 1689–1694.

Charlwood, B.V. and Charlwood, K.A. (1991) Terpenoid production in plant cell cultures. In: *Ecological Chemistry and Biochemistry of Plant Terpenoids* (Eds. Harborne, J.B. and Tomas-Barberan, F.A.), pp. 95–132. Clarendon Press, Oxford.

Chew, F.S. (1988) Searching for defensive chemistry in the Cruciferae, or do glucosinolates always control interactions of Cruciferae with their potential herbivores and symbionts? In: *Chemical Mediation of Coevolution* (Ed. Spencer, K.C.), pp. 81–112. Academic Press, New York.

Chiu, S-F. (1984) The active principles and insecticidal properties of some Chinese plants, with special reference to Meliaceae. In: *Natural Pesticides from the Neem tree and Other Tropical Plants* (Eds. Schmutterer, H. and Ascher, K.R.S.), p. 255. GTZ Press, Eschborn, West Germany.

Chiu, S-F. (1989) Recent advances in research on botanical insecticides in China. In: *Insecticides of Plant Origin* (Eds. Arnason, J.T., Philogene, B.J.R. and Morand, P.), pp. 69–77. ACS 387, Washington.

Chou, W.S. and Lu, H-S. (1980) Growth regulation and silk production in *Bombyx mori* L. from phytogeneous ecdysteroids. In: *Progress in Ecdysone Research* (Ed. Hoffman, J.A.), pp. 281–297. Elsevier North Holland, Biomedical Press, Amsterdam.

Cole, M.D., Anderson, J.C., Blaney, W.M., Ley, S.V. and Simmonds, M.S.J. (1990) Novel neo-clerodane insect antifeedants from *Scutellaria galericulata*. *Phytochemistry* 29, 1793–1796.

Dawson, G.W., Griffiths, D.C., Merritt, L.A., Mudd, A. and Pickett, J.A. (1990) Aphid semiochemicals—a review, and recent advances on the sex pheromone. *Journal of Chemical Ecology* 16, 3019–3030.

Dekker, J. (1963) Antibiotics in the control of plant diseases. *Annual Review of Microbiology* 17, 243–262.

Delle Monache, F., Marini Bettola, G.B. and Bernays, E.A. (1984) Isolation of insect antifeedant alkaloids from *Maytenus rigida* (Celastraceae). *Zeitschrift für angewandte Entomologie* 97, 406–414.

Dreyer, M. (1987) Field and laboratory trials with simple neem products as protectants against pests of vegetable and field crops in Togo. In: *Proceedings of the 3rd Neem*

Conference, Nairobi, Kenya 1986 (Eds. Schmutterer, H. and Ascher, K.R.S.), p. 431. GTZ Press, Eschborn, West Germany.

Elliger, C.A. and Waiss, A.C. Jr. (1989) Insect growth inhibitors from petunia and other solanaceous plants. In: *Insecticides of Plant Origin* (Eds. Arnason, J.T., Philogene, B.J.R. and Morand, P.), pp. 188–205. ACS 387. Washington.

Elliger, C.A., Zinkel, D.F., Chan, B.G. and Waiss, A.C. (1976) Diperpene acids as larval growth inhibitors. *Experientia* 32, 1364–1365.

Elliott, M. (1977) Synthetic insecticides from the natural pyrethrins. In: *Natural Products and the Protection of Plants* (Ed. Marini-Bettolo, G.B.), pp. 157–176. Pontificia Academia Scientiarvm, Citta del Vaticano.

Ellison, C.A. and Evans, H.C. (1990). Preliminary assessment of fungal pathogens as biological control agents for *Rottboellia cochinchinensis* (Gramineae). *Proceedings of the VII International Symposium on Biological Control of Weeds* (Ed. Delfosse, E.S.), pp. 477–482. Instituto Sperimentale per la Patologia Vegetale Ministero dell 'Agricultura e delle Foreste: Rome, Italy.

Engel, J.F., Harnish, W.N. and Staetz, C.A. (1990) Challenges: The Industrial Viewpoint. In: *Safer Insecticides* (Eds. Hodgson, E. and Kuhr, R.J.), pp. 551–573. Marcel Dekker, inc, New York.

Fagoone, I. (1987) Use of neem in vegetable crop protection in Mauritius. *Proceedings of the 3rd Neem Conference, Nairobi, Kenja, 1986* (Eds. Schmutterer, H. and Ascher, K.R.S.), p. 109. GTZ Press, Eschborn, West Germany.

Fellows, L.E., Kite, G.C., Nash, R.J., Simmonds, M.S.J. and Scofield, A.M. (1989) Castanospermine, swainsonine and related polyhydroxy alkaloids: structure, distribution and biological activity. In: *Plant Nitrogen Metabolism* (Eds. Poulton, J.E., Romeo, J.T. and Conn, E.E.), pp. 395–427. Plenum Press, New York.

Fisher, D., Kogan, M. and J. Paxton, J. (1990) Effect of glyceollin, a soybean phytoalexin, on feeding by three phytophagous beetles (Coleoptera: Coccinellidae and Chrysomelidae): Dose versus response. *Environmental Entomology* 19, 1278–1282.

Gershenzon, J., Rossiter, M., Mabry, T.J., Rogers, C.E., Blust, M.H. and Hopkins, T.L. (1985) Insect antifeedant terpenoids in wild sunflower: A possible source of resistance to the sunflower moth. In: *Bioregulators for Pest Control* (Ed. Hedin, P.A.), pp. 433–446. ACS 276, Washington.

Gomes, C.M.R., Gottlib, O.R., Marini Bettolo, G.B., Delle Monache, F. and Polhill, R.M. (1981) Systematic significance of flavonoids in *Derris* and *Lonchocarpus*. *Biochemical Systematics and Ecology* 9, 129–147.

Grainge, M. and Ahmed, S. (1988) *Handbook of Plants with Pest-Control Properties,* 470 pp. John Wiley & Sons, New York.

Harborne, J.B. (1991) Recent advances in the ecological chemistry of plant terpenoids. In: *Ecological Chemistry and Biochemistry of Plant Terpenoids* (Eds. Harborne, J.B. and Tomas-Barberan, F.A.), pp. 399–426. Clarendon Press, Oxford.

Harmatha, J., Mancham, B., Arnault, C. and Slama, K. (1987) Identification of a spirostane-type saponin in the flowers of leek with inhibitory effects on growth of leek-moth larvae. *Biochemical Systematics and Ecology* 15, 113–116.

Harwood, S.H., Moldenke A.F. and Berry, R.E. (1990) Toxicity of peppermint monoterpenes to the variegated cutworm. *Journal of Economic Entomology* 83, 1761–1767.

Hassanali, A. and Lwande, W. (1989) Antipest secondary metabolites from African plants. In: *Insecticides of Plant Origin* (Eds. Arnason, J.T., Philogene, B.J.R. and Morand,

P.), pp. 78–94. ACS 387, Washington.

Hawksworth, D.L., Sutton, B.C. and Ainsworth, G.C. (1983) *Ainsworth and Bisby's Dictionary of the Fungi*. CAB International Mycological Institute, Kew.

Hilder, V.A., Gatehouse, A.M.R. and Boulter, D. (1990) Genetic engineering of crops for insect resistance using genes of plant origin. In: *Genetic Engineering of Crop Plants* (Eds. Lycett, G.W. and Grierson, D.), pp. 51–66. Butterworths, London.

Hongo, H. and Karel, A.K. (1986) Effect of plant extracts on insect pests of common beans. *Journal of Applied Entomology* 102, 164.

Imai, S., Toyosata, T., Sakai, M., Sato, Y., Fujioka, S., Murata, E. and Goto, M. (1969) Screening results for phytoecdysones. *Chemical Pharmacology Bulletin* 17, 335–339.

Isman, M.B. (1989) Toxicity and fate of acetylchromenes in Pest Insects. In: *Insecticides of Plant Origin* (Eds. Arnason, J.T., Philogene, B.J.R. and Morand, P.), pp. 44–58. ACS 387, Washington.

Jacobson, M. (1989) Botanical Pesticides Past, Present and future. In: *Insecticides of Plant Origin* (Eds. Arnason, J.T., Philogene, B.J.R. and Morand, P.), pp. 1–10. ACS 387, Washington.

Jacobson, M., Reed, D.K., Crystal, M.M., Moreno, D.S. and Soderstrom, E.L. (1978) Chemistry and biological activity of insect feeding deterrents from certain weed and crop plants. *Entomologia experimentalis et applicata* 24, 448.

Jermy, T. (1966) Feeding inhibitors and food preference in chewing phytophagous insects. *Entomologia experimentalis et applicata* 9, 1–12.

Kawazu, K., Nakajima, S. and Ariwa, M. (1979) Xanthumin and 8 epixanthumin as insect development inhibitors from *Xanthium canadense* leaves. *Experientia* 35, 1294–1295.

Kioy, D., Gray, A.I. and Waterman, P.G. (1990) A comparative study of the stem-bark drimane sesquiterpenes and leaf volatile oils of *Warburgia ugandensis* and *W. stuhlmannii*. *Phytochemistry* 29, 3535–3538.

Kirsh, K. (1987) Studies on the efficacy of neem extracts in controlling major pests in tobacco and cabbage. *Proceedings of the 3rd Neem Conference, Nairobi, Kenya, 1986* (Eds. Schmutterer, H. and Ascher, K.R.S.), pp. 495. GTZ Press, Eschborn, West Germany.

Klocke, J.A., Balandrin, M.F., Barnby M.A. and Yamasaki, R.B. (1989) Limonoids, phenolics and furancocoumarins as insect antifeedants, repellents and growth inhibitory compounds. In. *Insecticides of Plant Origin* (Eds. Arnason, J.T., Philogene, B.J.R. and Morand, P.), pp. 136–149. ACS 387, Washington.

Koul, O., Smirle, M.J. and Isman, M.B. (1990) Asarones from *Acorus calamus* L. oil. Their effect on feeding behaviour and dietary utilization in *Peridroma sucia*. *Journal of Chemical Ecology* 16, 1911–1925.

Kubo, I., Asaka, Y., Stout, M.J. and Nakatsu, T. (1990) Structure of a novel phytoecdysteroid vitexirone, and efficient isolation of phytoecysteroids from *Vitex fisherii*. *Journal of Chemical Ecology* 16, 2581–2588.

Lange, W. (1985) Experiments on synergistic and residual effects of neem seed extracts. Ph.D thesis, University of Giessen, West Germany.

Langenheim, J.H., Convis, C.L., Macedo, C.A. and Stubblebine, W.H. (1986) *Hymenaea* and *Copaifera* leaf sesquiterpenes in relation to lepidopteran herbivory in southeastern Brazil. *Biochemical Systematics and Ecology* 14, 41–49.

Ley, S.V., Anderson, J.C., Blaney, W.M., Lidert, Z., Morgan. E.D., Robinson, N.G. and

Simmonds, M.S.J. (1988) Chemistry of insect antifeedants from *Azadirachta indica* (Part 3): Reactions on the C-22, 23 enol ether double bond of azadirachtin and conversion to 22, 23-dihydro-23-beta-methoxyazadirachtin. *Tetrahedron Letters* 29, 5433–5436.

Ley, S.V., Anderson, J.C., Blaney, W.M. and Simmonds, S.M.J. (1989) Insect antifeedants from *Azadirachta indica* (Part 5): chemical modification and structure activity relationships of azadirachtin and some related limonoids. *Tetrahedron Letters* 45, 5175–5192.

Luthria, D.L., Ramakrishnan, V., Verma, G.S., Prabhu, B.R. and Banerji, A. (1989) Insect antifeedants from *Atalantia racemosa*. *Journal of Agricultural Food Chemistry*, 37, 1435–1437.

Macleod, J.K., Moeller, P.D.R., Molinski, T.F. and Koul, P. (1990) Antifeedant activity against *Spodoptera litura* larvae and (13C) NMR spectral assignments of the meliatoxins. *Journal of Chemical Ecology* 16, 2511–2518.

Makanjuola, W.A. (1989) Evaluation of extracts of neem (*Azadirachta indica* A. Juss) for the control of some stored product pests. *Journal of Stored Product Research* 25, 231–237.

Matsuoka, T., Imai, S., Sakai, M. and Kamada, M. (1969) Studies on phytoecdysones: a review of our works. *Annual Reports Takada Research Laboratories* 28, 221–271.

Mikolajczak, K.L, Zilkowski, B.W. and Bartelt, R.J. (1989) Effect of meliaceous seed extracts on growth and survival of *Spodoptera frugiperda* (J.E. Smith). *Journal of Chemical Ecology* 15, 121–128.

Miles, D.H., Hankinson, B.L. and Randle, S.A. (1985) Insect antifeedants from the Peruvian Plant *Alchornea triplinervia*. In: *Bioregulators for Pest Control* (Ed. Hedin, P.A.), pp. 469–476. ACS 276, Washington.

Milner, R.J. (1987) From Chinese medicine to mycoinsecticides: man's exploitation of entomogenous fungi. *The Mycologist* 21, 147–151.

Misato, T. and Yamaguchi, I. (1984) Pesticides of microbial origin. *Outlook on Agriculture* 13, 136–139.

Miyakado, M., Nakayama, I. and Onho, N. (1989) Insecticidal unsaturated isobutylamides: from natural products to agrochemical leads. In: *Insecticides of Plant Origin* (Eds. Arnason, J.T., Philogene, B.J.R. and Morand, P.), pp. 173–187. ACS 387, Washington.

Murdock, L.L., Brookhart, G., Edgecomb, R.S., Long, T.F. and Sudlow, L. (1985) Do plants "Psychomanipulate" insects? In: *Bioregulators for Pest Control* (Ed. Hedin, P.A.), pp. 337–351. ACS 276, Washington.

Nakatani, K., James, J.C. and Nakanishi, K. (1981) Isolation and stuctures of trichilins, antifeedants against the southern armyworm. *Journal of the American Chemical Society* 103, 1228–1230.

Norris, D.M. (1986) Anti-feeding compounds. In: *Chemistry of Plant Protection Vol 1*, (Eds. Haug, G. and Hoffman, H.), pp. 97–146. Springer-Verlag, Berlin.

Parmar, B.S. and Srivastava, K.P (1987) Development of some neem formulations and their evaluation for the control of *Spilosoma obliqua* (Wlk) in the laboratory and *Euchrysops cnejus* (F.) in the field. *Proceedings of the 3rd Neem Conference, Nairobi, Kenya, 1986* (Eds. Schmutterer, H. and Ascher, K.R.S.), pp. 205. GTZ Press, Eschborn, West Germany.

Pickett, J.A. (1991) Lower terpenoids as natural insect control agents. In: *Ecological Chemistry and Biochemistry of Plant Terpenoids* (Eds. Harborne, J.B. and Tomas-

Barberan, F.A.), pp. 297–313. Clarendon Press, Oxford.

Polonsky, J., Bhatnagar, S.C., Griffith, D.C., Pickett, J.A. and Woodcock, C.M. (1989) Activity of quassinoids as antifeedants against aphids. *Journal of Chemical Ecology* 15, 993–998.

Powell, R.G., Mikolajczak, K.L., Zilkowski, B.W., Lu. H.S.M., Mantus E.K. and Clardy, J. (1991) Dithyreanitrile: an unusual insect antifeedant from *Dithyrea wislizenii. Experientia* 47, 304–306.

Proksch, P., Wray, V., Isman, M.B. and Rahaus, I. (1990) Ontogenetic variation of biologically active natural products in *Ageratina adenophora. Phytochemistry* 29, 453–458.

Reed, D.K. and Reed, G.L. (1986) Control of vegetable insects with neem seed extracts. *Proceedings Indiana Academy of Science* 94, 335.

Rees, S.B. and Harborne, J.B. (1985) The role of sesquiterpene lactones and phenolics in the chemical defence of the chicory plant. *Phytochemistry* 24, 2225–2231.

Roberts, D.W. (1981) Toxins of entomopathogenic fungi. In: *Microbial Control of Pests and Plant Diseases, 1970–1980* (Ed. Burgess, H.D.), pp. 441–464. Academic Press, London & New York.

Rodriguez, E. (1985) Insect feeding deterrents from semi-arid and arid land plants. In: *Bioregulators for Pest Control* (Ed. Hedin, P.A.), pp. 447–453, ACS 276, Washington.

Romeo, J. and Simmonds, M.S.J. (1989) Nonprotein amino acid feeding deterrents from *Calliandra.* In: *Insecticides of Plant Origin* (Eds. Arnason, J.T., Philogene, B.J.R. and Morand, P.), pp. 59–68. ACS 387, Washington.

Samson, R.A., Evans, H.C. and Large, J.P. (1988) *Atlas of Entomopathogenic Fungi*, Springer-Verlag, Berlin. 192pp.

Saxena, R.C. (1988) Insecticides from Neem. In: *Insecticides of Plant Origin* (Eds. Arnason, J.T., Philogene, B.J.R. and Morand, P.), pp. 110–135. ACS 387, Washington.

Simmonds, M.S.J. and Blaney, W.M. (in press) Insect antifeedants from the Labiates In: *The Labiates* (Eds. Harley, R. and Reynolds, T.)

Simmonds, M.S.J., Blaney, W.M., Delle Monache, G., Marquina MacQuhae, M. and Marini Bettolo, G.B. (1985) Insect antifeedant properties of anthranoids from the genus *Vismia. Journal of Chemical Ecology* 11, 1593–1599

Simmonds, M.S.J., Blaney, W.M., Ley, S.V., Savona, G., Bruno, M. and Rodriguez, B. (1989) The antifeedant activity of clerodane diterpenoids from *Teucrium. Phytochemistry* 28, 1069–1071.

Simmonds, M.S.J., Blaney, W.M. and Fellows, L.E. (1990a) Behavioural and electrophysiological study of antifeedant mechanisms associated with polyhydroxy alkaloids. *Journal of Chemical Ecology* 16, 3167–3196.

Simmonds, M.S.J., Blaney, W.M., Delle Monache, F. and Marini Bettolo, G.B. (1990b) Insect antifeedant activity associated with compounds isolated from species of *Lonchocarpus* and *Tephrosia. Journal of Chemical Ecology* 16, 365–379.

Simmonds, M.S.J., Blaney, W.M., Ley, S.V. Anderson, J.C. and Toogood, P.L. (1990c) Azadirachtin: structural requirements for reducing growth and mortality in lepidopterous larvae. *Entomologia experimentalis et applicata* 55, 169–181.

Sombatsiri, K. and Tigvattanont, S. (1987) Effects of neem extract on some insect pests of economic importance in Thailand. In: *Natural Pesticides from the Neem Tree and Other Tropical Trees* (Eds. Schmutterer, H. and Ascher, K.R.S.), GTZ Press, Eschborn, West Germany.

Srivastava, R.P., Proksch, P. and Wray, V. (1990) Toxicity and antifeedant of a sesquiterpene lactone from *Encelia* against *Spodoptera littoralis*. *Phytochemistry* 29, 3445–3448.

Sugawara, F., Strobel, G. Fisher, L.E. Van Duyne, G.D. and Clardy, J. (1985) Bipolaroxin, a selective phytotoxin produced by *Bipolararis cynodontis*. *Proceedings of the National Academy of Science* (USA) 82, 8291–8294.

Sutherland, O.R.W., Russell, G.B., Biggs, D.R. and Lane, G.A. (1980) Insect feeding deterrent activity of phytoalexin isoflavonoids. *Biochemical Systematics and Ecology* 8, 73–75.

Swingle, W.T., Haller, H.L., Siegler, E.H. and Swingle, M.C. (1941). A Chinese insecticidal plant, *Tripterygium wilfordii*, introduced into the United States. *Science* 93, 60–61.

Takemoto, T., Ogawa, S. and Nishimoto, N. (1967) Isolation of the moulting hormones of insects from *Achyranthes radix*. *Yakugaku Zasshi* 87, 325—327.

Thappa, R.K., Tikku, K., Saxena, B.P., Vaid, R.M. and Bhutani, K.K. (1989) Conessine as a larval growth inhibitor, sterilant, and antifeedant from *Holarrhena antidysenterica* Wall. *Insect Science its Application* 10, 149–155.

Wright, V.F., Vesonder, R.F. and Ciegler, A. (1982) Mycotoxins and other fungal metabolites as insecticides. In: *Microbial and Viral Pesticides* (Ed. Kurstak, E.), 720pp. Marcel Dekker Inc., New York & Basle.

Yano, K. (1987) Minor components from growing buds of *Atemisia capillaris* that act as insect antifeedants. *Journal of Agricultural Food Chemistry* 35, 889–891.

Yokota, T., Arima, M. and Takahashi, N. (1982) Castasterone, a new phytosterol with plant-hormone potency, from chestnut gall. *Tetrahedron Letters* 23, 1275–1278.

Chapter eleven:
Pesticide Use in Malaysia in the Year 2000

S.H. Tan, H.F. Yeoh, S. Ramasamy and A.M. Salleh
Pesticides Section, Crop Protection Branch, Department of Agriculture, Kuala Lumpur

Summary

Agriculture has for a long time been the mainstay of Malaysia's economy. Despite the stronger performance of the manufacturing and service sectors in the country in recent years, the contribution of agriculture to the GDP is still substantial. The population is expected to increase by 25% to 22.4 million in year 2000. Therefore, the importance of the agricultural sector in the national economy as well as its role in providing food and employment cannot be overemphasised. Hence, inputs such as pesticides are expected to continue to play an important role in the agricultural sector.

The pesticide industry in Malaysia is characterised by a heavy dependence on imported technical active ingredients and foreign product technology. It is projected that there will be an annual growth rate of 6.5% for the period 1990–95 and this is expected to be maintained for the remaining five years of the decade. It is anticipated that herbicides will remain the largest group of pesticides used.

Pesticides will continue to play an important role in crop protection for the foreseeable future as there are no practical alternatives at the moment. However, there have been tremendous changes in some of the newer pesticides introduced. One of these is the significant reduction in dosage rates. Another is their reduced persistence in the environment compared to the earlier pesticides.

It is envisaged that although the total value of pesticides used annually will increase, the volume used will not. This is because of the trend towards the use of more highly active and selective pesticides.

Introduction

The early methods of pest control in agriculture in Malaysia were not based on chemical pesticides but largely on mysticism, superstition and cultural practices (Lim *et al.*, 1983). In the intervening years until after the Second

World War some chemical pesticides, mainly inorganic chemicals, were used. However, it was only following the discovery of the insecticidal properties of DDT and its introduction in the mid-1940s that crop protection practices have been centred around the use of pesticides.

There is no denying that synthetic pesticides have contributed greatly towards higher agricultural productivity in the country, but in recent years they have come under criticism because of problems related to their use. There is increasing evidence that contamination and deterioration of the environment are indeed taking place, as a result of neglect, ignorance, accidents and in many cases a total disregard by man to preserve and protect it. The use of pesticides has been identified as one of the causes of environmental contamination.

Despite the stronger performance of the manufacturing and service sectors, the contribution of agriculture to the Gross Domestic Product (GDP) in the country, in 1986–88 averaged 22%, employment 32% and its share in total export earnings, was 31% (Mid-Term Review of the Fifth Malaysia Plan 1986–90). The population is expected to increase by 25% to 22.4 million in the year 2000 (Department of Statistics Malaysia, 1987). Therefore, the importance of the agricultural sector in the national economy as well as its role in providing food and employment cannot be overemphasised. Hence, inputs such as pesticides are expected to continue to play an important role in the agricultural sector.

Current Trends in Agriculture and Pest Control in Malaysia

While agriculture continues to make significant contributions to the gross domestic product, this trend has been changing since the 1980s with the manufacturing sector making increasingly higher contributions. Rubber, oil palm, cocoa, coconut, paddy, fruits, vegetables, pepper, tobacco, coffee and tea, presently the main crops in the country, will continue to be so even by the year 2000. However, there will be a gradual conversion of some rubber, cocoa/coconut and marginal paddy lands for industrial and urban development. Fruits and high-value food production are expected to become more important. The areas under the various major crops over the last few years are given in Table 1.

From the above, there appears to be a slowing down in the rate of development of agricultural land under the major crops. In fact, there has been a decrease in the hectarage for rubber and paddy in the last few years. This trend is expected to continue. Although the rate of increase will be slower, productivity is expected to increase as a result of improved planting material and better management including the area of pest control.

Pest control practices in rubber, oil palm and cocoa are focused mostly on weed control. With the shortage of labour in the plantation industry, it is not surprising that the use of herbicides is a very important and integral part

Table 1. Malaysia: crops by hectarage, 1985–90

Crop	1985	1986	1987	1988	1989	1990*
Rubber	1,950,400	1,916,053	1,903,453	1,890,853	1,878,253	1,865,653
Oil palm	1,468,214	1,599,311	1,685,581	1,745,581	1,805,600	1,865,600
Cocoa	303,897	322,344	341,000	396,600	453,300	518,200
Paddy	661,443	627,565	623,828	647,439	642,413	638,693
Coconut	224,100	219,771	215,525	211,362	207,279	203,279
Pepper	5,413	6,170	8,900	9,000	10,000	11,200
Pine-apple	10,336	10,400	7,690	8,700	10,501	10,000
Vege-tables	9,772	11,972	10,797	9,946	9,236	8,661
Orchards	106,545	110,269	115,070	121,174	129,914	138,811
Tobacco	16,180	15,822	12,314	9,600	11,556	12,667

Source: Mid-term Review of the Fifth Malaysia Plan 1986–90.
* estimate

of the production of the above commodities. This trend will continue for the coming years, though the introduction of sheep rearing under these crops has helped to reduce the use of herbicides in some situations. Since these crops will continue to be the main contributors to the agricultural sector, it is anticipated that herbicides will remain the largest group of pesticides used in the country in the year 2000 and after.

In oil palm, besides weeds, insect pests such as the leaf-feeding nettle caterpillars, bagworms, the bunch moth (*Tirathaba mundella*), the rhinoceros beetle (*Oryctes* sp.) and grasshoppers (*Valanga* sp.) can cause serious damage. Therefore insecticides will continue to play an important role in their control.

Direct seeding in rice has also resulted in weeds becoming a more serious problem. This has forced many farmers who are also facing labour shortages to resort to herbicides to control weeds such as *Echinochloa crus-galli, E. colonum, Leptochloa chinensis* and other grasses. In Japan it has been reported (Ozawa, 1990) that before herbicides were introduced in 1949, rice growers spent 506 hours out of 2162 hours per hectare, weeding. Today weed control is reduced to 32 hours per hectare.

Insect pests such as the brown plant hopper, the green leaf hopper, stem borers and rice bugs have been perennial pests of rice. It is likely that they will continue to be the major pests, thus requiring the continued use of insecticides for their control. However, with the adoption of Integrated Pest Management (IPM) by rice farmers, pesticides should be used only when necessary.

In cocoa, pests such as the tingid bug, *Helopeltis theobromae* and the pod borer, *Conopomorpha cramerella*, are likely to remain serious pests of cocoa and therefore will require the continued use of insecticides for their control.

The decline in vegetable areas during the last few years is of concern. This has been attributed partly to farmers changing to fruit cultivation and growing of ornamentals, which are more profitable. Insect pests such as the diamondback moth, *Plutella xylostella*, army worm, *Prodenia* sp., fruit borers and leaf feeders are the major pests of vegetables. The use of broad spectrum non-selective pesticides in the past has reduced the population of natural predators of these pests and it will be a long time before the natural balance can be regained. Insecticides will continue to be an important agricultural input. However, the use of microbial pesticides that are more compatible with integrated pest management now being introduced in vegetable production, will become more important. The use of fungicides on vegetables, fruits and ornamentals will continue.

World Pesticides Market

Bernet (1989) stated that the world agrochemical market is now approaching maturity and only moderate to low growth is anticipated especially for the developed countries. However, the industry is expected to experience greater growth in the developing countries with the market for herbicides showing a marked increase due to labour shortage caused by rural migration. For 1989 the world agrochemical market was valued at US $21,500 million. The breakdown for the different categories of pesticides is as shown in Table 2.

Table 2. World agrochemical market (1989)

	Sales (US $ million)		
Product category	1988	% change	1989
Herbicides	8,896	+ 6.8	9,503
Insecticides	6,135	+ 0.9	6,192
Fungicides	4,192	+ 7.2	4,494
Others	1,227	+ 6.8	1,311
Total	20,450	+ 5.1	21,500

Source: Agrow (1990a).

Pesticides Market in Malaysia

Malaysia is basically an importer of pesticides. There is not likely to be any change in the foreseeable future. The pesticide industry is characterised by heavy dependence on imported technical active ingredients and foreign product technology. The main manufacturing activity is formulation from imported technical materials although a few active ingredients are being manufactured locally. This is likely to continue with a few additional active ingredients manufactured by the turn of the century. The supply and demand projections for pesticides are as shown in Table 3.

Table 3. Supply and demand projections for pesticides

	1991	1992	1993	1994	1995	Ave. annual growth rate % 1991–95
	(million ringgit in 1981 constant prices)					
Domestic consumption	494.1	526.6	561.0	594.8	632.6	6.4
Export	35.8	38.9	42.3	45.9	49.9	8.7
Total	529.9	665.5	603.3	640.7	682.5	6.5

Source: Malaysian Industrial Development Authority (1985).

The Medium and Long Term Industrial Master Plan of Malaysia on the Chemical and Petro-Chemical Industry (1985) projected an annual growth rate of 6.5% for pesticides for the period 1990–95. It is anticipated that this trend of growth will be maintained for the remaining five years of the decade.

Estimates by the Malaysian Agricultural Chemicals Association (MACA) on the amount of pesticides used from 1986 to 1990 are as shown in Table 4. This table excludes the household pesticide market which is valued at 130 million ringgit for 1990. From this table it is clear that as elsewhere in the world, Malaysia is also a herbicide market.

Table 4. Malaysia: Estimates of agrochemical dollar (ringgit) value

	(million ringgit — end-users' level)				
	1986	1987	1988	1989	1990
Herbicides	171.8	210.0	230.0	240.0	250.0
Insecticides	38.0	42.0	43.5	36.0	41.0
Fungicides	15.0	16.0	16.5	12.0	14.0
Rodenticides	8.0	9.0	9.5	9.0	10.0
Total	232.8	277.0	299.5	297.0	315.0

Source: Malaysian Agricultural Chemicals Association (1991).

Current Trends in Pesticides

In the last decade, a number of environmental pollution problems as well as food contamination incidents by pesticides have been reported. Such problems are partly due to the very characteristics of the pesticides that have been used. Pest control using such chemicals often results in loading the environment with large amounts of the chemicals that in many cases are also persistent. There have, however, been tremendous changes in some of the newer pesticides introduced. One of the features of this new generation of pesticides is the significant reduction in dosage rates; smaller amounts of pesticides are needed to treat the same area of cultivated land. In addition, the newer pesticides are also less persistent in the environment. Examples of some of these new pesticides in comparison with some older ones are as shown in Table 5.

Table 5. Comparison of field application rates of some pesticides

Pesticide group	Pesticide	Year introduced	Field rate g/ha
Organochlorine	DDT	1944	1000–3000
	dieldrin	1949	1000–3000
Organophosphorus	diazinon	1952	500–1500
	dimethoate	1956	300–700
Carbamate	carbaryl	1956	250–2000
	carbofuran	1965	250–1000
Synthetic pyrethroid	cypermethrin	1976	20–50
	alphacypermethrin	1983	7–15

Warrell (1990) reported that records of the British Agrochemicals Association for the UK showed a 41% reduction in weight of active ingredients sold over the past eight years, from 35,000 to 23,000 tonnes per annum, principally through the introduction of active ingredients that are used at lower dosages.

Similarly, although the value of US pesticide production between 1977 and 1988 increased at an average annual rate of 8% reaching $6600 million in 1988, the volume of production declined by 30% during the same period. This increase in value and reduction in volume are again attributed to the reduced application rates. It is projected that there will be no increase in volume during the period 1988–2000 (Agrow, 1990b).

Taking into cognisance the trends in the developed countries and that it takes some time for these pesticides to be introduced into the Malaysian market, it is envisaged that Malaysia will experience a similar trend: the value of pesticides used will increase while the volume will decrease during the remaining years of this century. In fact this trend has already started as some of these highly active pesticides have already been introduced in the local market.

Pesticide Legislation

Most countries have regulations to control pesticides. In Malaysia, the principal legislation is the Pesticides Act 1974. The main aspect of this Act is the control of the manufacture and import of pesticides through registration. Registration of pesticides is a dynamic process as pesticides are reviewed from time to time with the availability of new information. For pesticides that have fallen into disuse under local conditions and where risks outweigh benefits, the registration of such products will be withdrawn.

As problems related to pesticides are being highlighted and as information becomes available, many regulatory authorities especially in developed countries are reviewing both the pesticides and the regulations that control them. In Denmark, legislation was enacted some time ago calling for a 50% reduction in the use of pesticides by 1997. Sweden also adopted a similar legislation but the target date for the 50% reduction is 1993. More recently, the Netherlands proposed that the use of pesticides should be reduced by 50% by the year 2000. In the US, particularly in the State of California, it was reported (Maddy, 1990) that many pesticides have had their registrations severely restricted, withdrawn or suspended. Many pesticides are also under review and data are being requested to fill "data-gaps". In Malaysia, there is at present no policy for setting target reductions in the use of pesticides but a review of the registration status of some pesticides is in progress. This is particularly important as new information becomes available and as risk–benefit considerations of certain pesticides have also undergone changes.

Malaysia, like many other countries, has also accepted the FAO International Code of Conduct on the Distribution and Use of Pesticides, which has among other things set standards to:

- promote practices which encourage the safe and efficient use of pesticides, including minimising adverse effects on humans and the environment and preventing accidental poisoning from improper handling;
- ensure that pesticides are used effectively for the improvement of agricultural production and the protection of human and plant health.

The Code is particularly useful for countries without pesticide control schemes. For countries that have established such schemes such as Malaysia, this Code serves as a useful supplementary guide.

The Future

Finney (1990) stated that as a result of mergers and acquisitions among pesticide companies and pressures facing the industry, the percentage of sales invested in research and development has started to decline and is expected to decline at an even faster rate in the next few years. A consequence of this is the decline in the introduction of new active ingredient products in the market. Table 6 shows the number of new products that were introduced between 1984 and 1989. As this trend is expected to continue in the coming years, product development through innovative formulations and application technology will become more important. Biotechnology will play an increasingly important role in crop protection in the future but it is doubtful that it will create a major impact by the year 2000.

Table 6. New product introductions 1984–89

Year	Number of new product introductions
1984	11
1985	20
1986	27
1987	22
1988	20
1989	16

Source: Finney (1990).

The need to produce more food to feed the increasing population will continue. The systems and technologies that are being used now may have to be modified and new technologies adopted in order to ensure that this goal can be achieved. Pesticides will continue to play an important role in crop protection for the foreseeable future as there are no practical alternatives at the moment. However, of late there have been many encouraging developments that give hope for the future.

The old concept of "immediate kill" or "kill all" chemicals will have to give way to the concept of chemicals that are less hazardous to the natural parasites and predators and keep pests at manageable levels. Hence, there is a need to consider future pesticides as bioregulators which should have the following features:

- Selectivity for the target pests and safety to non-target organisms
- Effectiveness at very low doses to keep the input into the environment as low as possible

- Biodegradability
- Compatibility with IPM

Pesticides introduced more recently are more specific in their use and this may help to minimise some of the problems caused by pesticides of the past, but pest control strategies will have to be improved. In Malaysia, Integrated Pest Management (IPM), which was first introduced in 1980 for paddy and is now being introduced to vegetables, will have to be incorporated for pest control in other crops.

Farmer education and transfer of technology on pesticides to end-users will have to be intensified. In this context the Department of Agriculture has already started a pesticide extension programme to train farmers on the judicious use of pesticides. With the introduction of the new generation of pesticides that are not only highly active and selective but are highly dependent on the timing and method of application, farmers will also have to be informed to ensure that such chemicals are used properly. The use of IPM alone in the country may not be sufficient, but a more holistic approach that takes into consideration cultural practices such as crop rotation, use of resistant varieties, accurate matching of fertiliser inputs to crop requirements and, where possible, use of biological control, may be the long-term solution to the problems faced by the agricultural sector.

The FAO in its *State of Food and Agriculture Report* (1989) stated that "All nations must search for ways of attaining developmental goals without causing irrevocable environmental damage: international organizations must do their part, but so must governments and the people they represent." It goes on further to identify inappropriate agricultural technologies that are not sustainable as one of the three major problems faced by mankind today. The indiscriminate use of pesticides has caused disruption of ecosystems because of the death of non-target species, the accumulation of pesticide residues in the environment and in food and the build-up of pesticide resistance in the target species. Therefore, it is important for all sectors, be it the pesticide industry, the regulatory bodies, public interest groups or pesticide users, to ensure that pesticides are used judiciously with minimal deleterious effects on non-target organisms and the environment.

Acknowledgement

The authors wish to thank the Director-General of Agriculture for permission to present this paper.

References

Agrow (1990a) Some Growth in World Market in 1989. No. 118, 16.

Agrow (1990b) US pesticide exports to exceed domestic use by 2000? No. 105: 18–19.

Bernet, E.E. (1989) The pesticide industry: Future trends and responsibilities. In: *Report of the Workshop and Symposium on Strengthening Pesticide Regulation.* Food and Agriculture Organization & Asian Development Bank, Manila, pp. 65–70.

Department of Statistics, Malaysia (1987) Population Projections, Malaysia 1980–2000, Kuala Lumpur, p. 113

FAO (1989) *The State of Food and Agriculture.* Food and Agriculture Organization, Rome, 171 pp.

Finney, J. (1990) World Crop Protection Prospect: Where Do We Stand? Where Do We Go? *Seventh International Conference of Pesticide Chemistry*, International Union of Pure and Applied Chemistry, Hamburg.

Lim, G.S., Hussein, M.Y., Ooi, A.C.P. and Zain, M.B.A.R. (1983) Pesticide application technology in annual crops in Malaysia. In: *Pesticide Application Technology* (Eds. Lim, G.S. and Ramasamy A.), pp. 13–41, Malaysian Plant Prot. Soc. (MAPPS), Kuala Lumpur.

Maddy, T.K. (1990) The Fate of Herbicides as a Result of Registration Process in the United States and the State of California and the Increasing Level of Public Concern About Exposure to any Pesticide. *Third Tropical Weed Science Conference*, Kuala Lumpur.

Malaysian Agricultural Chemical Association (1991) Annual Report 1990/91 and Directory, Petaling Jaya.

Malaysian Industrial Development Authority (1985) Medium and Long Term Industrial Master Plan. *Chemical and Petro-Chemical Industry.* Vol II part 5, p. 124.

Mid-Term Review of the Fifth Malaysia Plan (1986–1990) 1989. National Printing Department, Malaysia Kuala Lumpur. 295 pp.

Ozawa, K. (1990) Rice farmers choosing one-shot herbicides. *Farm Chemicals International*, pp. 68–69.

Warrell, E. (1990) Reducing pesticide use: The Danish experience. Shell Agriculture No. 8, pp. 18–20.

Chapter twelve:
Pesticides
Discussion

Chairman: Dato' Dr Hashim Abdul Wahab

Dr K. Gallagher (FAO Indonesia) "Botanicals" or "natural" insecticides imply soft, ecologically safe insecticides, but actually are also quite toxic and disruptive (nicotine, pyrethrins, etc.). Aren't you misleading the general public (farmers, consumers, etc.) about the toxicity and biological activities of these compounds?

Dr M.S.J. Simmonds (Royal Botanic Gardens, Kew) Yes, I take your point about the term "botanical pesticides". This can imply a "safe" pesticide. We have tried to use other terms but without much success. However, we usually try and get the word "biodegradable" into our title. But this still doesn't help us to imply a "safe pesticide". Many of the chemicals that we are dealing with have gone through toxicological tests and have been shown to be inactive. Problems associated with toxicity of "botanicals" is something that we are very aware of. If you can come up with a term to replace "botanical insecticide" that implies safe and natural, I'd be very interested.

Dr K. Gallagher Your presentation claimed increased use of herbicides in direct seeded rice in the future. I wonder if you have considered research underway on rice varieties producing root exudates which prevent weed germination (a common characteristics of many plants)?

Dr S.H. Tan (DOA, Malaysia) Our projection is based on the existing situation and trends in the country. We have not considered the research you mentioned. Once there are proven results from varieties of rice producing such substances, the use of herbicides will certainly be reduced. Herbicide usage is in any case very labour intensive, at a time when agricultural labour in Malaysia is in short supply.

Dr Jeff Waage (IIBC, UK) Are there real prospects for pesticide rotation as a means of managing pesticide resistance? And might it apply to the Malaysian situation where there are so few key crops and these are plantation crops where enforced or agreed rotation may be more easily implemented than with a mixed system of small farm field crops?

Dr W.P. Davies (Royal Agricultural College, UK) There is no doubt that reducing selection pressure for resistance to a particular pesticide, through rotation of products with different modes of action will be helpful. In the United Kingdom, fungicides used on cereal crops are grouped on the basis of their mode of action and chemical rotation is encouraged. It is very difficult to test experimentally hypotheses such as chemical rotation on the development of resistance. Large areas are required, usually making it a difficult experimental system. I would support your comments with reference to Malaysia and the plantation sector, which may prove easier to regulate.

Dr R.J. Williams (Ciba-Geigy, UK) What is the adaptation (resistance) potential of insects when faced with a whole crop uniformly producing the repellent?

Dr M.S.J. Simmonds So far there is very little evidence of adaptation to those compounds that act by suppressing the response to sugars. We found that when chemicals just stimulate "deterrent" receptors, adaptation can occur quickly, within a few hours. However, adaptation does not occur so quickly to those chemicals that appear to bind with sugar receptors. These chemicals could inhibit the enzymes that may be involved in releasing chemicals from their binding sites. Another factor that can occur is that insects can avoid being stimulated by the chemicals. For example, locusts can lift their legs off the surface of a leaf treated with an antifeedant and thus the receptors on the tarsi are not stimulated. Thus we need to have compounds that can combine antifeedant activity with toxic activity to control a range of insects.

Dr R.J. Williams It is very difficult to enforce an alteration of pesticides in practice. But one indirect example is the withdrawal of metalaxyl in some countries where the potato late blight pathogen has developed resistance to this fungicide and within a few years the product was successfully reintroduced (in mixture with other fungicides).

Dr W.P. Davies That same situation seems to have prevailed with the withdrawal of ethirimol for powdery mildew control on barley. When released in mixed products, the pathogen population was found to be very sensitive. This has been described as negative cross-resistance, and may possibly be attributable to the commoner availability and use of other products (e.g. Triazoles).

Dr R.J. Williams I would like to correct Dr Davies's comment that a new international body called IOPRM will replace the Industry Resistance Action Committe (HRAC, IRAC, FRAC). There is no such intention.

Dr W.P. Davies I accept the comments made by Dr Williams. A committee and a working group structure has been set up by pesticide scientists from industry, academia and government together with ecologists, regulators, conservationists, and growers' representatives to form an international organisation for pest resistance management (IOPRM) to co-ordinate worldwide efforts of pesticide resistance management.

Incorporation of IOPRM as a legal entity and formal agreement of its charter, purpose, organisation, activities and financial resources are anticipated to be completed at the general assembly of IOPRM at the First International Pest Resistance Management Congress planned for 1992 in Washington DC.

Three technical working groups are reported to have been set in place covering insect pathogen and weed resistance management. To what extent these groups will collaborate with, or substitute for IRAC, FRAC and HRAC is still not clear.

Dr Peter Dart (University of Queensland) You mentioned decrease in fitness as a corollary of development of resistance of pests to chemical pesticides. Does this decrease their effect on plant productivity such that they are no longer serious pests?

Dr W.P. Davies There are certainly reports, in the literature, of a loss of "fitness" in insects of weed populations resistant to certain insecticides and particular herbicides. As mentioned in my talk there are also cases of a reduction in pathogenicity of certain fungal pathogens which have developed resistance to fungicides.

In other cases the development of resistance to particular pesticides has reduced the capacity of certain pests to survive. In the UK reduced sensitivity to dicarboximide fungicide by *Botrytis cinerea* on strawberries did not allow the fungicide-resistant strain to survive between seasons, through the winter period.

One imagines that removal of a pesticide selection pressure would result in a poorer ability of pesticide-resistant organisms to compete as effectively with their wild type relatives, where their fitness has been reduced.

Prof Dr George Varghese (Plantex Bhd.) Mr Chairman in response to your comment on local experience in delaying resistance build up of pests to pesticides and use of rotation of pesticides, I would like to add that one of the popular strategies now adopted by pesticide manufacturers in the case of systemic fungicides which develop early resistance is to mix a new systemic fungicide with a coventional fungicide. For example metalaxyl mixed with mancozeb (EBDC) sold as Ridomil (MZ). In this way, the build-up of resistant populations of the target pathogen against the fungicide has been considerably delayed and so far it has proved to be a successful practical method. This type of strategy is now also widely adopted in cases of other systemic fungicides such as benomyl and some of the triazoles.

Dr W.P. Davies I agree with Dr Varghese's comments. In my talk I described the benefit of combining active ingredients with different modes of action in a pesticide product. I referred specifically to improved control of powdery mildew on barley in Europe by combining propiconazole with tridemorph. The combination of different active ingredients is a common strategy adopted by agrochemical manufacturers and may be adopted by growers where compatible products can be tank-mixed.

Dr Peter Dart The apparent effect of crop rotation in Egypt in decreasing pest incidence on cotton and other crops was mentioned by Dr Mochida and Dr Davies. Is there any explanation of the reason for this effect? Would not this be an ecological system to examine in more detail to elucidate the factors controlling pest populations, with the possibility that they can be added to IPM programmes in other countries? Where is this phenomenon in Egypt best documented?

Dr W.P. Davies My comments are related to cotton in the Sudan, Thailand, Australia and the United States. I mentioned the high requirement for pesticide inputs in cotton, making it an inevitable crop for the development of pesticide resistance problems and the upsurge of secondary pests with injudicious use of chemicals.

The agroecology of crop systems certainly requires more study, and integrated resistance management (IRM) must require other crop enterprises in the rotation in addition to the target crop.

Dr Peter A.C. Ooi (IIBC, Malaysia) Is it possible that the failure in using pheromones for the control of the cocoa pod borer (CPB) is due to resistance? What is the possibility of development of resistance to the baits for fruit flies?

Dr W.H. Loke (MARDI, Malaysia) The "failure" (maybe not the right word) of using the CPB sex pheromone for mass trapping on a large scale is perhaps due to the issue of costs and the possibility of the existence of CPB on non-cocoa hosts serving as a continuous source of infestation. The current pheromone blend available may not be the ultimate/optimal combination.

The possibility of resistance development to baits certainly cannot be ruled out as the evolutionary capability of insects is quite phenomenal. A parallel can be drawn using resistance development to hormone analogues as an example.

PART 4
Management of Pests and Diseases of Plantation Crops

Chapter thirteen: **Management of Rubber Diseases in Malaysia in 2000**

Ismail bin Hashim
Rubber Research Institute of Malaysia
P. O. Box 10150
50908 Kuala Lumpur

Abstract

Natural rubber will remain an important commodity to the economy of Malaysia in the future. As a measure to increase natural rubber production without further expansion in rubber hectarage, yield losses due to pests and diseases need to be minimised.

Rubber plants in Malaysia suffer from several diseases. Root diseases which kill the plants remain and will continue to pose a challenge especially during the immature stage of rubber. Leaf diseases, especially secondary leaf fall (SLF) will continue to be important. Diseases are continuously undergoing dynamic changes determined by climatic conditions, agronomic practices and host genotypes. *Hevea* clones GT 1 and to a lesser extent RRIM 600 were previously rated as fairly resistant to Colletotrichum SLF, but are currently severely infected. Corynespora leaf fall which was insignificant five years ago currently infects several important clones especially RRIM 600. Since a large planted hectarage of this popular clone is still immature, SLF and Corynespora leaf fall control will be the challenge in the next two decades or so. In addition, several leaf diseases (*Fusicoccum*, *Cylindrocladium*, *Guignardia* and algal spot) which were absent or obscure in the past have caused sporadic and localised problems. Pink disease (*Corticium salmonicolor*) and black stripe (*Phytophthora palmivora*), the two important diseases of the stem and the panel, though they can be controlled, need continuous attention.

Rubber cultivation in Malaysia is undergoing changes especially in the agronomic practices dictated by the need to reduce the immaturity period as well as the shortage of labour and land which will be more acute in the future. The nursery practices such as green budding and core stump production alter the disease scene in the nursery. The expected future reduction in incision tapping practices will reduce the importance of tapping panel diseases. The increasing cultivation of rubber on hilly and steep terrain and the increasing shortage of labour will influence disease management practices.

The traditional approaches of managing rubber diseases, through the integration of good husbandry, genetic resistance, cultural practices, plant quarantine and disease avoidance will continue, especially against leaf diseases. Stem and root diseases will still depend on chemical control, which is currently not popular as a method of controlling leaf diseases due to the unavailability of spraying applicators suitable for tall trees, especially those planted on hilly terrain. Aerial application is viewed to be too costly as multiple applications are necessary to treat leaf diseases. Fortunately, several sprayers are found suitable to treat leaf diseases of mature trees from the ground. The future utilisation of higher yielding clones will influence the economics of leaf disease control. The traditional laborious procedure of treating root diseases with collar protectant dressings needs modification or replacement. Fungicide soil drenching which is giving promising results offers the advantage for the future. In addition, the integration of chemical with biological control with the sole purpose of enhancing control and reducing cost is the other future alternative.

Introduction

Natural rubber is an important commodity to Malaysia, currently the world's greatest producer, and to several other countries in south east Asia. Many people speculate that natural rubber will continue to be an important contributor to the economies of these countries in the future. The process of crop diversification in Malaysia sees the expansion in areas of oil palm and cocoa, sometimes at the expense of flat rubber land. New rubber plantings increasingly occupy land which is marginal both in soil properties and topograhy. Since land is limited, there is little hope of a drastic increase in rubber areas in the next decade. As a measure to increase natural rubber production, the only strategy available is to increase tree productivity and reduce yield losses incurred by pests and diseases.

Important Diseases of Rubber Plants

Wastie (1975) listed and classified the various diseases of rubber based on their severity. This list was updated by Chee (1990). Over about a decade, there have been noticeable changes in that new diseases appear or a disease has lost or gained in severity (Table 1). Among the root diseases, *Rigidoporus lignosus* (Klotzsch) Imazeki was the most important, followed by *Ganoderma philippii* (Bres. & Henn.) Bres. and *Phellinus noxius* (Corner) Cunn. This phenomenon will continue to the year 2000 and probably beyond. Pink disease (*Corticium salmonicolor* Berk. & Br.) is an important disease of the stem and this status will be maintained. Black stripe (*Phytophthora palmivora* Butl.) was important in the 1980s. However, currently the disease can be easily controlled.

Table 1. Severity of *Hevea* diseases in Malaysia

Disease	Pathogen	Severity rating*		
		1975†	1991	2000
LEAF DISEASES				
Oidium secondary leaf fall	*Oidium heveae*	+++	+++	+++
Colletotrichum secondary leaf fall	*Colletotrichum gloeosporioides*	+++	+++	+++
Phytophthora abnormal leaf fall	*Phytophthora botryosa*	++	++	++
Corynespora leaf fall	*Corynespora cassiicola*	+	+ (+)	++ (+)
Bird's eye spot	*Bipolaris heveae*	++	++	+
Cylindrocladium leaf spot	*Cylindrocladium quinqueseptatum*	+	+	+?
Guignardia leaf disease	*Guignardia heveae*	–	+	+?
Algal spot	*Cephaleuros mycoidea*	(+)	+	+?
Fusicoccum leaf fall	*Fusicoccum* sp.	–	+?	+?
TAPPING PANEL AND STEM DISEASES				
Black stripe	*Phytophthora palmivora*	++	++	+ (+)
Mouldy rot	*Ceratocystis fimbriata*	+	– (+)	– (+)
Pink disease	*Corticium salmonicolor*	+++	+++	+++
ROOT DISEASES				
White root disease	*Rigidoporus lignosus*	+++	+++	+++
Red root disease	*Ganoderma philippii*	++	++	++

Table 1. *Continued*

Disease	Pathogen	Severity rating*		
		1975†	1991	2000
Brown root disease	*Phelinus noxius*	+	+	+
Ustulina root rot	*Ustulina deusta*	+	– (+)	– (+)
Poria root rot	*Poria hypobrunnea*	+	– (+)	– (+)
Stinking root rot	*Sphaerostilbe repens*	+	– (+)	– (+)

* –, absent; +, minor : ++, moderate ; +++ major.
† See Wastie (1975).

Similarly mouldy rot (*Ceratocystis fimbriata* Ell. & Holst.) has declined in importance. Secondary leaf fall (SLF) caused by either *Oidium heveae* Steinm. or *Colletotrichum gloeosporioides* Pens. will remain an important disease beyond 2000. Abnormal leaf fall (*Phytophthora botryosa* Chee) has increased in geographical area but remains less important because of its occasional outbreaks and low economic impact so far recorded.

The dynamic changes of *Hevea* diseases are noticeably demonstrated by the gain in prominence of several leaf diseases. *Corynespora cassiicola* (Berk. & Curt) Wei was first recorded in 1960 (Newsam, 1961) confined to nursery plants of South American origin. In 1975, an outbreak of Corynespora leaf fall (CLF) caused defoliations of immature clone RRIM 725, also of South American parentage (Rubber Research Institute Malaysia, 1975). Subsequently, severe infections of clone RRIC 103, a clone introduced from Sri Lanka caused growth retardation and even death of the trees. A survey conducted in 1990 indicated that CLF is widespread throughout Peninsular Malaysia affecting among others the popular clones such as RRIM 600, GT 1 and also newer clones (Tan, 1990b). Apart from a few localities, CLF is light in severity. However, CLF may be an important disease in the year 2000 as a large planted hectarage of the susceptible clone such as RRIM 600 and GT 1 is still immature. Morever the newer recommended clones are also susceptible to CLF (Tan, 1990b).

Other leaf diseases which are caused by *Cylindrocladium quinqueseptatum* Boedijn & Rietsma (Radziah & Chee, 1988), *Guignardia heveae* Sydow (Tan, 1985) and the algal spot caused by *Cephaleuros mycoidea* Karst which were obscure in the past are now noted to cause defoliation of plants in certain localities. These disease if not closely monitored and properly managed may gain prominence beyond the year 2000. In addition, a new disease of rubber caused by a fungus *Fusicoccum* sp. was detected in 1987 (Radziah & Chee, 1989a, b).

Rubber is a perennial crop with more than 20 years' economic life. The long period permits changes in climatic conditions as well as fungal pathogenicity to take effect. An exemplary case is the increased susceptibility of a previously resistant clone, GT 1 to *Colletotrichum* SLF (Ismail, 1984). Similar situations may develop with the other clonal–disease interactions.

Influence of New Technologies

Rubber experiences about six years of unproductive life during the immature period exclusive of the time spent in the nursery to prepare the planting materials. Green budding techniques, core-stump planting materials and adoption of proper pruning practices will shorten the immaturity period. These practices also influence disease: for example, the production of green budding reduces exposure time of stock seedlings to diseases. On the other hand, nursery diseases are important in the preparation of core-stump planting materials as the plants remain for several years in the nursery. Changes in the latex exploitation systems such as adoption of lower frequency tapping and use of tapping aids such as rain-guards reduce the impact of tapping panel diseases. In addition, tapping panel diseases would be insignificant if the incision method of tapping is replaced with puncture tapping as expected with the current encouraging breakthrough in the puncture tapping techniques (Sivakumaran, 1991, pers. comm.).

The yield per hectare of a smallholding is generally lower than that of an estate and since smallholdings constitute about 80% of the total rubber areas, special efforts are being made to increase the productivity per hectare as well as the household income. One of the technologies tried is increased density of planting. The effects of high planting density on diseases have not been quantified. However, it is expected that the microclimate under dense planting would be conducive to leaf and stem diseases. The closer planting distance between trees also facilitates easy and faster spread of root diseases.

In the past, stumpy high yielding clones were ideal for high density planting and also facilitated easy access for leaf disease control. The current increased awareness of the need to conserve tropical hardwood increases the value of rubber timber. It has been suggested that emphasis is given to breeding for tall, robust high yielding trees, and this will cause nightmares to the crop protectionists. The planting of intercrops along the inter-rows of young rubber supplements the income of the smallholder. In addition, crop mixing, i.e. combining rubber with other crops such as fruit trees, cocoa and coffee is being practised. Though the true effects of crop mixing on diseases are not known, it is a fact that intercropping of cocoa and coffee under rubber hinders ground spraying operations especially when tractor-mounted machines are used. The more humid environment created would be ideal for panel diseases such as mouldy rot. Organisms like *R. lignosus, G. pseudoferreum, P. palmivora* and

C. salmonicolor are common pathogens in these crops.

The incidence of root diseases in the new plantings is influenced by the method and quality of clearing the land and disposing of the old timber (Newsam, 1967). Currently the useless timber is burnt. It is anticipated that open burning will be restricted and new methodologies will need to be designed to dispose of the old timber. This new development will influence either positively or negatively the incidence and control of root diseases.

Some of these technologies are still on the drawing board or being practised only on a very limited scale. Some of them will be more widely accepted and practised in the next decade. In anticipation, crop protectionists have to devise suitable solutions to complement these developments.

Management of *Hevea* Diseases: Current Approaches

Exclusion of Exotic Diseases

Worldwide there are about 40 diseases of rubber and only a dozen of them occur in Malaysia (Chee, 1990). Several important diseases of rubber are exotic to Malaysia (Table 2), and strict quarantine regulations had and will ensure the continuation of this status. Special efforts have been established to exclude South American leaf blight (SALB) which is the most feared disease of *Hevea* rubber. Through the Association of Natural Rubber Producing Countries (ANRPC), an agreement was established with the sole purpose of excluding SALB through a concerted regional effort (Association of Natural Rubber Producing Countries, 1983). Malaysia, in addition has established a contingency plan, outlining the strategies to detect an outbreak of SALB and actions to be taken during the occurrence (Department of Agriculture and Rubber Research Institute of Malaysia, 1988).

Table 2. Diseases absent from Malaysia

Diseases	Pathogen	Endemic areas
South American leaf blight	*Microcyclus ulei*	South and Central America
Target leaf spot	*Thanatephorus cucumeris*	South America
Tar spot	*Catacauma huberi*	South America
Periconia leaf fall	*Periconia heveae*	China
Armillaria root rot	*Armillaria mellea*	Africa
Purple root rot	*Helicobasidium purpureum*	China

Genetic Resistance

Hevea diseases are currently being managed through an integration of genetic resistance, cultural practices and use of fungicides. The control of leaf diseases and to a lesser extent diseases of the stem and tapping panel, depends heavily, if not solely, on genetic resistance. The concept of enviromax planting (Ho *et al.*, 1974) recommends only resistant clones for planting in predemarcated disease endemic areas and this strategy has proven successful. Subsequently, an expert system was devised to assist in the choice of clones to be planted in an environment with their inherent soil, wind and disease limitations (Leong, 1988, pers. comm.). In the case of root diseases, apart from few initial reports (Liyanage, 1979), there are no known *Hevea* clones which are resistant to root diseases.

Cultural Practices

Adoption of suitable cultural practices is an important strategy to reduce the incidence and severity of SLF and especially root diseases. On mature trees, SLF is severe during the refoliation period after the annual wintering season, i.e. the period during which the deciduous rubber plants change their leaves. Wet rainy periods are conducive to SLF and near total defoliation of young susceptible leaves often occurs if refoliation is accompanied with rains. Mature rubber leaves, however, are immune to SLF. This phenomenon has been used to advantage to avoid severe infection of SLF. The application of desiccant defoliants to mature rubber leaves well ahead of the normal wintering period induces refoliation which can be timed to occur during the expected drier months. This technique has been useful in overcoming SLF (Rao & Azaldin, 1973; Azaldin & Rao, 1974).

The control of root diseases has been accorded special attention during the immature phase of a rubber holding. Research on root rubber diseases has resulted in an effective control method which advocates an integration of cultural strategies with chemical treatments. Pieces of infected roots and tree stumps are the main sources of infection and spread of root diseases. The incidence of root disease is reduced when the inoculum source is effectively removed during clearing of the old trees. Uprooting the trees, collecting the roots and ploughing the land is very effective (Newsam, 1967; Azaldin, 1979). However, this method is expensive as compared to the alternative method of felling the trees and poisoning the stumps which is more appropriate to the small farmers.

Horticultural and Agronomic Practices

A three part tree is produced when a high yielding trunk is top-budded with another clone possessing the desired characteristics to provide the crown (Yoon, 1976). When crowns of resistant clones are used, the technique is effective in controlling leaf diseases. In Brazil, the clone PA 31 is being crown budded on high yielding trunk clones to control SALB (Chee, 1990). The technique has also been tried to overcome *Phytophthora* leaf fall in India (Sarma, 1985).

Extra nitrogen manuring when applied at proper times enhances refoliation and reduces the severity of Oidium SLF (Lim, 1974). Planting of creeping legume cover is also effective in reducing the incidence of root disease (Newsam, 1963).

Chemical Control

Chemical fungicides are more commonly used to treat diseases of the tapping panels, stems and roots and only rarely are leaf diseases of mature plants chemically treated except for the sporadic treatment of Oidium SLF with sulphur dusting. Various fungicides effective against leaf diseases of rubber have been identified: however, their use is limited to nursery plants. There is a lack of spraying machinery suitable to spray effectively trees, often 20–30 m tall, and especially those planted on hilly terrain. In the past, mistblowers used to treat Colletotrichum SLF on mature plants did not produce excellent results (Wastie, 1971, 1975). Subsequently, experiments with fogging machines were undertaken to treat *Hevea* leaf diseases, producing some degree of control (Lim *et al.*, 1981; Ismail & Radziah, 1987). However, this technique has not been adopted by the planters, probably due to the high cost of the machines. Also the technique sometimes produces unsatisfactory results. The results in terms of increased latex production following treatment of leaf disease of mature trees with chemical fungicides may not be attractive enough to cover the high costs of the machine and chemicals especially when repeated spraying rounds, often not less than four, are necessary. The other challenges of chemical control are related to high rainfall which requires special rainfast formulations as developed for treating pink (Yeoh & Tan, 1974; Tan & Yeoh, 1976) and black stripe diseases (Yeoh & Tan, 1979; Tan, 1989).

New Challenges in Pest Management

The cost of production of rubber in Malaysia is higher than in the other neighbouring countries. In addition, the industry is also saddled with labour insufficiency. Mechanisation of farming operations especially in latex extraction and collection will reduce the chronic dependence on the decreasing labour force. Higher crop productivity will keep the plantation industry viable and attractive. Reducing crop loss due to pests and diseases contributes to increased crop productivity. However, it is pertinent that the procedures of reducing crop loss are rewarded with substantial economic gains. Less laborious, economical and environmentally safer techniques are desired.

In the year 2000 and beyond, several great challenges will confront the managers on diseases of rubber. The main hurdle is the lackadaisical attitude towards disease control especially for leaf diseases and also the slow acceptance of newly developed technologies. Disease management has to be viewed in the light of long-term strategies with less emphasis on short-term corrective measures. Better yield increases were only obtained following treatment of SLF disease carried out repeatedly for several years (Azaldin & Rao, 1974; Ismail & Radziah, 1987).

The other challenges are related to difficulties anticipated with the current increasing trends of planting rubber on hilly and steep terrain. The road systems for such terrains should cater for crop protection activities. Construction of enough tractor paths has been suggested (Lim *et al.*, 1981).

Future Prospects

Leaf disease will continue to demand special attention in view of the susceptibility of the clones presently planted and those recommended for the future (Rubber Research Institute of Malaysia, 1989). The economic gains resulting from control of leaf diseases range from marginal (Wastie & Mainstone, 1969; Ismail & Radziah, 1987; Tan & John, 1985) for one season disease treatment of Phytophthora leaf fall or SLF in Malaysia to greater yield increases obtained with control of SLF repeated for many years (Azaldin & Rao, 1974), severe Phytophthora leaf fall in India (Ramakrishnan, 1960) or SALB in Brazil (Mainstone *et al.*, 1977). In the year 2000, increased hectarages of the new clones such as the RRIM 900 series which possess potential for higher yield production will reach the tappable maturity. These higher yielding clones may offer the economic attractiveness to induce better acceptance and practice of disease control.

The strategies to manage rubber leaf diseases in the next 10 years will not differ drastically from the present, whereby greater emphasis is given to breeding for disease resistant clones which will continue to be planted based on the enviromax concept. Only when this strategy fails, application of chemical fungicides will be resorted to. These chemicals are likely to be sprayed or dusted from the ground unless there is a tremendous development in the local agricultural aviation industry which can provide the spraying service which is currently viewed as expensive. Several newer chemicals such as the triazoles are effective against the leaf diseases; however they are not expected to replace the non-proprietary and cheaper chemicals such as copper, chlorothalonil, sulphur, etc. A portable mistblower currently being tested is found suitable for treating leaf diseases affecting trees of medium height, and this machine is suitable for trees planted on hilly or flat terrain. Another bigger machine which needs to be mounted on a tractor has better vertical coverage. These sprayers offer the interim solution for treatment of diseases of mature trees. The current progress, if there is any, in developing spraying machines for tall perennial crops is not quick enough.

Stem Diseases

Chemical fungicides are more widely used to control pink and black stripe diseases and this practice will continue. Other chemicals, e.g. metalaxyl (Tan, 1989) will replace captafol which is found not safe to humans. Rainfast formulations which reduce the frequency of treatments are more appropriate when cheap labour is limiting. It is difficult to replace Bordeaux mixture in the treatment of pink disease as the chemical is cheap and the method of application is rapid and simple in comparison to the laborious brushing treatment. Other treatment methods which are simple, quick to apply and requiring only one application are still desired.

Root Diseases

Following identification of a diseased tree by foliage symptoms, the roots of the infected and neighbouring trees are exposed for inspection. Severely infected trees are uprooted and disposed of, while trees with lighter infections as well as the immediate neighbouring trees are painted with collar protectant (pentachloronitrobenzene or tridemorph) following surgical removal of rotted tissues. This procedure is laborious and cannot be carried out when there is no labour. Drenching of fungicides around lightly infected trees has been shown to cure white root disease affecting young plants (Canh, 1985; Canh *et al.*, 1990; Tan, 1990a). Multiple soil drenchings of infected trees with fungicides propiconazole and triadimefon are not cheaper than digging and applying collar protectant fungicides. However, the method is attractive and permits treatment of root disease quickly with minimum labour. What is required for the future is a cheaper and effective fungicide.

Sulphur applied around the collars of plants at the time of planting reduces the incidence of white root disease (Basuki, 1985; Peries & Liyanage, 1980; Ismail & Azaldin, 1985). The acceptance of this technique is gaining momentum and it is projected that it will be more widely used in the next decade.

The presence or increased population of micro-organisms especially *Trichoderma* in the soils under legume cover (Fox, 1965) and in soils treated with sulphur (Azaldin, 1979; Peries, 1966) has also contributed to the reduction in root disease incidence. *Trichoderma* spp. are antagonistic to *Rigidoporus lignosus* (Basuki, 1985; Ismail & Azaldin, 1985). The encouraging results obtained following integration of chemical fungicides with *Trichoderma* (Ismail,

1990) and the use of *Hansfordia pulvinata* to control SALB (Juncqueira *et al.*, 1988) indicated that biological control using micro-organisms provides a promising scope for the future. Micro-organisms if not alone may enhance chemical control in an integrated biological–chemical control system. It would be most attractive if the combination can contribute to reducing the cost of treating root diseases with fungicide applied by drenching.

Conclusion

Managing diseases of rubber will continue to be an important component of rubber cultivation. The increasing labour shortage and expanding rubber hectarages occupying difficult terrain demand preparations such as enough tractor paths to facilitate pest and disease control activities. Suitable spraying machines are desired especially suited for tall trees planted on hilly terrain. The adoption of control technologies will depend on the final economic benefit.

Genetic resistance will continue to be the main strategy to manage *Hevea* leaf diseases. More clones, especially those resistant to the newer diseases especially Corynespora will be introduced. The long breeding cycle of *Hevea* slows down the flow of resistant clones. Current developments exhibit the potential and successes of creating resistant plants through modern techniques in biotechnology. In the case of rubber, worldwide research using these breeding techniques is still at the infant stage and their produce may be realised only after the year 2000. Nevertheless, the present basket of available clones properly screened against the major leaf and stem diseases and planted according to an enviromax planting concept will continue to provide satisfactory benefits.

Acknowledgements

The author thanks the Director of the Rubber Research Institute of Malaysia for the encouragement and permission to write and present this paper at the Conference.

References

Association of Natural Rubber Producing Countries (1983) ANRPC Agreement on South American Leaf Blight (SALB), 30 pp.

Azaldin, M.Y. (1979) Effect of clearing methods and soil amendment on root disease incidence. *Planters' Bulletin Rubber Research Institute Malaysia* 158, 25–27.

Azaldin, M.Y. and Rao, B.S. (1974) Practicability and economics of large scale artificial defoliation for avoiding secondary leaf fall. *Proceedings Rubber Research Institute of Malaysia Planters Conference*, 1974, Kuala Lumpur, pp. 161–170.

Basuki (1985) Efforts in the control of white root disease (*Rigidoporus lignosus*) of rubber in Indonesia. *Proceedings International Rubber Conference*, 1985. Kuala Lumpur, 3, 209–221.

Canh, T.V. (1985) Use of Calixin and Sandofan F against white root disease and black stripe of *Hevea brasiliensis*. *Proceedings International Rubber Conference*, 1985, Kuala Lumpur, 3, 222–237.

Canh, T.V., Gohet, E., Louanchi, M. and Despreaux, D. (1990) Root rot disease of *Hevea brasiliensis* in Africa: Improvement of chemical control. *Proceedings of the IRRDB Symposium*, 1990, China, pp. 19–30.

Chee, K.H. (1990) Present status of rubber diseases and their control. *Review Plant Pathology* 69, 423–430.

Department of Agriculture Malaysia and Rubber Research Institute of Malaysia (1988) Contingency plan for the eradication and control of a dangerous pest: South American leaf blight of rubber (*Microcyclus ulei*), 55 pp.

Fox, R.A. (1965) The role of biological disease control in replantings of *Hevea brasiliensis*. In: *Ecology of Soil Borne Plant Pathogens* (Eds. K.F. Baker and W.C. Snyder,), pp. 348–362, University of California Press.

Ho, C.Y., Chan. H.Y. and Lim, T.M. (1974) Enviromax planting recommendations. A new concept in choice of clones. *Proceedings Rubber Research Institute of Malaysia Planters Conference*, 1974, Kuala Lumpur, pp. 293–310.

Ismail Hashim (1984) The severity of *Colletotrichum* on GT 1 — A Survey. *Planters Bulletin Rubber Research Institute of Malaysia* 177, 124–127.

Ismail Hashim (1990) Possible integration of *Trichoderma* with fungicides for the control of white root disease of rubber. *Proceedings of the IRRDB Symposium*, 1990. China, pp. 1–8.

Ismail Hashim and Azaldin, M.Y. (1985) Interaction of sulphur with soil pH and root diseases of *Hevea* rubber. *Journal Rubber Research Institute of Malaysia* 33, 59–69.

Ismail Hashim and Radziah, N.Z. (1987) Fogging with fungicides to control *Colletotrichum* secondary leaf fall of *Hevea* rubber in Malaysia. *Journal Natural Rubber Research* 2, 231–245.

Juncqueira, N.T.V., Silva, S.E.L., Silva, H.M. and Silva, M.A.M. (1986) Controle biologico do mal das folhas da seringueira por *Hansfordia pulvinata*. *Pesquisa em Andamento*, EMBRAPA-CNPSD, 5 pp.

Lim, T.M. (1974) Enhancing post-wintering tree vigour for avoiding *Oidium* secondary leaf fall. *Proceedings Rubber Research Institute of Malaysia Planters Conference*, Kuala Lumpur, 1974, pp. 178–187.

Lim, T.M., Radziah, N.Z. and Abdul Aziz, S.A.K. (1981) Rubber leaf disease control — A case for mechanisation. *Proceedings of Rubber Research Institute of Malaysia Planters Conference*, 1981, Kuala Lumpur, pp. 311–326.

Liyanage, A. de S. (1979) Review of Plant Pathology Department. *Annual Review Rubber Research Institute Sri Lanka* 1978, pp. 54–87.

Mainstone, B.J., Mcmanaman, G. and Begeer, J.J. (1977) Aerial spraying against South American leaf blight of rubber. *Planters Bulletin Rubber Research Institute of Malaysia* 148, 15–26.

Newsam, A. (1961) Pathology division. *Report Rubber Research Institute Malaya*, 1960, pp. 63–70.

Newsam, A. (1963) Covers and root disease. *Planters Bulletin Rubber Research Institute of Malaysia* 68, 177–181.

Newsam, A. (1967) Clearing methods and root disease control. *Planters Bulletin Rubber Research Institute of Malaysia* 92, 176–182.

Peries, O.S. (1966) The economics of disease control in *Hevea. Bulletin Rubber Research Institute Ceylon* 1, 27–33.

Peries O.S. and Liyanage, N.I.S. (1980) The use of sulphur for the control of white root disease caused by *Rigidoporus lignosus. Proceedings International Rubber Conference*, India, 1980.

Radziah, N.Z. and Chee, K.H. (1988) Cylindrocladium leaf spot. *Planters Bulletin Rubber Research Institute of Malaysia* 197, 146–149.

Radziah, N.Z. and Chee, K.H. (1989a) Fusicoccum leafblight — a new disease of rubber. *Proceedings Rubber Growers Conference*, 1989, pp. 471–484.

Radziah, N.Z. and Chee, K.H. (1989b) A new anthracnose disease of rubber. *Plant Pathology* 38, 293–296.

Ramakrishnan, T.S. (1960) Experiments on the control of abnormal leaf fall of *Hevea* caused by *Phytophthora palmivora* in South India. Proceedings of the Natural Rubber Conference, 1960, Kuala Lumpur, pp. 454–466.

Rao, B.S. and Azaldin M.Y. (1973) Progress towards recommending artificial defoliation for avoiding secondary leaf fall. *Proceedings Rubber Research Institute of Malaysia Planters Conference*, Kuala Lumpur, 1973, pp. 267–280.

Rubber Research Institute of Malaysia (1989) RRIM Planting recommendations 1989–91. *Planters Bulletin Rubber Research Institute of Malaysia* 198, 3–23.

Rubber Research Institute of Malaysia (1975) *Corynespora* leaf spot. *Planters Bulletin Rubber Research Institute of Malaysia* 139, 84–86.

Sarma, S.R. (1985) Some aspects of abnormal leaf fall of *Hevea* caused by *Phytophthora palmivora* in South India. *Proceedings International Rubber Conference*, 1985, Kuala Lumpur, 3, 238–251.

Tan, A.M. (1985) *Guignardia* leaf disease. *Planters Bulletin Rubber Research Institute of Malaysia* 182, 11–12.

Tan, A.M. (1989) New fungicide and formulations for the control of blackstripe. *Proceedings Rubber Growers Conference* 1989, pp. 463–470.

Tan, A.M. (1990a) The present status of fungicide drenching on the control of white root disease of rubber. In: *Root Diseases of Hevea brasiliensis*, *Proceedings of IRRDB Symposium,* Kunming, China, 1990, pp. 47–54.

Tan, A.M. (1990b) Survey on *Corynespora* leaf fall disease. *Planters Bulletin Rubber Research Institute Malaysia* 204, 80–85.

Tan, A.M. and John, C.K. (1985) Economic benefits of disease control in rubber — Malaysian experiences. *Proceedings International Rubber Conference*, 1985, Kuala Lumpur, 3, 270–279.

Tan, A.M. and Yeoh, C.S. (1976) A spray-on natural rubber latex formulation for controlling pink disease. *Proceedings of Rubber Research Institute of Malaysia Planters Conference* 1976, Kuala Lumpur, pp. 243–249.

Wastie, R.L. (1971) The control of secondary leaf fall of *Hevea brasiliensis*. In *Crop Protection in Malaysia* (Eds. Wastie, R.L. and Woods, B.J.), pp. 197–203, Incorporated Society of Planters, Kuala Lumpur.

Wastie, R.L. (1975) Diseases of rubber and their control. *PANS* 21, 268–288.

Wastie, R.L. (1986) Disease resistance in rubber. *FAO Plant Protection Bulletin* 34, 193–199.

Wastie, R.L. and Mainstone, B.J. (1969) Economics of controlling secondary leaf fall of

Hevea caused by *Oidium heveae*. *Journal Rubber Research Institute Malaysia* 21, 64–72.

Yeoh, C.S. and Tan, A.M. (1974) Natural rubber latex formulation for controlling pink disease. *Proceedings Rubber Research Institute of Malaysia Planters Conference*, Kuala Lumpur, 1974, 171–177.

Yeoh, C.S. and Tan, A.M. (1979) A new formulation for controlling blackstripe. *Proceedings of the Rubber Research Institute of Malaysia Planters Conference*, 1979. Kuala Lumpur, pp. 400–409.

Yoon, P.K. (1976) Technique of crown budding. Rubber Research Institute of Malaysia, 27 pp.

Chapter fourteen : **Management of Oil Palm Pests and Diseases in Malaysia in 2000**

Gurmit Singh
United Plantations Berhad
Teluk Intan, Malaysia

Summary

The major pests of oil palms in Malaysia are bagworms, nettle caterpillars, rhinoceros beetle, bunch moth, termites and rats, while the most important disease is *Ganoderma* basal stem rot and, to a lesser extent, *Marasmius* bunch rot. Insecticides will remain the main weapon against insect pests in the foreseeable future, but integrated control involving regular monitoring, suitable agronomic practices, conservation, supplementation and utilisation of natural enemies, and the judicious use of chemicals, will become increasingly more important. Chemicals with a broad spectrum of activity, long and persistent residues, and high mammalian toxicity will continue to be replaced by safer and more target-specific products that are also softer to the environment.

Rat control, which constitutes an important item of cost in crop protection, will continue to rely on rodenticides, but biological control using the barn owl, which is proving to be successful in some estates, will become increasingly more popular.

Labour shortages presently affecting the industry, will become more pronounced in the future. Besides directly affecting productivity, they will also have a bearing on the manifestation of certain pests and diseases. *Tirathaba* bunch moth and *Marasmius* bunch rot, which are usually kept in check by good palm sanitation via regular rounds of harvesting and pruning, will become more prevalent, as workers increasingly shy away from these physically demanding jobs, especially in tall palm areas. Similarly, rhinoceros beetle, which is presently more a problem in the coastal districts, will become more widespread in distribution as large areas of inland palms, which are coming up for replanting, are felled and replanted. The high cost of replanting will futher aggravate the situation as some estates and smallholders may be forced to sacrifice on hygiene and sanitation.

The accelerated pace of replanting will also have a deleterious effect on *Ganoderma* basal stem rot. The disease, which is on the rise in the coastal areas, will become more rampant as the inoculum potential increases from one generation to another. Clean clearing of the old stand of oil palms or coconuts, which is now commonly practised, is only partially effective as replants continue to succumb to the disorder. Research into the disease has in recent years been intensified and satisfactory progress is being made. It is envisaged that by the year 2000, a significant breakthrough will be made in its management.

The basic spray equipment currently in use for the different age groups of palms will remain, but improved models will continue to be evaluated and introduced. The conventional knapsack sprayer will continue to be used in the nurseries, power sprayers and mistblowers in young field plantings and tractor-drawn or tractor-mounted airblast sprayers in the older areas. The demand for aerial spraying is expected to increase, especially for the treatment of large pest outbreaks. The problem of insufficient jobs to keep aerial operators in business will, in part, be overcome as more hectarage becomes available for aerial manuring. Trunk injection, a useful tool for the treatment of leaf-eating pests, will undergo refinement. The drill and drench system will be replaced by a one-step automated gun. Work along these lines is already in progress in some quarters and others may follow suit. Operator safety will take on greater priority and more suitable protective gear will be developed and introduced.

Finally, as international traffic and cargo increase, greater demands will be placed on plant quarantine support services to prevent the inadvertent introduction of exotic pests and potentially lethal diseases, several of which are known to plague the African and South American oil palm industry.

Introduction

The oil palm industry in Malaysia has undergone unparalleled expansion in the last three decades to emerge as the largest producer of palm oil in the world. In 1989, there were 1.95 million hectares of oil palm in the country producing 6.06 million tonnes of crude palm oil (PORLA Review, 1989). By the year 2000, this is projected to increase to 2.67 million hectares, with a total production of 8.59 million tonnes of oil (Chow, 1990). Being a monoculture with a narrow genetic base (Rajanaidu *et al.*, 1982), and considering the favourability of the hot and humid climate for pest and disease development, it is fortunate that, contrary to the potential, there have been few problems of any major consequence. Apart from the occasional outbreak of leaf-eating caterpillars and the looming threat of *Ganoderm*a basal stem rot, the industry has remained relatively trouble-free. As for the future, many challenges lie ahead, and much will depend on man's continued efforts to maintain the delicate balance between pests and their natural enemies, and on inoculum management in the case of basal stem rot.

Major Pests and Diseases

Insect Pests

Of the more than sixty insect pests that have been recorded on the oil palm in Malaysia (Wood, 1968), only around ten are of economic importance. Foremost among these are the leaf-eating caterpillars. They comprise two major lepidopteran groups, namely bagworms (Psychidae) and nettle caterpillars (Limacodidae). The larvae of both these groups feed on the leaves and when present in large numbers, they can cause extensive defoliation, adversely affecting yield. Outbreaks are usually associated with weather changes or the use of broad-spectrum, long-residual life chemicals, disrupting natural balance (Wood, 1971).

The species of bagworms commonly encountered are *Metisa plana, Pteroma pendula* and *Mahasena corbetti*. As the name implies, the caterpillars live in cases made of silk, to which fragments of plant tissues are often attached. Pupation occurs within the cases and the cocoons are distinct for each species. The nettle caterpillars, on the other hand, are brightly coloured larvae armed with stinging bristles. A number of species are known to attack oil palms, but the most important are *Setora nitens, Darna trima, Setothosea asigna, Darna diducta* and *Birthosea bisura*. The other important insect pests are rhinoceros beetle (*Oryctes rhinoceros*), the adults of which bore into the base of the growing shoot causing spear collapse and frond damage, bunch moth (*Tirathaba rufivena*), larvae of which feed on the developing fruits, and termites (*Coptotermes curvignathus*). The latter is especially problematic on peat soils.

Mammalian Pests

Rats are the most serious pests in this group. They feed on the fruits on developing bunches, causing substantial losses, but can also attack young nursery and field palms. The species commonly found in oil palm estates are *Rattus argentiventer* (rice field rat), *R. tiomanicus* (Malayan wood rat) and *R. diardii* (house rat).

Diseases

More than twenty diseases are known to affect oil palms in Malaysia (Turner, 1981), but the most important are basal stem rot caused by *Ganoderma* sp. and, to a lesser extent, *Marasmius* bunch rot caused by *Marasmius palmivorus*. Others like anthracnose, leaf spots, blast and spear rot can sometimes occur on seedlings in the nursery, but are usually of minor importance.

Future Pest and Disease Trends

Leaf-eating Caterpillars

A nationwide survey of bagworms conducted by PORIM in 1985 showed that this pest is on the rise, with outbreaks occurring more frequently and covering a larger hectarage (Wahid *et al.*, 1988). Based on the size of the outbreaks, the most damaging species in Peninsular Malaysia was *M. plana*, while in Sabah and Sarawak it was *M. corbetti*. It was further noted that outbreaks could occur more than once within the same area and in some instances the problem had become chronic. Although nettle caterpillars were not covered by this survey, there is reason to believe that the situation with regards to this pest would not be very different. Based on the above observations, it can generally be expected that leaf-eating caterpillars will become increasingly more problematic in the future.

Tirathaba Bunch Moth and *Marasmius* Bunch Rot

Outbreaks of *Tirathaba* bunch moth and *Marasmius* bunch rot usually occur in young plantings where pollination and fruit set can sometimes be poor and where bunches are left to rot on the palms. In older areas, both these problems are usually kept in check by good palm sanitation involving regular rounds of harvesting and pruning, and are seldom seen to be of any great importance. With the migration of workers from the rural to the urban sectors of the economy, labour is becoming increasingly short in the plantation industry (UPAM, 1989). The position in some estates is so serious that even the key operations of harvesting and pruning are sometimes compromised (Chew, 1990). If the situation is allowed to continue, and indications are it will further deteriorate, then the poor palm sanitary conditions, especially in tall palm areas, will create conditions that are conducive to the development of *Tirathaba* and *Marasmius*. Such a situation has in fact recently occurred in an estate in Lower Perak, where the bunch moth, triggered by the above events, has become firmly established in 200 hectares of tall palms, and is proving difficult to control.

Rhinoceros Beetle

The rhinoceros beetle is basically a pest of coconuts that has become adapted to oil palms, on which it can cause serious damage. Palms of all ages are attacked, but damage to young palms, up to two years old, has the greatest effect on growth. The problem is usually more serious in replants where logs and stumps of the former oil palm or coconut stand are windrowed and allowed to rot, providing abundant sites for breeding. The adults which emerge in large number, six to nine months later, attack the young palms, many of which, due to recurring attacks, can be very badly damaged. In Peninsular Malaysia, the beetle is presently more a problem in the coastal plantings along the west coast, where, due to the long history of coconut cultivation and the large expanse of oil palms, it has become endemic (Barlow and Chew, 1971; Gurmit Singh, 1987). The problem, however, is expected to become more widespread as large areas of the inland plantings, planted in the 1960s and 1970s, come up for replanting. The high cost of replanting will further aggravate the situation, as some estates and smallholders may be forced to sacrifice on hygiene and sanitation.

Basal Stem Rot

This is the most important disease of oil palms in Malaysia. The fungus initially invades one or more roots and then moves into the bole and stem tissues, causing a dry rot which eventually leads to death of the palm. Primary infection occurs through tissues of the old stand left buried in the soil. Secondary spread is through root contact between healthy and diseased palms. The disease is currently confined to the coastal areas, where incidence can reach as high as 85% by the time palms are replanted (Turner, 1981; Gurmit Singh, 1991). In plantings from jungle or rubber, the disease usually develops when palms are 10–12 years old, but with replants from coconuts or oil palms, the disease can set in as early as 12–24 months, but more usually when palms are 4–5 years old (Gurmit Singh, 1991).

The disease is on the increase in the coastal areas and will become more rampant as old stands, due for replanting, are felled and replanted. In recent years, the disease has also spread into the peat soil areas which, till a few years ago, were considered clean. The plantings on inland soils, besides some isolated pockets (PORIM, 1989, 1990a), are the only areas that are currently free of the problem, but indications are that the disease will also make inroads into these areas. In Indonesia, where oil palms have been cultivated over a much longer period, *Ganoderma*, according to Turner (pers. comm., 1991) is already a problem on these soils.

Management Strategies in 2000

Insect Pests and Insecticides

Insecticides will, for the foreseeable future, remain the main weapon against insect pests, but broad-spectrum, long-residual (BSLR) contact products, which can do more harm than good, will continue to be replaced by more target-specific compounds with fast-fading residues and products that are generally softer to the environment. From the classical studies of Wood (1971) in the

1960s, it is clear that the use of BSLR compounds like endrin, aldrin, dieldrin and DDT can lead to explosive outbreaks of bagworms and other leaf-eating caterpillars. These insecticides are now no longer in use and in their place, safer products like the synthethic pyrethroids and the bio-insecticide *Bacillus thuringiensis* are becoming more popular. These, together with other products like trichlorophon, chlorpyrifos, endosulfan, carbaryl, monocrotophos, acephate and methamidophos will remain the mainstay of the industry for some time to come. New compounds will continue to be synthesised and introduced. One such group of products that are showing good promise are the insect growth regulators. In trials Ho and Sidhu (1986) and Ng *et al.* (1986) reported good control of nettle caterpillars using the chitin synthesis inhibitors, triflumuron and diflubenzuron. The latter has also been found to be effective against the bunch moth (UPRD, 1988).

Integrated Control

There is a growing awareness among growers of the importance of integrated pest management (IPM) and there is good potential for further development in this area. As pointed out by Cock *et al.* (1987), IPM considerations are especially important in tropical plantation crops where, unlike annual crops, biological interactions have a chance to reach equilibrium or near equilibrium conditions. Allowing a natural balance to develop between pests and their natural enemies may reduce the need for chemical intervention.

Essentials of IPM in Oil Palms

Pest monitoring

This is an important tool in pest management. It helps pinpoint early outbreaks and the progress of the infestation. It also enables decisions to be made on insecticide application and the efficiency of the treatment. The application of chemicals at the right time results in their efficient use and reduces the frequency of spraying.

Realising the importance of early detection, several plantations have adopted pest surveillance as an intergral component of estate operations, regularly monitoring the pest status in the estate, using the censusing system of Wood (1968) or Syed & Speldewinde (1974).

Natural control agents

The natural enemies of oil palm pests include parasitoids, predators and pathogens, all of which can exert good control in nature. The key to their efficient utilisation lies in the conservation and enhancement of their activity (Wood, 1979).

In the Limacodidae alone more than eighty parasitoids, fifteen predators, ten fungal pathogens and six families of viruses have been recorded in south-east Asia (Holloway *et al.*, 1987), of which several occur in Malaysia (Table 1).

Bagworms, like the nettle caterpillars, are also attacked by a number of natural enemies which can cause considerable mortality of larvae and pupae (Table 2). PORIM's ongoing studies with this pest have shown that parasitism varies with the season and location, and can be as high as 56% (PORIM, 1990b).

The most important natural enemies of the rhinoceros beetle are the virus, *Baculovirus oryctes* and the entomophagous fungus *Metarrhizium anisopliae*. Both these pathogens are present in Malaysia, but very little is known of their

Table 1. Natural enemies of some common nettle caterpillars in Malaysia*

	Natural enemies		
Pest species	Parasitoids	Predators	Pathogens
Setora nitens	Braconidae: *Spinaria spinator*	Reduviidae : *Sycanus leucomesus*	Fungus *Cordyceps* sp. nr. *pruinosa*
	Eulophidae: *Euplectromorpha malayensis* *Neoplectrus areolatus*		Virus Non-occluded virus
	Eurytomidae: *Eurytoma sp.*		
	Ichneumonidae: *Buysmania oxymora* *Goryphys mesoxanthus* *Theronia orientalis*		
	Sarcophagidae: *Seniorwhitea orientalis*		
	Tachinidae: *Chaetexorista javana* *Exorista sorbillans*		

Table 1. *Continued*

	Trichogrammatidae: *Trichogrammatoidea thoseae* Bombyliidae: *Systropus roepkei*		
Setothosea asigna	Ichneumonidae: *Chlorocryptus purpuratus*	Pentatomidae: *Cantheconidea furcellata* *Cantheconidea gaugleri*	Fungus *Cordyceps* sp. nr. *militaris*
	Trichogrammatidae: *Trichogrammatoidea thoseae*	Pyrrhocoridae: *Dindymus rubiginosus*	Virus *Picorna & Nudaurelia* β *virus*
		Reduviidae: *Cosmolestes picticeps* *Isyndus heros* *Sycanus dichotomus* *Sycanus leucomesus* *Sycanus macracanthus*	
Darna trima	Bombyliidae: *Systropus roepkei*	Pentatomidae: *Cantheconidea furcellata*	Virus *Granulosis virus & Nudaurelia* β *virus*
	Braconidae: *Apanteles aluella* *Fornicia penanga*	Reduviidae: *Sycanus dichotomus*	
	Chalcididae: *Brachymeria lasus*		

* Recorded by Wood (1968, 1971), Tiong (1979), Tiong and Munroe (1977), Entwistle (1987), Holloway *et al.* (1987).

Table 2. Natural enemies of bagworms in Malaysia

Pest species	Natural enemies		
	Parasitoids	Predators	Pathogens
Metisa plana	Ichneumonidae: *Goryphus inferus*	Cleridae: *Callimerus arcufer*	Fungus *Aspergillus* sp. *Beauveria* sp.
	Braconidae: *Aulosaphes psychidivorus* *Dolichogenedia metesae* *Apanteles* nr. *parasae*	Reduviidae: *Sycanus dichotomus*	
	Elasmidae: *Elasmus* sp.		
	Eupelmidae: *Eupelmus catoxanthae*		
	Eulophidae: *Pediobius imbreus* *Pediobius anomalous imbreus* *Tetrastichus* nr. *coeruleus*		
	Eurytomidae: *Eurytoma* sp.		
Mahasena corbetti	Ichneumonidae: *Echthromorpha agrestoria* *Xanthopiapla* sp.	Reduviidae: *Cosmolestes picticeps* *Isyndus heros* *Sycanus macracanthus* *Sycanus* sp.	Fungus *Paecilomyces fumosoroseus*
	Sarcophagidae: *Sarcophaga* sp.		Virus Non-occluded RNA virus

Table 2. *Continued*

	Tachinidae: *Palexorista solennis* *Eozenillia equatorialis* *Eozenillia psychidarum* *Eozenillia* sp. *Tricholyga aberrans* Braconidae: *Aulosaphes* sp. *Apanteles* sp. Salticidae: *Myrmarachne* sp. Chalcididae: *Brachymeria lugubris* *Brachymeria ghani* *Brachymeria* sp.	Cleridae: *Callimerus* sp. Formicidae: *Oecophylla smaragdina*
Pteroma pendula	Chalcididae: *Brachymeria* sp. Eulophidae: *Pediobius imbreus*	Reduviidae: *Sycanus dichotomus*

* Recorded by Wood (1968, 1971, 1976); Syed *et al.* (1973); Tiong (1979); PORIM (1990b).

effects in regulating beetle numbers. In the Pacific islands, where the beetle is a major problem on coconuts, the introduction of the virus has given good control of the pest (Marschall, 1978), and is now regularly released together with *Metarrhizium* to keep the population in check.

The natural enemy complex of *Tirathaba* on coconuts in Indonesia consists of several hymenopteran and dipteran parasites (Kalshoven & van der Laan, 1981), but in Malaysia only two parasitoids have been recorded on oil palm, namely *Antrocephalus* "sp. nr. *hakonensis*" and *Venturia palmaris* (Ng, 1981), with the latter being the more important.

Suitable Cultural Practices

Cultural practices can directly or indirectly affect pest populations. Weeds, while generally recognised as undesirable plants, can be a useful source of food, perch or pupation sites for parasitoids and predators of oil palm pests, and their indiscriminate destruction can lead to pest outbreaks (Syed *et al.*, 1973; Syed & Shah, 1977; Tiong, 1982). Not all weeds, however, are useful. The beneficial ones are mainly the broadleaf flowering plants that produce abundant nectar and pollen. Included in this category are *Euphorbia geniculata, Euphorbia prunifolium, Ageratum conyzoides, Cleome aciliata, Cleome rutidosperma, Hyptis brevipes, Phyllanthus niruri, Erechtites valerianifolia, Borreria* spp., *Eupatorium odoratum, Euphorbia heterophylla.*

Planters are becoming increasingly aware of the importance of these weeds and attempts are now being made in some estates to encourage their growth along road edges, non-harvesting avenues and in vacant points within a field.

Ground cover is also useful in the control of rhinoceros beetle. The maintenance of a dense leguminous or natural vegetation cover over the breeding sites of the pest can significantly reduce beetle numbers (Wood, 1969). This is now commonly practised in areas where clean clearing, due to high cost, is not done during replanting and instead the logs and stumps of the former stand are windrowed in the avenues and allowed to rot.

Rat Control

Rats are insidious pests of oil palm and need to be controlled from time to time to keep their numbers down. Baiting is the standard method of control, but biological control using the barn owl *Tyto alba*, which is proving to be effective in some estates (Duckett, 1976, 1981; Smal, 1989; Duckett & Karuppiah, 1989), will become more popular in the future. The barn owl is a very efficient predator of rats, with the latter making up virtually all its diet requirements. Nevertheless, it is unlikely that total control could be achieved by this method; periodic baiting would still be required.

Warfarin will remain the rodenticide of choice, but resistance to this anti-coagulant, first noticed in *R. tiomanicus* in the Lower Perak District in 1983 (Lee *et al.*, 1983), is on the rise along the west coast of Peninsular Malaysia (Wood *et al.*, 1989). In such situations, the more potent second generation rodenticides, brodifacoum, bromadiolone and flocoumafen are required to control the population. The problem, however, becomes more complex when the barn owl is included in the integrated control programme. Warfarin is reported to be safe to the owl, but the second generation products, especially brodifacoum, can cause secondary poisoning (Duckett, 1984). Alternatives are required and studies by Mendenhall and Pank (1980) have shown that diphacinone, chlorophacinone and fumarin are relatively safe and could be used as substitutes for warfarin.

Disease Control

Basal Stem Rot

Sanitation during replanting

Since tissues of the former stand of oil palms or coconuts are the primary source of infection in a replant, disease avoidance through clean clearing of the old stand during replanting is currently the main method of control practised in the coastal areas and will remain an important aspect of disease management in the future. A hydraulic excavator with a modified cutter bucket is used for the operation (Gurmit Singh, 1991). The palms are felled, windrowed and shredded into small fragments. The bole and adjoining matt of roots are extracted by digging a 1.5 m × 1.5 m × 1.0 m pit. When sufficiently dry all the tissues are completely burnt and the field ploughed and rotavated.

The operation, however, is only partially effective, but it is the best that is available under the circumstances.

Treatment of existing stands

The main emphasis here is to prevent secondary spread of the disease, and where possible, save palms that can be surgically treated. Surgery is only possible where the decay is not too advanced. The recent developments in early detection of the disease and mechanisation of the surgical procedure (Gurmit Singh, 1991), are important modifications that will improve the overall efficiency of the operation and give it a wider appeal. This, together with the removal of badly infected palms, will help minimise secondary infection and spread of the disease. At present this aspect of disease management is only practised on a limited scale, but with the satisfactory results being obtained (Gurmit Singh, 1991), it will become more important in the future.

Other approaches under study

Other aspects of disease management currently under study include screening of systemic funcidies for their prophylactic and curative properties (Chung, 1991; Khairuddin, 1991; Gurmit Singh, 1991), trunk injection of fumigants (Darus, 1991), biological control through soil amendments (Gurmit Singh, 1991), flood-fallow (Gurmit Singh, 1991), and crop rotation with padi and bananas (UPRD, 1991). Satisfactory progress is being made in some of these areas, and it is envisaged that by the year 2000, a significant breakthrough will be made in the management of the disease.

Application Technology

Ground Spray Equipment

Height, terrain and the dense umbrella-like canopy of the palms, can be a major constraint to efficient spray coverage and deposition. Machines selected should have adequate capabilities to do a satisfactory job. A variety of equipment is currently available for the different age groups of palms. These machines will remain the mainstay of the industry, although new models will continue to be evaluated and introduced. The conventional knapsack sprayer with hydraulic nozzles, will continue to serve the nurseries and young field plantings; portable power sprayers and mistblowers for palms up to six years old and tractor-drawn or tractor-mounted airblast sprayers for the older areas (Ho, 1988; Chung, 1988).

Aerial Application

Aerial spraying is the most suitable method of treating large outbreaks of leaf-eating caterpillars and the demand for this service is expected to increase by the year 2000. The problem of insufficient jobs to keep aerial operators in operation, would, in part, be overcome as more hectarage becomes available for aerial manuring. The application cost of M$55 per hectare, inclusive of chemicals (trichlorophon 1.75 kg/ha), is comparable to trunk injection with monocrotophos, but twice as expensive as ground application with tractor-drawn airblast sprayers. Exemption of surcharge and duties on Avgas would make the service more competitive, and appeals have been made to government to consider this waiver.

Trunk Injection

This drill and drench method of applying systemic insecticides for the control of leaf-eating caterpillars is ideal for tall palms, where ground spraying can sometimes be a constraint. The technique, apart from some minor modifications (Sarjit Singh, 1986), has remained largely unchanged since its introduction in the mid-1970s (Wood *et al.*, 1974). There are, however, good prospects for further refinement. Considering the fact that the stem tissues of the oil palm are much softer than those of dicotyledonous trees, it may be possible to develop a one-step pressure injector gun-type device that would not only simplify the operation but also minimise injury to the palms. There is already some interest in the development of such a device and a prototype may soon be available for testing (Nordmeyer, pers. comm., 1991).

Operator Safety

Pesticides are toxic substances and when wrongly handled, can cause serious health hazards to the user. Lee (1991), in his studies on pesticide contamination, showed that hazards can occur at all levels of pesticide handling, including mixing, filling/loading and spraying. He further reported that with insecticides and fungicides which, in the case of oil palms, are usually applied overhead, the greatest risk of contamination occurs on the face, neck, chest, shoulders and hands.

The incidence of pesticide poisoning among estate workers is one of the highest in the country (Ramasamy and Nursiah, 1988) and there is concern that with the increasing trend in pesticide usage, the incidence will increase further, unless remedial measures are taken to counteract this. More training and education programmes on the safe handling of pesticides are required, and there is also an urgent need to develop protective clothing suitable for tropical climates. Operator safety is the combined responsibility of the agrochemical companies, the plantation industry and the government, and greater efforts are required on the part of all concerned to remedy the problem.

Table 3. Important exotic quarantine pests and diseases of oil palm

Pests/diseases	Species	Distribution
Pests		
Leaf miner	*Coelaenomenodora elaeidis*	South America
Long-horned grasshopper	*Segestidea defoliana*	Papua New Guinea
Leaf and bunch stalk borer	*Eupalamides cyparissias*	South America
Palm weevil	*Rhynchophorus palmarum* *Rhynchophorus phoenicis*	All tropical America West Africa
Nettle caterpillars	*Acharia fusca, Euclea diversa* *Acharia intensa, Euprosterna elaeasa, Natada pucara*	South America
Leaf-eating caterpillars	*Stenoma cecropia* *Opsiphanes cassina* *Pimelephiea ghesquierei*	South America South America West Africa
Rooting feeding caterpillars	*Sagalassa valida*	South America
Dynastid beetle	*Strategus aloeus* *Scapanes australis*	South America Papua New Guinea

Table 3. *Continued*

	Augosoma centaurus	West Africa
	Oryctes monoceros	West Africa
Tissue boring weevil	*Temnoschoita quadripustulata*	West Africa
Mites	*Retracrus elaeis*	South America
Nematodes		
Red ring disease	*Radinaphelenchus cocophilus*	South America
Diseases		
Fusarium wilt	*Fusarium oxysporum f.* sp. *elaeidis*	Africa, Brazil
Freckle	*Cercocospora elaeidis*	Africa
Sudden withers/ marchitez	*Phytomonas* sp.	South America
Patchy yellow	*Fusarium oxysporum*	West Africa
Ring spot/ Leaf mottle	*Unknown etiology*	South America
Fatal yellowing	*Unknown etiology*	South America

Plant Quarantine

In its native Africa and in South America, the oil palm is host to a number of serious pests and potentially lethal diseases which pose a grave threat to the industry in the respective regions. With the increase in international traffic and movement of plant materials, the role of plant quarantine will become increasingly more important in the future and greater demands will be placed on its services to keep out exotic pests and diseases (Table 3).

Acknowledgements

The author would like to thank United Plantations Berhad for permission to present this paper.

References

Barlow, H.S. and Chew, P.S. (1971) The rhinoceros beetle, *Oryctes rhinoceros*, in young oil palms planted after rubber on some estates in West Malaysia, pp. 133–145. In:

Crop Protection in Malaysia (Eds. Wastie, R.L. and Wood, B.J.). Incorporated Society of Planters, Kuala Lumpur, 261 pp.

Chew, P.S. (1990) Research and development on mechanisation in oil palm estates. Guest editorial. *The Planter*, 66, 384–385.

Chow, C.S. (1990) Forecast of Malaysian palm oil production up to year 2000 (revised). PORIM Report, unpublished. Palm Oil Research Institute of Malaysia.

Chung, G.F. (1988) Spraying and trunk injection of oil palm. MOPGC-ARC Workshop on Mechanisation in Oil Palm Estates, April 1988, Kuala Lumpur. Malaysian Oil Palm Growers Council Agricultural Research Committee (abstract).

Chung, G.F. (1991) Preliminary results on trunk injection of fungicides against *Ganoderma* basal stem rot in oil palm. *Ganoderma* Workshop, September 1990, Kuala Lumpur. Palm Oil Research Institute of Malaysia.

Cock, M.J.W., Godfray, H.C.J. and Holloway, J.D. (1987) *Slug and Nettle Caterpillars*. CAB International, UK 270 pp.

Darus, A. (1991) Progress on *Ganoderma* research at PORIM. *Ganoderma* Workshop, September 1990, Kuala Lumpur. Palm Oil Research Institute of Malaysia.

Duckett, J.E. (1976) Owls as major predators of rats in oil palm estates with particular reference to the barn owl (*Tyto alba*). *The Planter* 52, 4–15.

Duckett, J.E. (1981) Barn owls (*Tyto alba*) — a proven predator of rats in oil palm, pp. 461–473. In: *The Oil Palm in Agriculture in the Eighties* (Eds. Pushparajah, E. and Chew, P.S.), Vol. II. Kuala Lumpur, Incorporated Society of Planters, 678 pp.

Duckett, J.E. (1984) Barn owls (*Tyto alba*) and the "second generation" rat-baits utilised in oil palm plantations in Peninsular Malaysia. *The Planter* 60, 3–11.

Duckett, J.E. and Karuppiah, S. (1989) A guide to the planter in utilising barn owls *Tyto alba* as an effective biological control of rats in mature oil palm plantations. pp. 357–372. In: *Proc. 1989 PORIM International Palm Oil Development Conference, Module II Agriculture*, Kuala Lumpur, Palm Oil Research Institute of Malaysia, 588 pp.

Entwistle, P.F. (1987) Virus diseases of Limacodidae, pp. 213–222. In: *Slug and Nettle Caterpillars* (Eds. Cock, M.J.W., Godfray, H.C.J. and Holloway, J.D.), CAB International, UK , 270 pp.

Gurmit Singh (1987) Naphthalene balls for the protection of coconuts and oil palm against *Oryctes rhinoceros* (L.). *The Planter* 63, 286–292.

Gurmit Singh (1991) *Ganoderma* — The scourge of oil palms in the coastal areas. *Ganoderma* Workshop, September 1990, Kuala Lumpur, Palm Oil Research Institute of Malaysia.

Ho, C.T. (1988) HMPB experience with the use of mechanised mistblowers for the treatment of oil palm leaf pests. MOPGC-ARC Workshop on Mechanisation in Oil Palm Estates, April 1988, Kuala Lumpur. Malaysian Oil Palm Growers Council Agricultural Research Committee (abstract).

Ho, C.T. and Sidhu, M. (1986) Some developments in chemical control of leaf pests in oil palms in East and Peninsular Malaysia. In Extended abstracts of 2nd Int. Conf. Pl. Prot. in the Tropics, March 1986, Kuala Lumpur, Malaysian Plant Protection Society, pp. 177–187.

Holloway, J.D., Cock, M.J.W. and Desmier de Chenon, R. (1987) Systematic account of South-east Asian Limacodidae, pp. 15–118. In: *Slug and Nettle Caterpillars* (Eds. Cock, M.J.W., Godfray, H.C.J. and Holloway, J.D.), CAB International, UK 270 pp.

Kalshoven, L.G.E. and van der Laan, P.A. (1981) *Pests of Crops in Indonesia.* Revised by van der Laan, P.A. and Rothschild, G.H.L. Jakarta, P.T. Ichtiar Baru, van Hoeve, 701 pp.

Khairuddin, H. (1991) Results of four trials on *Ganoderma* basal stem rot of oil palm in Golden Hope estates. *Ganoderma* Workshop, September 1990, Kuala Lumpur. Palm Oil Research Institute of Malaysia.

Lee, C.H., Mustapha, Md. D., Soh, K.G. and Mohan, E. (1983) Warfarin resistance in *Rattus tiomanicus* (Miller). *MARDI Res. Bull.* 11, 264–271.

Lee, S.A. (1991) Studies on contamination patterns, penetration of pesticides through clothing and spray operator protection. *The Planter* 67, 9–40.

Marschall, K.J. (1978) Biological control of rhinoceros beetles: Experiences from Samoa, pp. 658–672. In: *Proc. Int. Conf. on Cocoa and Coconuts*, June 1978, Kuala Lumpur, 766 pp.

Mendenhall, V.M. and Pank, L.F. (1980) Secondary poisoning of owls by anticoagulant rodenticides. *Wildlife Soc. Bull. (USA)* 8, 311–315.

Ng, K.Y. (1981) *Venturia palmaris* Wilkinson (Hymenoptera — Ichneumonidae) — A parasite of oil palm bunch moth *Tirathaba mundella* Walker (Lepidoptera — Pyralidae), pp. 519–528. In: *The Oil Palm in Agriculture in the Eighties* (Eds. Pushparajah, E. and Chew, P.S.), Vol. II. Kuala Lumpur, Incorporated Society of Planters, 678 pp.

Ng, K.Y., Schonlau, H.J. and Weber, P. (1986) Integrated pest control in oil palm (*Elaeis guineensis*) with Alsystin. *Pflanzenchutz-Nachrichten Bayer,* 39, 30–46.

PORIM (1989) Minutes of the *Ganoderma* Action Committee meeting, February 1989. Palm Oil Research Institute of Malaysia.

PORIM (1990a) Minutes of the *Ganoderma* Action Committee meeting, February 1990. Palm Oil Research Institute of Malaysia.

PORIM (1990b) Annual Research Report 1990, Biology Division. Palm Oil Research Institute of Malaysia, pp. 78–100.

PORLA (1989) Review of the Malaysian palm oil industry 1989. Palm Oil Registration and Licensing Authority, Malaysia, 22 pp.

Rajanaidu, N. Lawrence, M.J. and Ooi, S.C. (1982) Variation in Nigerian oil palm germplasm and its relevance to oil palm breeding, pp. 3–18. In: *The Oil Palm in Agriculture in the Eighties* (Eds. Pushparajah, E. and Chew, P.S.), Vol. I. Kuala Lumpur, Incorporated Society of Planters, 678 pp.

Ramasamy, S. and Nursiah, M.T.A. (1988) A survey of pesticide use and associated incidence of poisoning in Peninsular Malaysia. *J. Pl. Prot. Tropics* 5, 1–9.

Sarjit Singh (1986) The use of electrical drills for more efficient trunk injection against bagworms in oil palms. *The Planter* 62, 54–57.

Smal, C.M. (1989) Research on the use of barn owls *Tyto alba* for biological control of rats in oil palm plantations: 1986–1989, pp. 324–356. In: *Proc. 1989 PORIM International Palm Oil Development Conference, Module II Agriculture.* Kuala Lumpur, Palm Oil Research Institute of Malaysia, 588 pp.

Syed, R.A. and Shah, S. (1977) Some important aspects of insect pest management in oil palm estates in Sabah, Malaysia, pp. 577–590. In: *International Developments in Oil Palm* (Eds. Earp, D.A. and Newall, W.), Incorporated Society of Planters, Kuala Lumpur, 804 pp.

Syed, R.A., Shah, S. and Liew, K.T. (1973) Occurrence and Control of *Mahasena corbetti* Tams on oil palm in Sabah, pp. 402–413. In: *Advances in Oil Palm Cultivation* (Eds.

Wastie, R.L. and Earp, D.A.) Kuala Lumpur, Incorporated Society of Planters, 469 pp.

Syed, R.A. and Speldewinde, H.V. (1974) Pest detection and census on oil palms. *The Planter* 50, 230–233.

Tiong, R.H.C. (1979) Some predators and parasites of *Mahasena corbetti* (Tams) and *Thosea asigna* (Moore) in Sarawak. *The Planter* 55, 279–289.

Tiong, R.H.C. (1982) Oil palm pests in Sarawak and the use of natural enemies to control them. Pp. 363-372. In *Proc. Int. Conf. Pl. Prot. in Tropics*, Kuala Lumpur, 1982. Malaysian Plant Protection Society, 743 pp.

Tiong, R.H.C. and Munroe, D.D. (1977) Microbial control of an outbreak of *Darna trima* (Moore) on oil palm (Elaeis *guineensis* Jacq.) in Sarawak (Malaysian Borneo), pp. 624–639. In: *International Developments in Oil Palm* (Eds. Earp, D.A. and Newall), W. Incorporated Society of Planters, Kuala Lumpur, 804 pp.

Turner, P.D. (1981) *Oil Palm Diseases and Disorders*. Oxford University Press, Oxford 280 pp.

UPAM (1989) A survey of labour shortage in oil palm estates. United Planting Association of Malaysia, Kuala Lumpur.

UPRD (1988) Annual report for 1986. United Plantations Research Department, Teluk Intan, Malaysia, 86 pp.

UPRD (1991) Internal report. United Plantations Research Department, Teluk Intan, Malaysia.

Wahid, M.B., Halim, A.H. and Masijan, Z. (1988) Bagworms (Lepidoptera: Psychidae) of oil palm in Malaysia. PORIM Occ. Paper. Palm Oil Research Institute of Malaysia No. 23, 37 pp.

Wood, B.J. (1968) *Pests of Oil Palms in Malaysia and Their Control*. Incorporated Society of Planters, Kuala Lumpur, 204 pp.

Wood, B.J. (1969) Studies on the effect of ground vegetation in infestations of *Oryctes rhinoceros* (L.) (Col., Dynastidae) in young oil palm replantings in Malaysia. *Bull. Ent. Res.* 59, 85–96.

Wood, B.J. (1971) Development of integrated control programme for pests of tropical perennial crops in Malaysia, pp. 422–457. In: *Biological Control* (Ed. Huffaker, C.B.), Plenum Press, N.Y. 511 pp.

Wood, B.J. (1976) Insect pests in South-east Asia, pp. 347–367. In: *Oil Palm Research* (Eds. Corley, R.H.V., Hardon, J.J. and Wood, B.J.), Elsevier Scientific Publishing Company, Amsterdam, 532 pp.

Wood, B.J. (1979) Progress in the development of integrated pest control programmes in Malaysian Agriculture. *Proc. Pl. Prot. Sem.*, Kuala Lumpur, pp. 176–216.

Wood, B.J., Chung, G.F. and Sim, S.C. (1989) The occurrence and control of warfarin resistant rats in oil palm, pp. 317–341. In: *Proc. 1989 PORIM International Palm Oil Development Conference, Module II Agriculture*. Kuala Lumpur. Palm Oil Research Institute of Malaysia, 588 pp.

Wood, B.J., Liau, S.S. and Knecht, J.C.X. (1974) Trunk injection of systemic insecticides against the bagworm *Metisa plana* (Lepidoptera: Psychidae) on oil palm. *Oleagineux* 29, 499–505.

Chapter fifteen: **Management of Pests and Diseases of Vegetable Crops in Malaysia in 2000**

Syed Abdul Rahman
MARDI, 39007, Tanah Rata, Cameron Highlands Pahang DM, Malaysia

Abstract

The Malaysian Government has now placed high priority on the vegetable industry in its National Agricultural Policy. Vegetables represent an invaluable source of nutrients to the daily diet of Malaysians and form about 15% of their calorie intake. With the increasing demand, the per capita vegetable consumption between 1975 and 1985 has moved upwards from 36 kg to 48 kg, respectively. The Ministry of Agricultural has estimated that this increase will continue and production of vegetables will reach 922,000 metric tonnes by the year 2000.

A wide range of more than 50 types of vegetables, encompassing both tropical and temperate species are cultivated in a total area of 14,000 hectares, of which 2000 hectares are in the highland areas. The vegetable industry faces several major problems, among which pests and diseases are major factors. Approximately 70% of the growers depend solely on pesticides, and this syndrome has led to problems such as unacceptable levels of residues, pesticide resistance and upsurge in secondary pests.

Among the major pests of vegetables are the diamondback moth (DBM), *Plutella xylostella* L. and the cabbage webworm, *Hellula undalis*. The most serious diseases among the fruit vegetables are viruses on chilli, bacterial wilt, *Pseudomonas solanacearum* (Smith) and late blight, *Phythopthora infestans* (Mont.) on tomato. Weeds and nematodes are considered minor pests and they do not pose a serious constraint to vegetable production in Malaysia.

Although a wide range of control tactics are available, few are able to exploit and integrate such measures into a system. Nothwithstanding that the past research efforts were mostly confined to aspects not directly aimed at the conscious development of Integrated Pest Management (IPM), most of the knowledge generated will constitute the basis for IPM programmes for vegetables. One such programme already developed is the DBM–IPM model, which adopted biological control as the core component of the programme.

The DBM–IPM model has been tested and showed positive results in terms of yield and income to the farmers. A similar model has been developed for the management of virus diseases on chilli. The IPM model for chilli integrates cultural, physical and chemical methods of control.

In the future, IPM programmes will be developed for other minor pests and diseases of vegetables. The programmes will be developed and implemented by the National IPM Committee for vegetables. The committee will look into integrating components/techniques for successful IPM programmes. The latter mainly include harnessing key natural enemy species, using mircobials and applying relatively safer pesticides. Other useful but less commonly exploited elements include proper timing of planting, crop rotation, physical barriers and trapping.

With respect to promoting greater IPM implementation, special emphasis is necessary to generate increased awareness and transfer of available IPM programmes. The strategic steps in the future will include determination of farmers, pest management knowledge, attitude and practices, IPM trial demonstration, and appropriate training of extension personnel and farmers. Governmental support and policies will no doubt set the pace.

Introduction

Vegetables represent an invaluable source of nutrients to the daily diet of Malaysians and form about 15% of their calorie intake (Ding *et al.*, 1981). Although about 15% of the vegetables are exported annually to Singapore, the production of fresh vegetables in Malaysia is currently mainly for local markets. Also about MR$286 million worth of vegetables and vegetable preparations are imported annually, of which MR$160 million comprises fresh and frozen vegetables (Hanada, 1987).

It is recognised that vegetable cultivation can be highly lucrative to producers, provided it is properly managed. The industry has tremendous potential for expansion and further development, particularly with the anticipated increase in demand as the standard of living rises. Accordingly, with increasing demand, the per capita vegetable consumption between 1975 and 1985 has moved upwards from 36 kg to 48 kg (Hanada, 1987). The Ministry of Agriculture has estimated that this increase in demand will continue and will reach 722,000 metric tonnes by the year 2000. To accommodate the increasing demand, vegetable cultivation must be expanded by 7000–8000 hectares.

The vegetable industry now ranks seventh in terms of hectarage and production in Malaysia. A wide range of more than 58 types of vegetables, encompassing both tropical and temperate species, are planted in Malaysia (Ding *et al.*, 1981). These include cabbage (*Brassica oleracea capitata*), Chinese cabbage (*Brassica pekinensis cylindrica*), Chinese mustard (*Brassica rapa*), kai choy (*Brassica juncea*), chilli (*Capsicum annuum*), tomato (*Lycopersicum esculentum*), cucumber (*Cucumis sativus*), long bean (*Vigna sinensis*), French bean (*Phaeseolus vulgaris*) and eggplant (*Solanum melongena*). The main vegetable growing areas in Malaysia are in the states of Johore, Pahang, Kelantan, Perak, Selangor, Malacca and Penang. These contribute about 85% of the total area (ca. 14,000 hectares) in the country of which ca. 2000 hectares are in the highlands.

Realising the dietary importance, crop diversification, and export potential of vegetables, the Government has now placed high priority on the vegetable industry in its National Agricultural Policy (NAP). The policy stipulates that vegetable production will be intensified by encouraging the establishment of large-scale production units. A blueprint on the "National Agricultural Policy Programme for Development of Vegetable Commodities (1986–2000)" outlines the relevant strategies and provides semi-detailed programmes on production, marketing, research and other supporting services needed for the development of the vegetable industry.

Nonetheless, the vegetable industry faces several constraints which include (1) lack of quality, high-yielding and resistant varieties, (2) poor crop management practices, supply, (3) severe pest and disease problems, (4) limited seed supply, (5) land shortage, (6) unstable vegetable prices, (7) processing technology, (8) lack of post harvest handling, processing technology, (9) labour shortage, (10) marketing problems, (11) high production costs and (12) lack of support services (Ding *et al.*, 1981; Tengku Mohd. Ariff, 1985; Saharan, 1986; Chay, 1990). Among these, pest and disease emerge as the main crippling factor. In a nationwide survey carried out in 1981 (Ding *et al.*, 1981), at least 85% of the farmers reported pest and diseases to contribute a major problem, while 65% need to use pesticides substantially to control them. The unilateral approach of farmers to overcome the pest and disease problems with pesticides has itself generated other undesirable attributes. These include (i) development of pesticide resistance in the pest, (ii) presence of pesticide residue on marketed products, (iii) rising incidence of pesticide poisoning of farmers, (iv) hazards to non-target organisms in the environment and (v) increased cost of production.

Although a wide range of control tactics are available, few vegetable farmers are able to exploit and integrate such measures into a system. Clearly, breaking the vicious cycle of pesticide dependency is imperative. The solution to this lies in the Integrated Pest Management (IPM) strategy, realising the importance and immense value of IPM as the prime approach for crop protection practices in Malaysia (Min. of Agric., 1984). Presently, sufficient information has been generated to allow for the development of an IPM programme against the diamondback moth (DBM), *Plutella xylostella* L.

Thus, beginning with DBM as a base programme and a guiding model, it is envisaged that IPM for other pests will gradually grow and expand around it.

Major Pests and Diseases

Table 1 shows the major pests and diseases of economic importance and the associated problems of selected vegetables. Among the major pests of crucifers, the diamondback moth, *Plutella xylostella,* is the most destructive insect pest in both the highland and lowland areas. In addition to DBM, the cabbage webworm, *Hellula undalis,* is also a serious pest of crucifers, particularly cabbage, in the lowlands. The common diseases of cabbage, Chinese cabbage and cauliflower are soft rot, *Ervinia carotovora,* and black rot, *Xanthomonas campestris*. The flea beetle, *Phylotreta* spp., occasionally causes serious damage on the other leafy crucifers, e.g. Chinese mustard, Chinese kale and kai choy.

In the past, the control of DBM has mainly been with the use of insecticides (Lim, 1990), and this led to the development of insecticide resistance to almost all insecticides used by the farmers (Sudderuddin & Kok, 1978; Syed, 1990). However, with development and implementation of IPM on DBM in the late 1980s, the problem has virtually been solved, especially in the highland crucifer growing areas.

The most serious among the fruit vegetable pests are the chilli viruses and anthracnose, *Colletotrichum capsici*, on chilli, bacterial wilt. *Pseudomonas solanacearum* and the late blight disease, *Phytophthora infestans* on tomato. In the highland vegetable areas, the major pests of sugar peas are the leaf miner, *Chromatomyia horticola* and leaf blight, *Ascochyta* spp. Weeds and nematodes are considered minor pests and they do not pose a serious constraint to vegetable production.

Control Strategies

The common control strategies in pest and disease management of vegetables include: (i) resistant varieties, (ii) cultural control, (iii) chemical control, (iv) biological control and (v) integrated pest management.

Plant resistance to important pests and diseases is certainly a desired element in any variety of vegetable. However, most imported seeds and cultivars known to be resistant, in most cases succumb to diseases and pests, due to environmental factors, pathogen inoculum levels and development of biotypes. Breeding programmes have been initiated on tomato for bacterial wilt resistance, and chilli for resistance to chilli veinal mottle virus (CVMV), cucumber mosaic virus (CMV) and the anthracnose fruit rot disease. Although these programmes are still in the early stages, some progress has been achieved.

For bacterial wilt of tomato, variety MT-1 and MT-11 have been found to have good and moderate field resistance respectively (Ho, 1988). Among the many tomato lines from the Asian Vegetable Research and Development Centre (AVRDC) tested, two lines, *viz.* CL 5813-553 D4–3–0 and CL 5915-

Table 1. Important pests* of the major vegetable crops in Malaysia

Crop	Pests	Associated problems			Locally
		Effective control	Residue	Resistance	
Leafy vegetable					
Crucifers					
Brassica oleracea	*Plutella xylostella*	+	-	+	L & H
(Head and	*Hellula undalis*	+	-	+	L
Chinese cabbage)	*Erwinia carotovora*	-	-	-	L & H
	Xanthomonas campestris	-	-	-	L
Brassica rapa	*Plutella xylostella*	+	+	+	L
(Chinese mustard)	*Phyllotreta* spp.	+	-	+	L
Brassica alboglabra	*Plutella xylostella*	+	+	+	L
(Chinese kale)	*Phyllotreta* spp.	+	-	+	L
Brassica juncea	*Plutella xylostella*	+	-	+	L & H
(Kai choy)	*Phyllotreta* spp.	+	-	+	L
Fruit vegetables					
Capsicum annuum	*Dacus* spp.	+	+	-	L & H
(Chilli pepper)	*Heliothis* spp.	+	+	+?	L & H
	Viruses				

Table 1. *Continued*

Crop	Pests	Associated problems			Locally
		Effective control	Residue	Resistance	
Lycospersicum esculentum (Tomato)	*Heliothis* spp.	+	+	+?	L & H
	Phytophthora infestans	+	+	+?	H
	Pseudomonas solanacearum	-	-	-	L
Cucumis sativus (Cucumber)	*Pseudoperonospora cubensis*	-	-	-	L
Vigna sinensis (Long bean)	*Maruca testulalis*	-	-	-	L
	Ophiomyia phaseoli	-	-	-	L
	Pseudomonas phaseolicola	-	-	-	L
Solanum melongena (Eggplant)	*Leucinodes orbonalis*	+	-	+	L
	Pseudomonas solanacearum	-	-	-	L
Pisum sativum (Sugar peas)	*Chromatomyia horticola*	-	+	+	H
	Ascochyta spp.	-	+	+	H

* Pests refer to undesirable arthopods, pathogens, nematodes, weeds and molluscs.
+ Yes; - No; ? Strongly suspected present.
L, Lowland; H, Highlands.

93 D4-1-0 were found to be resistant to most virulent isolates of *Pseudomonas solanacearum* (Ho, 1989).

Studies on the breeding of chilli against viruses have shown that 34 and 27 genotypes are resistant or tolerant to CVMV and CMV respectively (Chew and Ong, 1990). Genetic studies on two CVMV resistant lines (local "cili padi" and Mexican Serrano Huasteco) revealed that the resistance is conferred by a pair of recessive genes which are on different loci. Currently, these genes have been incorporated into preferred local susceptible cultivars and screened for good agronomic characters (Chew and Ong, 1990). In vitro infection studies on anthracnose fruit rot of chilli showed the local cultivar "cili padi" and the Mexican Cultivar Serrano Tampiqueno as the most resistant. These cultivars are now used in the breeding programme (Mah, 1985).

A breeding programme for late blight resistance on tomato was initiated in 1983 in Cameron Highlands. In an effort to search for the source of late blight resistance for the breeding programme, 36 tomato lines were evaluated in the field. Two lines were found to have the resistance character, i.e USDA PI 162679 and West Virginia 700. To improve the fruit size, these resistant lines were crossed with the local susceptible variety using the inbred backcross line method. In the initial backcross studies, the segregating populations showed high intensity of resistance and improved fruit size (M. Shukor Nordin *et al.*, 1986).

Several cultural methods for pest and disease control have been used by the growers. Intercropping of tomato and cabbage was found to reduce DBM damage on cabbage (Sivapragasam, *et al.*, 1982). However, further studies are needed to determine its usefulness. Mulching is practised by about 19% of the vegetable growers, for various reasons such as conservation of moisture and controlling weeds (Ding *et al.*, 1981). However, recently the percentage of farmers using mulches have increased due to the use of plastic and aluminium foil mulches to repel alate aphids in chilli cultivation. Studies also show that maize planted as a barrier crop around chilli crops reduced alate aphid infestation by 49–65% and reduced chilli virus incidence with increased yields of 40% (Ho *et al.*, 1990)

Since the 1940s a wide range of pesticides have been used in the vegetable industry. These include a wide range of products comprising all groups of pesticides from organic to synthetic agrochemicals. Most of the pesticides are imported. In 1986, Malaysia imported herbicides, insecticides and fungicides valued at MR$171.8, MR$38.0 and MR$15.0 million respectively (MACA, 1987). It is also estimated that approximately 50% of the insecticides and fungicides are used in the vegetable industry.

Among the fungicides in particular, ethylenebisdithio-carbamate (EBDC) compounds are widely used on vegetables in Malaysia. These compounds includes the propineb, mancozeb, maneb and the mithirum group of fungicides. These fungicides are broad spectrum, protective and used against a wide range of foliar diseases including late blight, early blight of tomatoes and potatoes and mildews of the short-term leafy vegetables (Mah *et al.*, 1990). It is known that, under laboratory conditions EBDC compounds degrade to several compounds including ethylenethiourea (ETU), which is a weak carcinogen (Lyman, 1971). In 1987, large quantities of leafy vegetables exported to Singapore were rejected and destroyed due to high levels of EBDC. This has created awareness among consumers of the hazards of pesticides in food and the environment. In most of the vegetables tested, the level of EBDC was below the Marginal Residue Limit (MRL) only 14 days after application (Mah *et al.*, 1988). It was also suggested that application should stop at least 14 days before harvest, or that another fungicide, chlorothalonil, be used instead of EBDC compounds.

The common insecticides used in the vegetable industry include malathion, dimethoate, diazinon, trichlorfon, dichlorvos, aminocarb, quinalphos, methamidophos, cartap, and a wide range of synthetic pyrethroids, i.e. deltamethrin, fenvelerate, permethrin and cypermethrin. In the late 1980s insect growth regulators (IGRs) like diflubenzuron, teflubenzuron and chlorfluazuron were widely used by the vegetable growers in the effort to control lepidopterous pests. The commonly talked about "pesticide syndrome" is well documented in the case of insecticides used in the control of DBM on cabbage in Cameron Highlands (Ooi and Sudderuddin, 1978; Lim, 1974; Ooi, 1986). The history of insecticide usage for the control of DBM in Malaysia is presented in Table 2. In the last three decades the cabbage farmers replaced insecticides every two or three years due to serious DBM outbreaks. This is due to the development of insecticide resistance in DBM to a wide range of insecticides as reported by Sudderuddin and Kok (1978). Recently, the diamondback also was found to be resistant to the newly introduced IGRs, chlorfluazuron, teflubenzuron and diflubenzuron, and to *Bacillus thuringiensis kurstaki* (Syed, 1990).

It is realised now that pesticides not only are unable to provide long-term control but also have evoked undesirable effects. The major pests and diseases of vegetables are currently faced with combinations of three key issues, *viz.* no effective control, unacceptable levels of pesticide residues and resistance to pesticides. It can be said that the major pests and diseases are at the stage where conditions strongly favour development and implementation of IPM programmes. For most of the vegetables in Malaysia there are still no practical IPM programmes. However, there are two notable programmes developed that have shown some success. The first is the IPM programme for DBM on cabbage and the other for aphids/viruses on chilli.

IPM for DBM

The development of the IPM programme for DBM comprised several distinct stages. These ranged from initial appraisal of the problem and examination of

Table 2. History of insecticide usage for the control of DBM in Malaysia*

Insecticides	Year introduced	Status
Botanicals (Nicotine & Rotenone)	Before 1950	Replaced by synthetic organic insecticides
DDT, Gamma BCH Dieldrin	1950	DBM outbreaks in 1950 RF (DDT) = 530 RF (BHC) = 65
Malathion, Dimethoate, Diazinon	1956	Use in insecticide "cocktails" RF (malathion) = 2096
Trichlorfon	1964	Gave poor control in 1966
Dichlorvos	1966	Used in "cocktails"
Methomyl	1967	Mid-70s was not allowed to be used
Aminocarb	1964	Poor control by 1970
Methamidophos	1970	Used in high dosages RF = 6
Synthetic Pyrethroids	1974, 1975	Used in high dosages, not effective after 2–3 years
Bacillus thuringiensis	1979	Used for many years *B.t kurstaki* now resistant, RF = 120 *B.T. aizawai* now found to be effective
Insect Growth Regulators	Since 1984	Popular in 1984–87 DBM outbreak occurred in 1988. RF = 300–3000

* Modified from Lim, G.S. (1987)
RF = Resistance factor

strategic options, to component identification, development and integration, and eventual evaluation of the IPM package at the farm level. Several options were looked into, including resistant varieties, cultural management techniques, novel approaches like hormonal control and the use of antifeedants. However, these approaches were found to be largely exploratory in nature, and mainly confined to laboratory situations. Success in the suppression of DBM with the use of biological control has been reported elsewhere (Muggeridge, 1930; Hardy 1938;

Ullyett, 1947; Vos, 1953). The biological control approach, therefore, was given priority consideration in the DBM-IPM model.

A thorough search for parasitoids of DBM was initiated, and Lim and Ko (1975) uncovered *Apanteles plutellae*, an endolarval braconid parasitoid. Levels of parasitism averaged 26.6% to 38.8% in the highlands and 12.1% in the lowlands (Lim, 1982). In spite of its usefulness, *A.plutellae* was found to be inadequate in providing full suppression of DBM (Ooi, 1979; Lim 1982; Lim, *et al.*, 1986). Attempts were made in 1975–77 to introduce exotic parasitoids to complement *A.plutellae*. Four parasitoid species were brought in from Australia, India, Indonesia, and New Zealand, *viz. Diadegma semiclausum* (= *D.eucerophaga*), *Diadromus collaris* (= *Thyraeella collaris*), *Tetrastichus sokolowskii* and *Macromalon orientale*. These parasitoids were introduced into Cameron Highlands, and only *D. semiclausum* and *D.collaris* became established (Ooi, 1979; Ooi and Lim, 1983).

With the acceptance that parasitoids can play a significant role, an attempt was then made to begin formulating an IPM programme. Preliminary IPM approach was essentially treatment with insecticides when the number of DBM larvae exceeded the economic threshold of five larvae per ten plants. From the first six trials conducted in both farmers' and experimental farms, it was found that the economic returns in the IPM plots were marginally superior than the prophylactic farmers' plot (Sivapragasam *et al.*, 1985). Over the years as additional information became available the IPM package was improved by refining the adopted thresholds and incorporating the role of the parasitoids. The DBM-IPM package is shown in Table 3. Results of four trials in the highlands and four trials in the lowlands clearly established the superiority of IPM over prophylactic control practised by the cabbage growers. Marketable yields were 5–60% higher and up to six-fold increases in profits were obtained in IPM trial plots. The numbers of insecticide applications were also

Table 3. DBM-IPM package

No spray	-	If number of P. xylostella larvae < 4 per plant.
No spray	-	If number is >4 but <7 and level of parasitisation is > 40%.
Spray	-	Use *B. thuringiensis* if number is >4 but <7 and level of parasitisation is <40%. Example: Florbac at rate of 1 ml/litre.
Spray	-	Use insecticides if number is > 7 Example: Methamidophos at 10 ml/9 litres. Abemectin at 1 ml/litre.

DBM = Diamondback moth.
IPM = Integrated Pest Management.

Table 4. Summary of DBM-IPM trials in Cameron Highlands on cabbage crop* (after Loke *et al.*, 1990)

Mean values/crop/hectare	DBM-IPM package	Prophylactic
Cabbage yield (tonnes)	8.0	7.6
Net returns (MR$)	6610	4890
Spray frequency	1	7
Spray cost (MR$)	42	238

* Mean of four trials.

Table 5. Summary of DBM-IPM trials in Jalan Kebun (lowland) cabbage (after Loke *et al.*, 1990)

Mean values/crop/hectare	DBM-IPM package	Prophylactic
Cabbage yield (tonnes)	6.1	3.2
Net returns (MR$)	2417	307
Spray frequency	1	7
Spray cost (MR$)	186	446

* Mean of four trials.

significantly reduced from 7–9 times in prophylactic plots to a maximum of only 3 applications in IPM plots (Tables 4 and 5). Furthermore no insecticide residues were detected in the crop harvested from IPM plots (Loke *et al.*, 1990).

To implement DBM-IPM with the farmers, between 1988 and 1990 several training courses for extension agents, seminars, dialogue sessions with farmers and field demonstrations were held in Cameron Highlands. A recent survey in Cameron Highlands revealed that 54% of the farmers are already practising IPM and 86% are keen to implement the programme (Loke *et al.*, 1990). With the strong support from the research and extension services, DBM-IPM will be adopted area-wide in the very near future.

IPM for Aphids/Viruses on Chilli

Mosaic diseases caused by viruses have become the most devastating diseases on chilli. Reduction in chilli yield ranges from 10% to 15%, if infection comes in late and may reach 60% if plants were infected at an early stage (Ong *et al.*, 1980). The major viruses of chilli in Malaysia are chilli veinal mottle virus (CVMV) and cucumber mosaic virus (CMV). A survey conducted by Ong (1990) showed that CVMV was more common than CMV. About 72% of the isolates collected were CVMV while 20% of them were CMV. Mixed infection of the two viruses occurred in only 6% of the isolates. These two viruses are transmitted by several species of aphids namely, *Aphis craccivora, Toxoptera citricidus, Aphis spiraecola, Aphis gossypii, Myzus persicae, Rhopalosiphum maides* and *Hysteroneura setariae* (Ong *et al.*, 1978).

As mentioned earlier, cultural methods of control using reflective mulches and intercropping with maize have reduced virus diseases of chilli. In the IPM approach to control virus diseases on chilli these components were integrated with insecticides in proper agronomic practices. A complete IPM package (T1) was evaluated against two variations (T2 and T3) and a control (T4) (Table 6). The trial showed positive effects of the complete IPM package with good nursery protection, use of maize as a barrier crop and spraying with effective insecticides when vectors were first seen (Ho *et al.*, 1990). Yields of chilli were also significantly higher in complete IPM package (T1) plots (Table 7). The two criteria which were used to determine the initiation of chemical spraying were (i) when virus vectors were first seen (ii) when viral symptoms first appeared. Therefore, thorough surveillance and monitoring of virus vectors and virus symptoms is essential, as in any IPM programme. In addition the IPM package for chilli includes compulsory nursery protection of seedlings against virus infection.

The IPM programme for chilli is at the experimental stage and large-scale pilot projects in farmers' plots will be conducted before adoption of the programme.

Constraints to Adopting IPM

First, most pest management practitioners dealing with pest control have never needed or been encouraged to think other than of pesticides. Such a single-track approach came about from the burgeoning of cheap, simply-applied, and highly effective pesticides that fitted readily into the crop complex production system. Even today, very few are aware of the need for IPM, and this will delay the adoption of IPM programmes.

Another major constraint is that research into a particular pest is largely on component aspects rather than from a holistic approach. Furthermore, there is an overemphasis on the chemical component. Ecological studies which provide useful information to rational pest management are lacking. Of even greater importance, however, is the lack of true integration of research in developing the IPM package itself. Again, research programmes are conducted in isolation, away from the needs of the farmers, resulting in practical problems, when the results are delivered to the clientele.

Table 6. Effect of a complete integrated pest management package (T1) on the total number of aphids caught and its percentage reduction as compared to its variations (T2 and T3) and the control (T4) (after Ho *et al.*, 1990)

Treatment	Incidence of viral diseases	Mean total no. winged aphids caught	Mean % winged aphids reduction
T1	50.58a	38.3a	58.9a
T2	55.78a	44.0a	52.9a
T3	74.44b	55.3b	40.7b
T4	83.25c	93.3c	-
F-test **	**	*	
SE (mean)	1.3	2.25	2.18
CV (error)	12.0	6.76	7.27
DF (error)	39	6	4

T1	:	Nursery	-	Seedlings raised in insect proof cage + insecticide sprayings
		Field	-	Paddy straw mulching + maize as barrier + insecticide spraying (began when vector seen)
T2	:	Nursery	-	Seedling raised in open nursery + insecticides spraying
		Field	-	Paddy straw mulching + maize as barrier + insecticide spraying (began when symptoms seen)
T3	:	Nursery	-	Seedlings raised in open nursery + insecticide spraying
		Field	-	Maize as barrier + insecticide spraying (began when symptoms seen)
T4	:	Nursery	-	Seedlings raised in open nursery
		Field	-	No mulching, no barrier and no insecticide (control)

1 : Means of three replicates.

2 : Means and % reduction with the same letter are not significantly different at 5% level.

* : Significant at 5% level.

** : Significant at 1% level.

NB : % Aphid reduction was calculated using the following formula:

$$\% \text{ Aphid reduction} = (1 - \frac{\text{aphid catches in treatment plot}}{\text{aphid catches in control plot}} \times 100)$$

Differences in winged aphid population of the treatment plots is based on their reduction from total number of aphids in the control plot (T4).

Table 7. Effect of a complete integrated pest management package (T1) on the chilli as compared with its variations (T2 and T3) and the control (T4) (after Ho *et al.*, 1990)

Treatment	Mean total no of fruit/5 plant	Mean no. fruit/plant	Mean total fruit weight/5 plant (g)	Mean total fruit weight/plant (g)
T1	905.0	181.0a	8438.0	1687.6a
T2	898.7	179.7a	7795.0	1559.0a
T3	525.7	110.5b	4643.7	928.7b
T4	410.7	82.1b	3447.3	689.5b
Test of significance		**		**
SE (mean)		15.7		118.9
CV (%)		19.6		16.9
DF (error)		6		6

T1 : Nursery - Seedling raised in insect proof cage + insecticide sprayings
Field - Paddy straw mulching + maize as barrier + insecticide (began when vector seen)

T2 : Nursery - Seedlings raised in open nursery + insecticide spraying
Field - Paddy straw mulching + maize as barrier + insecticide sprayings (began when symptoms seen)

T3 : Nursery - Seedlings raised in open nursery + insecticide spraying
Field - Maize as barrier + insecticide spraying (began when symptoms seen)

T4 : Nursery - Seedlings raised in open nursery
Field - No mulching, no barrier and no insecticide (control)

1 : Mean of three replicates. Means with the same letter are not significantly different at 5% level.

** : Significant at 1% level

Communication of IPM technology to the farmers has met little success. This is especially so in most attempts since the IPM programmes are not simplified sufficiently, due to a poor understanding of farmers' behaviour and attitudes.

The complexity of the vegetable crop itself may be a constraint. The crop is generally short term with several major and minor pests, and development of an IPM programme may be demanding.

Future Strategies in IPM

Most of Malaysia's IPM research on vegetables began in the 1970s and was limited to DBM and chilli viruses. A more concerted effort is needed to expand the studies to other pests and diseases of vegetables. However, realising the need for this the Ministry of Agriculture wisely formed the National IPM Committee for vegetables in 1990. The goal of the committee will be to promote IPM practice by farmers for the control of pests and diseases in the production of vegetables in Malaysia. The Department of Agriculture (DOA) and the Malaysian Agricultural Research and Development Institute (MARDI) will be responsible for the overall implementation of IPM activities. Six working groups have been formed to assist the National Committee in planning and implementation of various aspects, *viz.* control tactics; pest surveillance and forecasting; implementation, demonstration and evaluation; training and

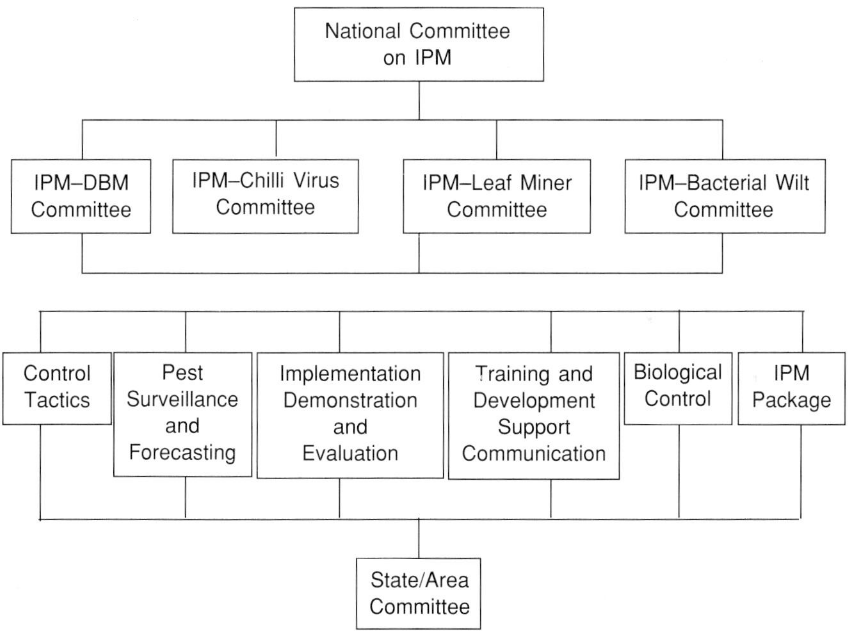

Fig. 1. Organisation for the development and implementation of IPM on vegetables

development of support communication; biological control and IPM packaging (Fig. 1). Besides implementing the ongoing IPM programmes, IPM of vegetables will be expanded to other major pests and diseases, e.g. bacterial wilt of tomatoes, leaf miner on sugar peas and thrips on vegetables.

Future strategies in IPM of vegetables will focus on the use of biological control agents and cultural methods of control, with little emphasis on the use of pesticides. Thus far, limited studies have been done on the conservation and augmentation of the natural enemies, and this will be given priority. Special emphasis will be given to generate increased awareness for the need of IPM in vegetables. The strategic steps taken will include IPM demonstration and appropriate training of extension personnel and the farmers. Particularly, in farmer training, the "field school" approach, as described by Lim (1990), should be adopted. In training, suitable development support communications will be utilised, encompassing pre-tested posters and pamphlets, and other audio-visual media.

Future research strategies on pest management of vegetables will be persued holistically. Component research will be conducted with the IPM package in mind and the practicality of implementation of the programme. All IPM research will be conducted in close collaboration with the farmers. Furthermore, studies on the farmers' knowledge, perception, attitude and practice relating to IPM will be initiated before implementation of such programmes. This, hopefully will improve the communicative processes crucial for translating IPM for adoption.

References

Chay, E.M. (1990) Farmers' perception and practices on DBM control. Paper presented in Sem. Management of *Plutella xylostella* : Perspective and Strategies, 24 Aug 1990, Kuala Lumpur.

Chew, B.H. and Ong, C.A. (1990) Genetics and breeding for chilli veinal mottle and cucumber mosaic virus resistance in hot pepper. Paper presented at the 3rd International Conference on Plant Protection in the Tropics. Genting Highlands Resort. 20–23 March 1990. Kuala Lumpur, Malaysia.

Ding, T.H., Vimala, P. and Yusoh Salleh (1981) *An Agroeconomic Survey of Vegetable Farming in Peninsular Malaysia*. MARDI. Serdang, Selangor. Malaysia. 139pp.

Hanada, T. (1987) *Promotion of Vegetable Production in the Tropics*. TARC. Japan. 86pp.

Hardy, J.E. (1938) *Plutella maculipennis* Curt., its natural and biological control in England. *Bull. Ent Res.* 29,343–372.

Ho, B.L. (1988) Performance of local tomato varieties against bacterial wilt disease. *Teknol. Sayur-Sayuran Jil.* 4,27–52.

Ho, B.L. (1989) Screening of tomato varieties for resistance to bacterial wilt. 12pp (unpublished).

Ho, B.L. Mohd. Roff, M.N., Abdul Aziz, A.M. and Ong, C.A. (1990) Integrated management of viral diseases of chilli in Malaysia. Paper presented at the 3rd International Conference on Plant Protection in the Tropics. Genting Highland Resort. 20–23 March 1990. Kuala Lumpur, Malaysia.

Lim, G.S. (1974) Integrated pest control in the developing countries of Aisa. In: *Environment and Development* (Ed. Dworkin, D.M.), pp. 47–76. SCOPE Misc. Publ., Indianapolis.

Lim, G.S. (1982) The biology and effects of parasites on the diamondback moth, *Plutella xylostella* (L) PhD thesis, Univ. of London. 317pp.

Lim, G.S. (1987) IPM of *Plutella xylostella* (L.) on vegetables in Southeast Asia: Rationale, needs and prospects. Paper presented to the Southeast Asian Pesticide Management Workshop, Febraury 23–27, 1987. Pattaya, Thailand. 14pp.

Lim, G.S. (1990) Integrated pest management of diamondback moth: Practical realities. Paper presented at the 2nd International Workshop on the Management of diamondback moth and other crucifer pests. 10–14 Dec. 1990. AVRDC, Shanhua, Taiwan (ROC).

Lim, G.S. and Ko, W.W. (1975) *Apanteles plutellae* Kurdj., a newly recorded parasite of *Plutella xylostella* (L) in Malaysia. *MARDI Res. Bull.* 3,94–95.

Lim, G.S., Sivapragasam, A. and Ruwaida, M. (1986) Impact assessment of *Apanteles plutellae* on diamondback moth using an insecticide-check method. In: *Diamondback Moth Management*. AVRDC. Shanhua, Taiwan (ROC) pp. 159–171.

Loke, W.H., Lim, G.S., Syed, A.R., Abdul Aziz, A.M., Rani, M.Y., Md. Jusoh, M. and Cheah, U.B. (1990) IPM on the diamondback moth in Malaysia: Development, implementation and impact. Paper presented at the 2nd International Workshop on the Management of diamondback moth and other crucifer pests. 10-14 Dec. 1990. AVRDC, Shanhua, Taiwan (ROC).

Lyman, W.R. (1971) The metabolic fate of Dithane-M45 (coodination product of zinc ion and manganous ethylenebisdithiocarbamate) In: *Pesticide Terminal Residues*, Invited papers from the International Symposium on Pesticide Terminal Residues, Tel-Aviv, Israel (Ed. Tahori, A.S.). Butterworth & Co. Ltd., London. pp. 243–256.

MACA (1987) Annual report and Directory, 1986/87. The Malaysian Agricultural Chemicals Association. 43pp.

Mah, S.Y. (1985) Anthracnose fruit rot (*Colletotrichum capsici*) of chilli (*Capsicum annuum*): causal pathogen, symptom expression and infection studies. *MARDI Teknol. Sayur-sayuran Jil.* 1, 35–40.

Mah, S.Y., Cheah, U.B. and Syed, A.R. 1988. Masalah sisa racun kulat pada sayur-sayuran di Semenanjung Malaysia. MARDI Teknol. *Sayur-sayuran Jld.* 4,39–45.

Mah, S.Y., Cheah, U.B and Syed, A.R. (1990) An overview of the fungicide residue problem on vegetable crops in Peninsular Malaysia, 3rd International Conference on Plant Protection in the Tropics. MAPPS. Genting Highlands. Kuala Lumpur. Malaysia. (Abstracts)

Ministry of Agriculture Malaysia (1984) *National Agriculture Policy-Development programme for vegetable commodities 1986–2000*. Min. of Agric. Kuala Lumpur. Malaysia. 192pp.

M. Shukor Nordin, Syed, A.R., Nishio, T., Mochzuki, T. and Kamimura, S. (1986) Breeding tomato for resistance against late blight (*Phytophthora infestans*) in Cameron Highlands. 2nd. Int. Conf. Pl. Prot. in the Tropics. MAPPS. Genting Highlands. Kuala Lumpur. Malaysia (extended abstracts).

Muggeridge, J. (1930) The diamond-back moth. Its occurrence and control in New Zealand. *New Zealand Journal of Agriculture* 18,235–264.

Ong, C.A. (1990) Studies of chilli viruses in Peninsular Malaysia. Progress report on the component studies of IPM in chilli. (Ed. Chew, B.H.) (MARDI).

Ong, C.A., Ting, W.P. and Varghese, G. (1978) Role of aphid vectors in the spread of chilli veinal mottle virus. *Proc. Pl. Prot. Conf.* 22–25 March 1978, 53–60pp., Kuala Lumpur, Malaysia.

Ong, C.A., Varughese, G., and Ting, W.P. (1980) The effect of chilli veinal mottle virus on yield of chilli (*Capsicum annuum*). *MARDI Res Bull.* 7,78–88.

Ooi, P.A.C. (1979) Incidence of *Plutella xylostella* (L.) (Lepidoptera: Yponomeutidae) and its parasite, *Apanteles plutellae* Kurdj. (Hymenoptera:Braconidae) in Cameron Highlands, Malaysia, *Malaysian Applied Biology* 8,131–143.

Ooi, P.A.C. (1986) Diamondback moth in Malaysia, pp. 25–34. In *Diamondback Moth Management. Proceedings of the First International Workshop.* AVRDC, Shanhua, Taiwan.

Ooi, P.A.C. and Lim, G.S. (1983) Introduction of exotic parasitoids to control the diamondback moth in Malaysia. Symposium on Res. in Biol. and Biotech. in Developing Countries: Singapore, 2–4 November 1983.

Ooi, P.A.C. and Sudderuddin, K.I. (1978) Control of diamondback moth in Cameron Highlands, Malaysia pp. 214–227. In *Proceedings of Plant Protection Conference 1978.* Rubber Research Institute of Malaysia, Kuala Lumpur.

Saharan, H.A. (1986) Vegetable production in Malaysia — current status and future prospects. Paper presented at the FAO Expert Consultation on Improved Production of Tropical Vegetables, Bangkok, Thailand. 16–20 Sept. 1986, 38pp.

Sivapragasam, A. and Lim, G.S. (1982) Control tactics for the diamondback moth, *Plutella xylostella* (L) in Malaysia. Paper presented in: National Symp. on Vegetables and Ornamentals in the Tropics, 7-8 Oct 1982. Univ. Pertanian Malaysia. Serdang.

Sivapragasam, A., Tee, S.P. and Ruwaida, M. (1982) Effect of intercropping cabbage with tomato on the incidence of *Plutella xylostella* (L). *MAPPS News.* 6,6–7.

Sivapragasam, A., Lim, G.S. and Ruwaida, M. (1985) Experimental trials on an integrated pest management programme for *Plutella xylostella* (L). In: *Integrated Pest Management in Malaysia* (Eds. Lee, B.S., Loke, W.H. and Heong, K.L.), pp. 191–207. MAPPS, Kuala Lumpur.

Sudderuddin, K.I. and Kok, P.F. (1978) Insecticide resistance in *Plutella xylostella* collected from Cameron Highlands of Malaysia. *FAO Pl. Prot. Bull.* 26,53–57.

Syed, A.R. (1990) Insecticide resistance in the diamondback *Plutella xylostella* L. in Malaysia. Paper presented at the 2nd International Workshop on the Management of Diamondback moth and other crucifer pests. 10–14 Dec. 1990, Tainan, Taiwan.

Tengku Mohd. Ariff, T.A. (1985) The status of the vegetable industry in Peninsular Malaysia and strategies for development, *MARDI Teknol. Sayur-sayuran Jld.* 1,1–6.

Ullyett, G.C. (1947) Mortality factors in population of *Plutella maculipennis* Curtis, and their relation to the problem of control. Dep. Agric, Forestry, Union S. Africa. *Entomol. Mem.* 2,77–202.

Vos, H.C.C.A.A. (1953) Introduction in Indonesia of *Angitia aerophaga* grav., a parasite of *Plutella maculipennis* Curt. Contrib. Gen. Agric. Research Station, Bogor 134, 1–32.

Chapter sixteen:
Management of Cocoa Pests and Diseases in Malaysia in the Year 2000

E.B. Tay[1] and C.L. Bong[2]

[1] *Malaysian Cocoa Board, Kota Kinabalu, Sabah, Malaysia*

[2] *Department of Agriculture, Kota Kinabalu, Sabah, Malaysia*

Introduction

The management of pests and diseases at a point in time is influenced by the cultivation practices and the developments in the related sectors of agriculture, with economics as the primary underlying consideration. Among them are the complexity and dynamics of development of pests and diseases of the crop, of the planting industry, of the agrochemical industry, advances in research on crop improvement, and the increasing global concern for the health of the earth. This paper discusses the likely developments in these areas and their applications on the management of pests and diseases of cocoa in the years to come.

Current Status and Relative Importance of Pests and Diseases of Cocoa in Malaysia

Cocoa is a relatively new crop to Malaysia. Historically, its presence was mentioned by Koenig (1894) in 1778 in Malacca in Peninsular Malaysia. In Sabah, it was first recorded by Von Donop in 1882 in his report on agriculture in British North Borneo. However, commercial cultivation and, therefore, its abundance as a host to local pests and pathogens did not begin until the mid-1950s. The area cultivated then was about 31 ha in Sabah and 400 ha in Peninsular Malaysia. By 1966, the area had increased to 3500 ha. During this short span of time, a total of 26 species of insects of importance were found to have caused damage to the growth and production of cocoa (Conway 1971).

When cocoa was first cultivated as a commercial crop, the insect pests of importance were the leaf feeders, branch and trunk borers, sap feeders including mirids, hoppers, aphids and mealy bugs. With the changes over the years in the environment surrounding cocoa plantations, only branch and trunk borers (*Zeuzera* spp., *Endoclita hosei* Tindale) and mirids (*Helopeltis* spp. and *Platyngomiriodes apiformis* Ghauri), remain as important pests. The branch and trunk borers are low density pests which can cause considerable losses of trees at times. The mirids are debilitating and contribute to the increase in brown pod rot. Currently the most important insect pest in terms of crop loss is undoubtedly the cocoa pod borer (CPB), *Conopomorpha cramerella* Snellen, a species that was not recorded on cocoa in Malaysia until August 1980. CPB infestation results in total loss of pods and where the infested pods are usable, there is a loss of quality in terms of smaller and clumped beans. If the pest is not controlled, the loss of production can be total.

With the best current management practice, it is estimated that there is still some 5% loss of the crop from CPB infestation. Another species of pod borer, *Cryptophlebia encarpa* Meyr was first recorded in Sarikei and Sibu in Sarawak in 1987. Currently the distribution of this pest is spreading slowly but is still confined to Sarawak.

Among the mammalian pests, squirrels (*Callosciurus notatus* and *C. prevostii*) appear to be the most important. On the other hand, rats, considered a serious problem in Peninsular Malaysia only occasionally cause problems in Sabah. At one time monkeys caused considerable damage in the early days of cocoa cultivation in Sabah where most plantations were surrounded by jungles. With the increase in the clearing of the jungles, monkeys cause fewer problems.

In terms of disease, presently in Malaysia, there are recorded on cocoa 14 diseases of known etiology, five diseases of partially understood and/or unestablished etiology, two virus diseases that had been eradicated from mid-1960s to early 1970s, a genetic disorder and several parasitic plants and epiphytes (Liu, 1979; Liu and Liew, 1975, 1979; Bong, 1982; Tay *et al.*, 1983; Wong, 1983; DOAS Ann. Res. Reps. for 1982 to 1989). Of these 21 or so recorded diseases, economically most important is the vascular streak dieback disease or VSD caused by *Oncobasidium theobromae*. National losses of the crop from VSD can be safely estimated to be in the range of 5% to 10% from year to year depending on the severity of the annual incidence of epiphytotic cases. Next in importance are the diseases caused by *Phytophthora* species, *P. palmivora*, the causal agent of black pod and canker in particular. Losses from black pod are generally below 5%. Because of the destructive potential, canker disease is accorded some importance despite the fact that losses from this disease are low and occasional. Although there is no estimate made on the economic importance of pink disease caused by *Corticium salmonicolor*, the disease is considered major and more important than canker because occasional epiphytotics do occur and because of its wide distribution and perennial presence. On the same par with pink disease and probably a greater threat in the near future are the root diseases caused by *Phellinus noxius* (brown root disease), *Rigidoporus lignosus* (white root disease), *Poria* and *Ganoderma* spp. (red root disease). At present root diseases, in particular the brown and red root diseases, are the major cause of loss of mature trees. Of lesser importance among the debilitating diseases are the thread blights caused by *Marasmius cyphella, M. neosessiles* and *M. scandens* (white thread blight),

M. byssicola (brown thread blight) and *M. crinisequi* (horsehair blight).

Diseases considered of minor importance now and likely to remain so are the brown (*Botryodiplodia theobromae*) and other *Phytophthora* pod rots (*P. parasitica, P. meadii* and *P. heveae*), fruit and leaf anthracnose (*Colletotrichum gleoesporioides*). Of the five diseases of lesser understood etiology, the most likely to constitute a significant threat because of its destructive potential is terminal wilt (likely causal pathogen, *Phialophora cyclamanis* or *Stagonospora* sp.). Massive branch dieback is a major concern on older stands and may be associated with wetwood disease. While the occurrence of wetwood disorder is less extensive than *Phytophthora* canker, the problem requires investigation.

Dynamics of the Development of Pests and Diseases and the Likely Challenges by the Year 2000

The importance of a species of insect or pathogen varies with time and cultural practices. Table 1 illustrates the changes in relative importance of the pests and diseases of cocoa over the last 30 years of cocoa cultivation. These changes are brought on by the spread of cocoa cultivation into different climatic zones, the changes of the cocoa environment and the cultivation practices.

The dynamics of the development of pests and diseases of cocoa is often more spectacular with insect pests, brought on by intensive application of insecticides and/or a sudden change in the environment. Outbreaks of pests following the application of a wide spectrum contact insecticide to control leaf

Table 1. The relative importance of pests and diseases of cocoa in the 1960s and 1980s

Pests and diseases	1960s	1980s
Leaf-feeding insects	+++	p
Trunk and branch borers	+++	+++
Mirids	++	++
CPB	–	+++
VSD	++	+++
Pink disease	++	++
Phytophthora diseases	p	++
Root diseases	p	++
Terminal wilt	–	+

– absent
p present but unimportant
\+ important only in some years
++ widespread and important
+++ very widespread and very important

feeding insects was reported by Conway (1971). The effect on the diseases is less apparent or dramatic in appearance.

The most recent example is the attempt to maximise yield by introducing high insolation through total removal of shade and increased fertiliser input, supported with a regular programme of insecticide application to control CPB. Temperature and transpiration increased as shade was removed and the plants were exposed to water stress. Weeds proliferated in the exposed areas and weedicide was applied to control the weeds. When weeds were controlled, more ground was exposed and was therefore less protected from erosion and moisture loss. During the first year when these measures were introduced and maintained, shoot feeding insects such as the leaf hoppers, mites, aphids and mealy bugs began to appear more abundant and shoot growth was inhibited or aborted which resulted in thinning of the canopy. As the canopy thinned, more ground was exposed and weeds sprung up in more areas. In the second year, the abundance of these pests and shoot damage increased. This necessitated another series of insecticide and weedicide applications. In so doing, the ecosystem was repeatedly upset and the plants were exposed to increasing stress from moisture deficit, heavy fruit load, pests and disease damage. By the third and the following years, canopy dieback occurred as a result of repeated aborted shoot growth, pest damage of the remaining young shoots and continuous senescence of older leaves. The deterioration continued: some of these trees never recovered; some died. The debilitated and dead trees showed extensive rotting of the root system. The pest damage on the canopy was conspicuous while the deterioration of the root system was not obvious during this period. Such was the apparent impact of imposing a sudden change in the environment on the development of the host, pests and pathogens. Here is a situation where a change in the cultural practice leads to a change in the host environment that in turn necessitates a series of measures to overcome the problems previously created. Although no data were collected on these observations to support the above hypothesis, such a sequence of events did not occur in fields where these measures were not introduced.

Another example of an apparent association between the occurrence of a disease and ecological disruption is the appearance of terminal wilt, a newly recorded disease that followed large-scale application of weedicides and use of 2, 4, 5–T in poisoning the *Gliricidia* shade trees. The widespread use of herbicide during the first two years of field planting also coincides with the period during which root rot complex is more prevalent. There is a need to investigate the apparent association of herbicide usage and root problems, and the effect of the herbicides on the changes in soil and rhizosphere microflora.

Therefore, changes in the relative importance of pests and diseases in the years to come will be determined by the ecosystem designed, built and maintained through cultural practices by the planting industry. The present development policy of the planting industry will give some indications of the likely changes in the status of the pests and diseases. The demand for a better quality product (higher butter fat content, bean uniformity, flavour), and the drive towards increasing productivity and profitability together push the industry towards looking for a highly productive and efficient system of cultivation. By the year 2000, a large proportion of the present hectarage of cocoa will be over 25 years old and will have exceeded its optimum economic life. It is anticipated that there will be replanting on a large scale with new planting materials. A major replanting problem is likely to be an upsurge in the occurrence and spread of root diseases if preventive or precautionary measures are not taken.

The trend is also towards clonal planting at a high density with irrigation for uniform yields and a consistent end-product, and to increase productivity per unit area. Clonal planting with a diverse gene base offers wider adaptability and better protection against explosive outbreaks of diseases or pests. Resistance to vascular streak dieback and to some degree black pod and canker is now among the selection criteria in crop improvement programmes. The threat from VSD will not be as great as in the past. Root rot complex and terminal wilt may remain problematic. The effect of irrigation on the epidemiology of root rot, black pod and canker is yet unknown. In high density planting, the spread of such diseases as root diseases and pink disease as well as a possible increase in pod damage by rodents is a concern to be watched for.

Crop diversification is now being seriously examined and practised by some plantations. The range of economic crops available for this purpose is at present limited. Crop diversification would offer an opportunity for the design of a stable and economically profitable agroecosystem consisting of complementary crops in terms of value of the crops and the mutual support of these crops for beneficial insects or biocontrol agents.

Major diseases which are not yet found in Malaysia are witches broom, *Moniliophthora* pod rot and cocoa swollen shoot diseases. With the increasing ease of air travel and the continuous quest for new planting materials by the planting community, it can be expected that such problems will in due course appear in Malaysia unless vigilance is maintained.

Developments in Crop Protection

Among the tools of crop protection are regulatory control, chemical control, biological control, cultural control and physical control. Developments in these areas will determine the range of choices for integration in the management of pests and diseases of cocoa in the years to come.

Regulatory Control

Regulatory control is particularly important and effective in excluding pests and pathogens which are not yet found in the country or in campaigns to eradicate pests which have been inadvertently introduced. Quarantine measures will remain the vanguard in the prevention of entry of pathogens of such destructive diseases as witches broom caused by *Crinipellis perniciosa*, now confined to South America and some Caribbean islands, frosty or watery pod rot caused by *Moniliophthora roreri* found in South America, and cocoa swollen shoot disease caused by a virus of which the most virulent strain is found in Ghana and Nigeria (Lass, 1985). On pests and pathogens already found in Malaysia, regulatory control should also be imposed to avoid the inadvertent introduction of races of pests or pathogens which may differ in virulence to those found in Malaysia. These include *Conopomorpha cramerella*, *Phytophthora* spp. and *Oncobasidium theobromae*.

Some form of new regulatory measures on planting materials like certification or registration is necessary. If the planting materials can be certified for a range of characters, including their resistance or susceptibility to specific pests and diseases, the planting industry can be protected and be warned of the crop protection measures needed.

Chemical Control

The term "chemical" as used here refers strictly to pesticides or synthetic chemicals which do not occur naturally. Chemical control is presently the main tool in the management of insect pests and weeds. The use of weedicides on exposed areas, particularly those on hill slopes is likely to be critically appraised and measured against soil cover or protection. Fungicides are used less frequently and less extensively than the insecticides or weedicides, generally on such diseases as black pod and canker, pink disease, VSD (first two years) and thread blights. There will be a search for protectant and eradicant fungicides for the management of root diseases caused by *P. noxius*, *R. lignosus*, *Ganoderma* and *Poria*.

The heaviest chemical use is on the control of the cocoa pod borer with pyrethroids as the most popular insecticide applied. The possible development of resistance among the pests, especially the pod borer to pyrethroids, is a matter for concern.

There will be a demand for a wider range of chemicals effective against such pests as *Conopomorpha cramerella* and *Oncobasidium theobromae*. As these two problems are not worldwide and therefore the market is limited, the agrochemical industry's response is not likely to be very enthusiastic. It is likely to be met from the existing array of chemicals developed or under development, as indicated by the gradual springing up of research stations owned by the multinational agrochemical companies in Malaysia and other Asian countries.

It does not appear likely that there will be any novel approach to rodent control. Shooting, poisoning and trapping of squirrels will remain the standard method used. In the case of poisoning, new rodenticides need to be found to replace current anticoagulants to which the pests are becoming increasingly resistant.

Increasing pressure from the global concern on environmental health is having an effect on the development policies of the agrochemical industry and these affect the choices available to us. On the development of pesticides, chemicals with higher specificity and lower persistence are likely to increase relative to those which are persistent, and formulations will be improved (for controlled release; for delivery in water as distinct from oil; dyed for easier monitoring of coverage and avoidance of abuse) (Spurrier, 1990; Gotsch & Rieder, 1990). Such chemicals will undoubtedly be more costly than existing compounds, which will in turn put pressure on the producers to attain higher productivity in order to maintain profitability. The agrochemical industry's newly developed products including genetically engineered herbicide-resistant plants or disease-resistant hosts will not be of direct benefit or applicability to cocoa crop protection in the immediate future. It is likely that in the management of pests and diseases of cocoa and other crops, managers will be encouraged to be more judicious or discriminating in exercising chemical control, particularly on the delivery system (sprayers, nozzles, formulations) (Spurrier, 1990). This, coupled with the development of accurate pest or disease warning systems, will reduce the usage of chemicals and thus present less hazard to the environment as well as prolonging the economic life of the chemical.

Although other control measures will be more intensively pursued to meet the challenges created by the use of chemicals, they will continue to have a definite role in cocoa crop protection.

Biological Control

Biological control as used here refers to the control of one organism by another, exclusive of man, as defined by Baker (1987). It includes the use of resistance in planting materials, natural enemies such as parasites, antagonists and suppressive organisms, and microbial associations.

Resistant planting materials

The use of resistant planting materials is the most effective way of controlling some pests and diseases. In countries where VSD is a threat as in Malaysia and Papua New Guinea, resistance to the disease is among the top selection criteria in crop improvement programmes. The present screening procedure used for VSD is a long process. There is so far no attempt to search for biochemical markers of resistance for VSD due to technical reasons such as the fastidiousness and obligate nature of the pathogen, *Oncobasidium theobromae*. It is likely that this line of research will be initiated to provide a faster route in breeding programmes and to try to define or identify the types of resistance mechanisms at molecular levels so that the different types of resistance if they exist can be more systematically and efficiently deployed.

Little has yet been done to identify materials resistant to insect pests. Only some observations have been carried out on the possible resistance of cocoa clones to CPB. Although nothing spectacular has so far been observed, there are indications that there exists resistance among the clones to the pest. Such characters as smooth skin and hard sclerotic layer of the pod husk constitute some structural resistance to the pest. This line of research is likely to continue. Resistance in planting materials to black pod and canker is present and is a selection criterion. The use of biochemical methods to screen for resistance to black pod among cocoa clonal materials is now under investigation and should prove to be a faster means of identifying resistance and thus quicken the pace of breeding (Blaha, 1987). Resistance to other diseases such as pink disease and root diseases has not been studied. The prospect of being able to employ resistance in the management of these diseases and pests in early 2000 is therefore rather mixed, with progress made in certain areas but not all.

Parasites and natural enemies

The use of natural enemies is more commonly seen so far in pest management. The control of *Conopomorpha cramerella* by *Trichogrammatoidea bactrae fumata* Nagaraja is effective but uneconomical at present. Research is likely to intensify in this area and with improvement in breeding methods for the parasitoid, it should prove to be an ideal alternative to chemical control or as a component in an integrated management scheme for the cocoa pod borer. Other parasitoids, indigenous or exotic, will continue to be investigated. Lim *et al.* (1982) had observed that ants and other predators played a significant role in the control of CPB.

In North Sumatra, Indonesia, the use of ants as biocontrol agents of *Helopeltis antonii* Signoret has been practised in the fields where artificial ant nests are placed to encourage the ants to stay. Further studies on the behaviour, searching, feeding and nesting habits of the ants will improve the effectiveness of the system. On the other hand, McGregor & Moxon (1985) reported that some tent-building ants are vectors of *P. palmivora* and their exclusion by the more dominant species, *Anoplolepis longipes* could help in the control of black pod.

The use of microbial pathogens of insect pests is quite limited, the most well known being *Bacillus thuringiensis*. Recently Utomo and Pardede (1991) successfully isolated a strain of *Beauveria* sp. and used it against the branch borer, *Zeuzera coffeae*.

The use of parasites and antagonists in cocoa disease management is still at the research stage. There is now under investigation a number of potential biocontrol agents of such pathogens as *Phytophthora palmivora* and *Crinipellis perniciosa*, the root pathogens and *Moniliophthora roreri* (Bastos *et al.*, 1981; Bastos and Figueiredo, 1982; Odigie & Ikotun, 1982; Odamtten and Clerk, 1984; Saedoval *et al.*, 1987; Jiminez *et al.*, 1987). The biocontrol agents being investigated include *Pseudomonas aeruginosa* M-50-1 and *Bacillus* species for the control of monilia pod rot; *Cladobotryum amazonense* on *Phytophthora* spp. and *C. perniciosa; Trichoderma viride* and *Aspergillus niger* on *Phytophthora* spp.; and a number of basidiomycetes on root pathogens such as *Phellinus noxius* and *Rigidoporus lignosus* (Jollands, 1984).

Much of this research is undertaken in other countries. As biological control is sensitive to ecological changes or variations, the methods developed elsewhere cannot be instantly adopted *in toto* without having been exhaustively tested and may not be suitable. There is thus a need to initiate research on biological control of cocoa disease in Malaysia. Root diseases will be an appropriate priority target. This is because these diseases are likely to be more prevalent in the near future and the control measures available at present for root diseases are time and labour intensive and therefore very costly.

Microbial associations

Other than parasitism, associations between organisms may be antagonistic or synergistic or competitive. These other associations have not been explored for the control of cocoa diseases or pests. There are soils which are suppressive to *Phytophthora* and those which are conducive. There is likely to be a potential in the exploitation of these associations as a complementary component in the integrated management of black pod, canker and root diseases.

Mycorrhizal associations play a definite role in disease development or host tolerance or susceptibility (Baker, 1987; Dehne, 1982). Ectomycorrhizae may provide a physical barrier or produce antibiotic, or enhance their host's production of inhibitory compounds to the pathogens. Ectomycorrhizae may act by decreasing the effect of environmental stress such as drought and enhance rooting or decrease host susceptibility to the root pathogens. Mycorrhizal association and its potential role in the control of root diseases is an area where investigations have begun in Malaysia on the control of *Ganoderma* infection of cocoa roots by prior inoculation of the roots with vesicular–arbuscular mycorrhizal (VAM) fungi (Chulan *et al.*, 1990). Not all mycorrhizal associations are effective in controlling soilborne pathogens of cocoa as indicated in a study where seedlings were more healthy in sterilised than in untreated soils where more VAM fungi were found (Hedger, 1987).

Pheromones and Other Semiochemicals

The use of pheromones and semiochemicals in insect pest management is relatively novel. Beevor *et al.* (1986) successfully identified the female pheromone of CPB. Although it was not as successful as anticipated in trapping the males for the control of CPB it has proven useful as a tool in monitoring the relative abundance of CPB. An analogue of the pheromone was tested for use in disruption of mating and this has shown promising results. Thistlewood *et al.* (1989) demonstrated the existence of a sex pheromone in a mirid, the mullein bug, *Canpylomma verbasei* (Meyer). Pheromones therefore hold potential which can be tapped in the management of other pests particularly of the lepidopterous insects such as the branch and trunk borers. Research on other semiochemicals including juvenile hormones has not yet started and it is very unlikely that any will be available for practical application within the next decade.

Cultural Control

Cultural practices are often adopted in the control of pests and diseases, particularly when the incidence is low and often in combination with other methods as the epidemic develops. Sanitation pruning of the cocoa canopy contributes to the control of thread blights and black pod and will continue to be practised in the future. Regular harvesting of ripened pods has proven to be effective in slowing down the build-up of CPB. Pruning to control the height and branching of the trees also helps in improving coverage of chemical application and the reduction of rodent damage. Plant architecture is currently receiving attention.

Manuring is an area that deserves more attention. The types and rates of fertilisers used may affect soil properties and thus the soil microflora and the susceptibility of the roots to root pathogens.

Shade tree types and density can affect the development of pests and diseases. Their influence on the cocoa environment will affect the interaction between the pest/pathogens, natural enemies and the crop. Certain shade trees may also serve as alternate hosts to cocoa pests and pathogens. Shade density and intensity is now being investigated for incorporation in the integrated management of VSD. For the present it would appear that choices are limited to the types of shade trees to be used. *Gliricidia* is an alternate host of some of the insect pests of cocoa, of major root pathogens of cocoa, terminal wilt and also pink disease. Thus under certain conditions, its use should be re-examined in replanting insofar as crop protection is concerned.

Irrigation to reduce water stress and thus enhance root growth is a study currently being undertaken. The types of irrigation (sprinklers, drip) chosen may affect such diseases as canker and root rots. The selection of areas to be planted with cocoa will be more critical and a set of guidelines (water table, climate, soil types or properties) may be developed.

Physical Control

Physical control measures are heavily relied on for the management of mammalian pests. Control of rodents is in general by means of trapping or shooting. In the management of other pests and diseases, they are applied to a limited extent. Nurseries with proper roofing will gain popularity as this provides good protection from VSD and seedlings are generally more healthy. Protective cover in the form of a cylinder made out of a fertiliser bag placed round transplanted seedlings for a period of about six months offers good protection against insect pests. It is a practice likely to be increasingly adopted. Pod sleeving is not as widely used as before for CPB control and its use may increase again if chemical control fails as a result of the development of resistance to insecticides. Pod sleeving may in some cases prove to be more economical or convenient to use by smallholders than other control measures. It is a device that excludes CPB and other pod attacking insect pests.

Developments in Crop Protection Aids

There is now available a wide range of communications and data processing technology, especially in computer science and infrared remote sensing systems that can be used in crop protection, the potential of which can be tapped for application in local plant protection. An infrared remote sensing system can be used in the large-scale assessment of pest and disease damage when an outbreak occurs. Advances made in electronics, especially for meteorological equipment, data loggers and accessories for monitoring the field environment can be exploited for epidemiological studies and the development of disease or pest warning systems necessary in integrated plant protection schemes.

A number of disease and pest warning systems have been developed and used for the management of potato blight and cereal diseases in the temperate countries, the most notable of which are BLITECAST (for potato late blight) and EPIPRE (for six fungal diseases and all aphids of wheat) (Zadoks, 1984). These systems operate on a central computer, and the individual farmer can phone in or access by other means (postcard for EPIPRE) the necessary inputs (crop stage, pest and disease incidence, weather, etc.) and receive the recommendations on the economics of applying or not applying the control. The use of these systems has helped to reduce the usage or wastage of chemicals, relieve the pressure of selection of pathogens/pests resistant to the chemicals, reduce unnecessary environmental pollution, save on costs and often makes economic sense. Similar systems can be developed for the management of cocoa pests and diseases. Initial steps have been taken along this direction on VSD (Bong, 1989) and *Helopeltis* (Wills, 1984).

Diversified Agroecosystem

As mentioned above, crop diversification is being seriously considered by the planting community. This, together with replanting and the anticipated problems, offers the industry an opportunity to redesign its planting system to include diversified cropping.

There are studies which support the concept that a diversified ecosystem is more stable than a monocrop system: for diversity guards against severe outbreaks of pests and diseases and offers wider adaptability to environmental changes (Altieri & Letourneau, 1982; Wolfe, 1985). The diversity in an agroecosystem also provides for the natural proliferation of beneficial insects, some of which may be natural enemies to the pests of the main and secondary crops. It is therefore worth while to conduct a careful selection of the crops to be interplanted based on economic and complementary values and their pest and disease complexity. This, together with a careful choice and placement of the plants including timber species for use as shade or windbreaks, will contribute towards better crop protection. At present there are some weak attempts in this direction as shown by the planting of banana among the cocoa trees. The decomposing banana leaves provide the food base for the proliferation of the insect pollinators. There will be some attempts towards evaluation of more varieties of economic crops for diversification in the agricultural producing sector. The development of long-term, crops easily requires more than ten years in duration. Diversification will thus be a gradual process.

Management of Major Pests and Diseases by the Year 2000

Taking into account the above developments, the management of the major pests and diseases of cocoa by the year 2000 or thereabout is projected as follows. There will be increasing integration of the various methods of control. But the use of biotechnology in commercial cocoa crop protection will perhaps not be realised so soon. Generally there will be incremental steps taken to adopt a multidisciplinary approach to the cultivation and protection of the crop.

Cocoa Pod Borer

The management of CPB has so far relied heavily on chemical control and frequent harvesting. There are several alternative control measures currently under investigation. Biological control using the egg parasitoid *T. bacterae fumata* is effective, particularly by inundated release (Lim & Chong, 1987). It has not been widely adopted primarily because of the technical constraints in large-scale production of the parasitoid. The technical difficulty of mass breeding of the parasitoid is not insurmountable. The effectiveness of the parasitoid can also be enhanced if the kairomones involved are known (Tan & Loke, 1984). So far only a weak attempt has been made at introducing exotic parasites of related species to evaluate their potential as biocontrol agents. It would be worth while to carry out further investigations along these lines.

Previous attempts to use synthetic CPB pheromone traps to control the pest have not proved to be successful. More recently encouraging results have been obtained with the use of synthetic female pheromones in mating disruption. Further work in this area including the use of male pheromone if any could enhance the effectiveness of the technique. Pheromones are species specific and therefore harmless to non-target organisms and ideal for incorporation in a management system that makes use of biocontrol agents or chemicals. Furthermore they are non-toxic and leave no residue. These features certainly make pheromones environmentally attractive to use.

As mentioned earlier, resistance among the planting materials to CPB exists and is being investigated. The next decade should perhaps see the introduction of resistant planting materials on a limited scale in the management of CPB.

It can thus be anticipated that the management of CPB by the year 2000 will be composed of a greater integration of the various methods of control as outlined above. There will be a reduction in the intensity of insecticide use while biological control including the use of pheromones will play an increasingly important role.

Mirids

Helopeltis is a major pest of cocoa in Peninsular Malaysia. In Sabah as in Peninsular Malaysia, the pest is chemically controlled and a warning system has been developed (Wills, 1984). Regular control of CPB using the wide spectrum insecticides in Sabah may also bring down the damage by mirids. It appears that the abundance of mirids is related to shade density. It has been observed in Sabah that when shade is correctly maintained, the pest damage to both the pods and shoots can be reduced. The potential of some ant species as biocontrol agents of the mirids is likely to be further exploited. Khoo & Chung (1988) demonstrated that the black ant, *Dolichoderus thoracicus* (syn. *D. bituberculatus*), was capable of reducing damage caused by *Helopeltis*, probably through deterring rather than active predation. Similarly, work in Indonesia (Surtiyati & Suhardjo, 1991) has shown that interplanting of coconut in immature cocoa area enhances the proliferation and activity of the ant.

If the sex pheromone of *Helopeltis* or other mirids can be identified, it could play an important role in the monitoring and management of this group of pests.

Branch and Trunk Borers

Branch and trunk borers are currently controlled by sanitation practices and injection with insecticides. Biocontrol of these pests is now being investigated and the fungal pathogen, *Beauveria* sp. is a potential candidate (Utomo & Pardede, 1991). Lim (pers. comm.) observed that braconids that attack larvae of *Corcyra cephalonica* would also attack larvae of the branch and trunk borers. It would also be of interest to find out whether pheromones play a part in the mating of the adult moths and whether they can be deployed as control agents.

Vascular Streak Dieback

Several methods are integrated in the management of vascular streak dieback with resistance of planting materials as the main tool to be supported by cultural practices and chemical application based on a disease warning system (Tay & Bong, 1989). As resistance to VSD is one of the top criteria in the present crop improvement programme in the country, there will be available a wider range of resistant hybrids and clones that are genetically more diverse than at present. There will be increasing use of rain-proof nurseries not only to reduce casualties due to VSD but also to provide a more uniform environment for the growth of the seedlings.

Chemical control will continue to be used to a limited extent particularly when the plants are young. It can also be applied selectively in plantings where certain less resistant cultivars are used because of the materials' better bean characteristics such as size, uniformity and butter fat content. For such plantings, chemical control will be more efficiently applied with the help of a disease warning system, improved formulations of the chemicals and application technology. Improved fertiliser programmes and approriate shade management which is less conducive to the development of the disease will be determined and maintained.

Future management of VSD will therefore employ a combination of cultural practices, resistant materials and limited usage of chemicals.

Root Diseases

Root diseases caused by *P. noxius, R. lignosus* and *Poria* or *G. phillipii* are expected to be the major replanting problem. Although biological control of the disease will be a major research programme in the years towards 2000, advances made by then may not reach the field application stage of testing. Many of the precautionary practices recommended for the control of these pathogens on rubber and oil palm are cultural measures which are costly because they are time and labour intensive. These include clean clearing after the trees are fast-killed by poisoning, intercropping with short-term leguminous crops to exhaust the remaining inoculum, regular inspection and immediate treatment of disease foci (uproot and burn dead trees, isolation by trenching, and treatment of infected roots by exposing and painting with fungicides). In Papua New Guinea, chemical control is being evaluated and it is a variation of the above in that trees are poisoned with 2, 4, 5-T and stumps treated with copper fungicides to prevent establishment of the root pathogens (Anon, 1983). New slow release techniques currently under development for the control of other diseases may improve the effectiveness of chemical application and its economy. Chemical control may therefore be resorted to as an alternative means of control if the economics are justifiable. Some of the triazole fungicides found to be effective against VSD show promising potential for the control of these root pathogens. If the mycorrhizal association is found to be effective, inoculation of the roots of seedlings with the VAM fungi before transplanting will be a practice to be adopted.

The management of root diseases by the year 2000 is likely to rely heavily on cultural and chemical control with the possible exploitation of mycorrhizal and other microbial associations. Economical biological control by inoculating shade or cocoa tree stumps with the pathogens' antagonists will be feasible, but may need well over a decade to develop.

Pink Disease

Chemical control will probably remain the main tool of management of the disease. There now exists a wide range of chemicals which are effective against the pathogen, *Corticium salmonicolor*. Among these are copper-based fungicides which are non-specific in action and some of the inhibitors of ergosteral biosynthesis such as carboxin and carbendazim. Another area that may change is the probable reduction in the use of *Gliricidia* as shade trees as it is a host of many pathogens of cocoa as mentioned above. Coconut palms, to provide shade and as an intercrop with cocoa in existing coconut farms, can be re-evaluated, particularly for the smallholders. Resistance to the disease is not uppermost in the selection criteria in the current crop improvement programmes and is unlikely to supersede chemical control in the near future in the management of the disease. Likewise there has not been any research undertaken so far on the biological control of the disease on cocoa.

Black Pod

Black pod is at present controlled chemically only during an unusually wet season and during the main peak crop period when the incidence is high. During other times, frequent harvesting has been found to be adequate. Chemical usage in Malaysia against the disease has never been as intensive as it has been in Africa, South America or Papua New Guinea. As researchers (Odigie & Ikotun, 1982) on this disease have shown, there exist natural antagonists such as *Aspergillus* and *Penicillium*. Resistance of the planting materials to the disease is also present. The low incidence and low damage of the disease here may have been the result of a combination of the two as fungicide application is in general infrequent in mature cocoa fields. Alternatively, it is possible that the local strains of *P. palmivora* are generally less virulent. Resistance to the disease is one of the selection criteria and the materials developed for commercial planting will have an acceptable degree of resistance to the disease.

Concluding Discussion

We do not anticipate a revolutionary change in the approach to the cultivation of cocoa and the management of its pests and diseases ten years from now. Instead there will be stepwise improvements in the integrated control of the pests and diseases as indicated above.

Chemicals will stay as a short-term control measure to be incorporated or combined with biological and cultural methods. There will be more advances made in application technology and more efficient product formulation which will in part reduce the amount of chemicals used per unit area. Together with an efficient disease and pest monitoring and warning system, its application can be more accurately timed. All these will reduce cost while at the same time reducing the risk to the environment.

The generalised use of weedicides will receive critical appraisal with regards to their effect on the crop, the soil, and the impact on the beneficial insects and micro-organisms that are normally harboured by these hosts or the environment they have created.

Biological controls including the use of disease-resistant materials will increase. There is great potential in the exploitation of beneficial soil and rhizosphere microbial associations in the management of root diseases. Measures towards augmenting them will gradually be identified and play an increasingly greater role in time. There will be a greater awareness of the importance of not disrupting the ecological balance established. In the short history of cocoa cultivation in the country, there are many examples of association between outbreaks of pests and diseases with disruptions of the cocoa ecosystem. The cocoa environment is by its very nature more complex than that of other tree crop environments, since to start with, it consists of at least two major tree species: the crop itself, and the shade crop. It therefore offers an opportunity to create a stable environment with the right mix of plant species. Properly designed diversity in the plants cultivated may reduce the number and severity of pest and disease outbreaks and economic losses. The latter is an attractive proposition that will not be realised soon, as the range of economic crops now available is limited and other potential plant species or crops have not been evaluated for their economic values and compatibility.

How dynamic the development of pests and diseases is depends to a great extent on how the cocoa environment is created and manipulated, or how the stability is maintained. The better the cocoa environment is understood, the more efficient we shall be in managing its pests and diseases. For the next decade a start should be made to integrate other disciplines of agriculture into the pest and disease management system. Pest and disease management can no longer depend solely on the entomologists and phytopathologists but must also involve the agronomists and others related to the cultivation of the crop.

In conclusion pest and disease management in cocoa in the year 2000 will involve a better understanding of the crop, its environment, the flora and fauna existing in such an environment and their interactions.

Acknowledgements

We wish to thank the Director-General of Malaysian Cocoa Board, Dato' Dr Hashim Abdul Wahab, and the Director of Agriculture, Sabah, Datuk Aripen Ampong for their encouragement and permission to present this paper. We would also like to thank Professor K. H. Tan of University Science Malaysia for discussion, Ms Rosie Liew and Mr Chong Tan Chun of the Department of Agriculture, Sabah, for the assistance given.

References

Anon (1983) Cocoa pathology. In: *Lowland Agricultural Experiment Station Annual Report*, 1982. Papua New Guinea.

Altieri, A.M., and Letourneau, D.K. (1982) Vegetation management and biological control in agroecosystems. *Crop Protection* 1, 405–430.

Baker, K. (1987) Evolving concepts of biological control of plant pathogens. *Ann. Rev. Phytopathol.* 25, 67–85.

Bastos, C.N. and Figueiredo, J.M.De. (1982) Effects of the culture filtrate of *Cladobotryum amazonense* Bastos, Evans & Samson on three *Phytophthora* species from cocoa. *Revista Theobroma* 13 (2), 113–117.

Bastos, C.N. Evans, H.C. and Samson, R.A. (1981). A new hyperparasitic fungus, *Cladobotryum amazonense*, with potential for control of fungal pathogens of cocoa. *Trans. Br. mycol. Soc.* 77, 273–278.

Beevor, P.S., Cork, A., Hall, D.R., Nesbitt, B.F., Day, R.K. and Mumford, J.D. (1986) Components of female sex hormones of cocoa pod borer moth, *Conopomorpha cramerella*. *J. Chem. Ecol.* 12, 1–23.

Blaha, G. (1987) Enzymatic polymorphism of the *Phytophthora* causal agents of black pod disease: research into variability linked to host-parasite interaction. *Proceedings of Tenth International Cocoa Research Conference, Santo Domingo, Dominican Republic*, pp. 397–406. Cocoa Producers Alliance, London, UK.

Bong, C.L. (1982) Some important diseases of cocoa in Sabah. Seminar Penyelidikan Koko Untuk Pembangunan Pertanian Negara, Stesyen Penyelidikan Koko Bukit Quoin, Tawau, Sabah, Malaysia.

Bong, C.L. (1989) VSD epidemic development and weather factors in Sandakan. Technical Bulletin No. 9, 50–56. Department of Agriculture, Sabah, Malaysia.

Chulan, A.H., Shaji, G.T. and Christine, Z.A. (1990) Interactions of mycorrhizal fungi with root pathogens of cocoa. (abs). In: *Current Trends in Mycorrhizal Research — Proceedings of National Conference on Mycorrhiza* (Eds. Jalali, B.L. and Chand, H.), pp. 78–79. Haryana Agricultural University, Hisar, India.

Conway, G.R. (1971) *Pests of Cocoa in Sabah and Their Control*. Ministry of Agriculture and Fisheries, Sabah, Malaysia.

Dehne, H.W. (1982) Interaction between vesicular-arbuscular mycorrhizal fungi and plant pathogens. *Phytopathology* 72, 1115–1118.

DOAS Ann. Reps. (Department of Agriculture Sabah, Annual Research Reports) for 1982 to 1990.

Gotsch, W. and Rieder, P. (1990) Forecasting future developments in crop protection. *Crop Protection* 9, 83–89.

Hedger, J.N. (1987) The effect of root infecting fungi on early seedling root growth of cocoa. *Proceedings of Tenth International Cocoa Research Conference, Santo Domingo, Dominican Republic*, pp. 503–507. Cocoa Producers Alliance, London, UK.

Jiminez, J.M., Galindo, J.J., Ramirez, C. and Enriquez, G.A. (1987) Evaluation of biological and chemical control to combat *Monilia* cocoa pod rot (*Moniliophthora roreri*) in Costa Rica. *Proceedings of Tenth International Cocoa Research Conference, Santo Domingo, Dominican Republic*, pp. 453–456. Cocoa Producers Alliance, London, UK.

Jollands, P. (1984) Potential for biological control of root diseases of cocoa in Papua New Guinea. *Proceedings of the International Conference on Cocoa and Coconuts*, Kuala Lumpur, Malaysia.

Khoo, K.C. and Chung, G.F. (1988) Use of the black cocoa ant to control mirid damage in cacao. *The Planter*, 370–383.

Koenig, J.C. (1894) Journal of a voyage from India to Siam and Malacca in 1779. *J. Str. Br. Roy. Asiat. Soc.* 26, 104.

Lass, R.A. (1985) Diseases. In *Cocoa*, 4th edn. (Eds. Wood, G.A.R. and Lass, R.A.). Longman, London.

Lim, G.T. and Chong, T.C. (1987) Biological control of cocoa pod borer by periodic release of *Tichrogrammatoidea bactrae fumata*. In: *Management of the Cocoa Pod Borer* (Eds. Ooi, P.A.C. *et al.*), pp. 71–80. Malaysian Plant Protection Society, Kuala Lumpur, Malaysia.

Lim, G.T., Tay, E.B., Pang, T.C. and Pan, K.Y. (1982) The biology of cocoa pod borer, *Acrocercops cramerella* Snellen and its control in Sabah, Malaysia. In: *Proceedings: International Conference on Plant Protection in the Tropics* (Eds. Heong, K.L. *et al.*), pp. 257–287. Malaysian Plant Protection Society, Kuala Lumpur, Malaysia.

Liu, P.S.W. (1979) "Watermark" — a new cocoa virus disease. *Technical Bulletin* No. 4, 19–20. Department of Agriculture, Sabah, Malaysia.

Liu, P.S.W. and Liew, P.S.C. (1975) Diseases of cocoa in Sabah. *Technical Bulletin* No. 1. 82 pp. Department of Agriculture, Sabah, Malaysia.

Liu, P.S.W. and Liew, P.S.C. (1979) Transmission studies of a cocoa virus disease (yellow-vein banding) in Sabah. *Technical Bulletin* No. 4, 11–15. Department of Agriculture, Sabah, Malaysia.

McGregor, A.J. and Moxon, J.E. (1985) Potential for biological control of tent building species of ants associated with *Phytophthora palmivora* pod rot of cocoa in Papua New Guinea. *Ann. appl. Biol.* 107, 271–277.

Odamtten, G.T. and Clerk, G.C. (1984) Studies on the possible use of two soil fungi, *Aspergillus niger* and *Trichoderma viride* as biological control agents of *Phytophthora palmivora* and the effect of their metabolites on cocoa seedlings. *Proceedings of the Ninth International Cocoa Research Conference*, 293–301.

Odigie, E.E. and Ikotun, T. (1982) *In vitro* and *in vivo* inhibition of growth of *Phytophthora palmivora* (Butl) Butl by antagonistic microorganisms. (abs). *Fitopatologia Brasileira* 7 (2), 157–167.

Saedoval, G.A. De, Armbrecht, H. and Granada, G.A. (1987) Possible biological control of *Monilia* pod rot of cocoa. *Proceedings of Tenth International Cocoa Research Conference, Santo Domingo, Dominican Republic*, 473–477. Santo Domingo, Dominican Republic. Cocoa Producers Alliance, London, UK.

Spurrier, E.C. (1990) Pesticides — there will be change. *Plant Disease* 74, 103–105.

Surtiyati, S. and Suhardjo (1991) Pengalaman praktek mengenai pengendalian nama *Helopeltis* di PT Perkebunan VI. Prosiding Konperensi National Kakao III, Medan, Indonesia, pp. 181–200.

Tan, K.H. and Loke, W.H. (1984) Insect growth regulators and semiochemicals for insect pest management. In *Proceedings of the Seminar on Integrated Pest Management* (Eds. Lee, B.S. *et al.*), pp. 105–131. MAPPS and MARDI, Kuala Lumpur, Malaysia.

Tay, E.B. and Bong, C.L. (1989) Research on the vascular streak dieback (VSD) of cocoa in Sabah. *Technical Bulletin* No. 9, 1–7. Department of Agriculture, Sabah, Malaysia.

Tay, E.B., Bong, C.L. and Lim, G.T. (1983) Pests and diseases in cocoa cultivation. *Seminar on Recent Developments and Future Prospects of the Cocoa Industry in Malaysia*, Kuala Lumpur, Malaysia.

Thistlewood, H. M. A., Borden, J. H., Smith, R. F., Pierce, H. D. Jr. and McMullen, R. D. (1989) Evidence for a sex pheromone in the mullein bug, *Campylomma verbasei* (Meyer) (Heteroptera: Miridae). *Can. Ent.* 121, 737–744.

Utomo, C. and Pardede, Dj. (1991) Pengendalian bilogis hama *Zeuzera coffeae* Nietn. pada kakao dengan jamur *Beauveria sp.* Prosiding Komperensi Nasional Kakao III, Medan, Indonesia, pp. 201–209.

Von Donop, L.B. (1882) Report on the Agricultural Prospects of British North Borneo. 33 pp.

Wills, G.A. (1984) Use of an early warning system for the control of *Helopeltis theobromae* in cocoa. *Proceedings of the International Conference on Cocoa and Coconuts.* Kuala Lumpur, Malaysia.

Wolfe, W.S. (1985) The current status and prospects of multiline cultivars and variety mixtures for disease resistance. *Ann. Rev. Phytopathol.* 23, 251–273.

Wong, P.H. (1983) Diseases of cocoa in Sabah with special reference to vascular streak dieback and stem canker. *Proceedings of the Seminar on Medium Size Agricultural Estate/Holding Development*, Sandakan, Sabah, Malaysia. SASS Publication, pp. 85–91.

Zadoks, J.C. (1984) A quarter century of disease warning, 1958–1983. *Plant Disease* 68, 352–355.

Chapter seventeen: *Management of Pests and Diseases of Plantation Crops Discussion*

Chairman: Dato' Dr Hj Abdul Halim Hassan

Mr Lakchai Meenakamit (Ministry of Agriculture, Bangkok) Are there any attempts to use a nylon net over vegetable fields and also using herbal or botanical pesticide such as neem, galanga, lemon grass, derris or tobacco as direct control and as an intercrop?

Are there any Non-Government Organisations (NGOs) involved such as vegetable dealers or consumers? These NGOs can put pressure on producers to produce vegetables free of pesticides.

Dr Syed A. Rahman Studies on production of pesticide-free vegetables in nylon net houses are now being undertaken by MARDI. We are working out the cost of producing these vegetables. Studies on botanical pesticides are at the early stages and will be emphasised in the future.

NGOs are not directly involved in our IPM programmes. However, we will include them in our task force in the future.

Dr Basuki (RRI, Indonesia) I have comments on the paper by Dr Ismail. Another approach in managing rubber diseases is through integration of chemical with biological control. Rubber root diseases can be controlled chemically and biologically. Leaf diseases can be controlled by cultural practices, such as by planting mixed, tolerant clones. Would Dr Ismail care to comment?

Dr Ismail I agree with your first comment. For the integration of biological control with chemical control for the control of root diseases, we have to select chemicals which are not toxic to the organism that we are using. This is of high priority. As for the second comment, so far we don't have many data on benefits in mixing of clones but we do have experience in glass-house experiments which shows mixing of clones might be of benefit in overcoming diseases such as South American Leaf Blight because of the induced resistance. But here we are dealing with clones with vertical resistance. We don't have many data yet on moderately resistant clones. For the smallholders, the mixing of clones is a normal practice because they are supplied with a basket of clones, not with one.

Dr O. Mochida The aerial spraying cost in Malaysia is M$55 per ha per round. What kind of application is this: ultra low, low or high volume?

Secondly, what is the percentage of popularity of the different kinds of aerial spraying?

Dr Gurmit Singh In aerial spraying of trichlorophon this is applied at 1.75 kg per hectare. The spray volume used is 18—23 litres per hectare. The aerial cost compares quite favourably with trunk injection with monocrotophos, which can cost about M$50 per hectare, but it is about 1.5 times more expensive than ground application with airblast spraying using trichlorophon. At present aerial application is limited to large outbreaks of leaf-eating caterpillars, usually covering more than 500 ha, or so.

Prof Varghese Can Dr Ismail give us some more information on the new disease of fusicoccum leaf blight of rubber? Is the disease still confined to the original location in Johore? Has the disease spread into a larger area? What is the present situation?

Dr Ismail Fusicoccum has only occurred within the last three years or so. So far it is still confined to Johore and we are constantly monitoring its spread. So far in the area where it was first detected the disease was minimal because the trees are undergoing normal wintering and being tapped. In fact before wintering the canopy was about 80%, as good as any other area. We are also screening our new clones including PC clones in the nursery located at the original site.

Prof Varghese Dr Gurmit Singh said that *Ganoderma* is now seen to be spreading into oil palm from ex-rubber areas. Is this new evidence of *G. philippii* from rubber now infecting oil palm which has never before been recorded?

Dr Gurmit Singh No, *Ganoderma philippii* from rubber does not infect oil palms. However, consolidated pieces of rubber tree tissues left buried in the soil can, in the coastal areas become colonised by *Ganoderma boninense* (as a saprophyte) and the fungus can then move on to oil palms that are growing in the vicinity.

Prof Varghese What is the Cocoa Board doing to breed lines of cocoa resistant to vascular streak dieback, as has happened in PNG?

Dr Tay There are at present many clones that are known to be resistant to VSD. Most major plantation companies have their own selection, e.g. the PBC materials and the KKM clones of MARDI. The Department of Agriculture Sabah's series III materials are also known to be more resistant to VSD than the earlier hybrids. Clonal selection for VSD resistance is also progressing well by that organisation. The Malaysian Cocoa Board has also placed high priority in this line of research. I do not wish to speculate on how Malaysian materials compare with PNG's. But certainly, we feel more comfortable with VSD now than we were say four or five years ago.

Dr C. Adalia (University of Philippines) We are currently experiencing a leafminer outbreak in string bean (*Vigna sesquipedalis*). We observed two groups — one causing a tiny tunnel which is a fly (Order Diptera) and another one causing a broader tunnel which is a microlepidoptera (Order Lepidoptera). May I know:

(a) What species or kind you have here (diptera or microlepidoptera)?
(b) Have you observed/recorded any effective parasitoids?
(c) Is any insecticide found effective to control this pest in Malaysia?

Dr Syed A. Rahman

(a) The species of leaf miner attacking the sugar peas is *Chromotomyia horticola*. In Malaysia, on vegetables we also have *Liriomysa* sp. These are dipteran leaf miners.
(b) We have recorded two parasitoids. They are *Hemiptarsenus* sp. and *Chrysocaris* sp., both hymenopterous parasitoids.
(c) Insecticides effective are deltamethrin and cyhalothrin.

Dr Peter Dart Would you please elaborate on the role of root diseases in cocoa root disease control. What kind of mycorrhiza is involved and how is the inoculum managed?

Dr Tay VAM experiments are carried out at UPM. It is cited as an example of having possibility of being utilised in the system against root diseases. VAM works conducted in UPM against Ganoderma pointed to this potential. Other experiments showed root disease fungi can be antagonised in their growth in vitro and their potential for biocontrol is being investigated.

It could possibly be used against root disease. VAM works conducted in UPM against *Ganoderma* pointed to this potential.

David Lim (BAL Plantations Sdn Bhd) It appears that in the management of P & D in plantation crops, there is a big gap or hardly any integration between the recommendation of the P & D man and that of the breeders/ agronomists. For example,

1. Dr Ismail's recommendation for "low" trees to facilitate spraying conflicts with current R & D advice for fast growing tall trees for later timber exploitation.
2. Likewise some advocate deep planting, while others advise against it because of problems with root disease.
3. Dr Tay recommends small canopy trees. This means closer spraying to achieve yields thus creating a denser canopy, which induces blight.

Dr Ismail In a way it is quite true that we are going for both short and tall trees. There is value in tall trees because of their wood. But tall trees will hinder chemical control from the ground. Although we can do aerial spraying the cost is prohibitive. So we have to know what we want. In my presentation, I stressed that genetic resistance is still the main method for controlling leaf diseases. Chemical control is the last resort. Deep planting is not really widely practised. To expose the roots of deep planted trees we have to do a lot of digging and this is going to be a big problem. We should be going for yield per hectare rather than yield per tree. The planters have to decide which technology they want to adopt. If deep planting is going to be a problem in your area due to root disease then do not accept this technology.

Dr Tay I agree entirely with Mr Lim's statement that there should be team work between the crop protectionists, breeders and agronomists. The slides that I have shown on the integrated management of VSD include agronomic practices such as spacing of shade trees and fertiliser application while the breeder contributed in supplying different genetic materials used in the experiment.

Dr Nathan (ASEAN Planti) In relation to Dr Syed's paper on IPM on vegetable crops, I would like to add that there is a urgent need to screen and certify seeds being imported into the country. This is vital so that new diseases and pests are not introduced. Screening and certification can be done by pooling resources and expertise of agencies like MARDI, DOA, ASEAN Planti, etc. Dr Gurmit Singh also highlighted in his paper the need for screening and effective plant quarantine of oil palm germplasm entering the country to prevent diseases such as the *Fusarium* wilt of oil palm (caused by *F. oxysporum* f. sp. *elaeidis*) which is present in Africa. We also should be on the look out for diseases like the South American Leaf Blight of rubber which is again confined only to the American tropics and not present in the Southeast Asian region where rubber is a major crop. If proper plant quarantine procedures are not enforced and followed and if diseases such as SALB are introduced in this region, it will have far reaching implications.

En Azizi (Kumpulan Guthrie) Mr Lim mentioned earlier the importance of integration between crop protectionists, breeders and agronomists. Please do not forget the planters. What you produce we are going to look after. Whether we go for short or tall trees, we are faced with an acute shortage of labour. To overcome this, we have to use aerial application techniques. Dr Gurmit Singh mentioned there are no takers for aerial application for fertilisers or chemicals. I think Kumpulan Guthrie can help because they have their own helicopter which can undertake such work. I am glad to advise that Guthries is extending its business here. The problem is that most planters are unduly conservative about the costs.

PART 5
Management of Weeds

Chapter eighteen: **Management of Weeds in the Year 2000**

R.J. Williams
CIBA-GEIGY Limited, Agricultural Division, CH-4002 Basle, Switzerland

Introduction

The control of weeds is an essential component of productive agriculture, and productive agriculture is essential to meet the food and fibre needs of the continuously growing world population.

By the year 2000, the world's human population will probably increase to about 6100 million persons, from ca. 5200 million in 1989. More than two-thirds of the extra 900 million will reside in what are now low-income, food-deficit nations in Africa, Asia and Latin America. The increases will not stop then, but go on to produce a population of about 8000 million by 2025, and will probably stabilise at about 11,000 million persons by the end of the twenty-first century. Of the 11,000 million, only 12% will be in what are currently the developed Western nations. Thus, the world and particularly the countries in the tropics and sub-tropics, will need to produce considerably more food (primarily cereal grains) than they do today (Brown *et al.*, 1990; Brown, 1988; Borlaug, 1990). To do this they will need to expand cropped area where this is possible and greatly to improve productivity on currently cropped lands, while maintaining the sustainability of that increased productivity (Barbier, 1989). The control of weeds, and the management of vegetation to prevent soil erosion and maintain soil fertility, will be essential components of the technology required to achieve and sustain the necessary productivity. Thus, it is important to understand the status and likely developments in weed control technology, and constraints in its deployment.

Technology Developments

Machinery and Tools

The age-old method of weed control is either to remove them from the crop by hand directly or with a tool such as a hoe, or a machine such as a harrow that is dragged along by humans, animals or oil-powered traction equipment. All of these methods are used around the world today to various degrees as part of our integrated crop management systems, and they will continue to be used, in situations where:

- there are no better alternatives,
- they are cost-effective and
- they provide a suitable component of an integrated weed-management programme.

Apart from efficiency improvements, with current systems there is no major breakthrough in sight for a completely new generation of weeding tools or machines using modern technology such as lasers or computers, as the costs involved, in terms of hardware and energy, prohibit their use in most agricultural systems. People and energy are key limitations to the degree of use of these manual and machinery systems. In addition, in certain situations, mechnical weed control can result in undesirable effects such as predisposition of soils to erosion and loss of needed soil moisture.

It is possible that computers will be used as tools by more advanced farmers to help in deciding whether or not it is desirable to weed or to apply herbicides, using weed threshold and economic information.

Chemicals for Weed and Vegetation Management

During the past 50 years, the development of modern chemical herbicides has been one of the key factors in the increased productivity of agriculture in North America and Western Europe, allowing large areas to be farmed by a small percentage of the population, and allowing new, more productive crop varieties to be able to respond to increased fertilisation, without the competition from the weed flora. Farmers today have a broad choice of herbicides, with varying selectivities and use timings to meet their various needs. The major developments of the 1980s include: the reduction of use rates (now many herbicides are available with use rates in the 20–100 g a.i./ha range); the greater availability of post-emergence products that allow farmers to wait and see the nature and extent of the developing weed problems; the introduction of safeners to act as specific antidotes and thus provide needed specificity to broad-spectrum products; and the introduction of novel packagings that make the products safer and easier to use and reduce packaging disposal problems (e.g. closed-filling-system, returnable re-fillable containers for liquid products; water dispersible granules in water-soluble film packs).

During the 1990s these developments will be continued, with further reduction in use rates (products that are effective in the rate ranges of 5–30 g a.i./ha will become widespread), with emphasis on post-emergence products which will be rapidly degraded in soil to avoid accumulation or carry-over problems. For soil-applied products, there is also likely to be the development of controlled-release formulations, which release active ingredients in the soil in response to environmental stimuli (e.g. moisture, temperature, etc.), to reduce the amount of active ingredient available to be leached at any one time. The variations possible in organic molecules are infinite, and provided the correct screens are employed in the search for new compounds, molecules with the required characteristics can and will be discovered and developed, providing farmers with an even greater range of weed management tools, with the environmental and toxicological characteristics needed to meet the demands of safe and sustainable agriculture.

Herbicide-Resistant Crops

The use of selective herbicides, i.e. herbicides that control the weeds but do not affect the crop plants, exploits naturally occurring resistance in certain species against certain herbicides. The use of "safeners", that is chemicals that activate or induce herbicide detoxification processes in the target crop plants but not in the target weeds, is another way of exploiting naturally occurring resistance to herbicides in plants. Both of these methods of using naturally occurring herbicide resistance have been widely practised. In the recent past, stimulated by the advances in biotechnology and particularly the possibilities for genetic engineering, attention has focused on the deliberate development of crop cultivars resistant to particular herbicides. Hence, efforts are underway to transfer into crop plants, genes that confer resistance to single herbicides (e.g. glyphosate) or herbicide families (sulphonylureas; imidazolinones) to produce herbicide-resistant cultivars for farmers to buy and use.

Two uses for herbicide-resistant crop cultivars are envisaged:

- to allow a previously unusable herbicide to be used in the crop;
- to allow sensitive crop species to follow in the rotation a crop in which a particular herbicide has a major advantage.

The advantages cited for the use of such engineered herbicide-resistant crop cultivars include:

- allow "wait and see" use of herbicides;
- potentially less herbicide used;
- replacement of older products by more modern (lower-rate, safer) products.

However, these advantages can also be achieved with modern selective post-emergence herbicides.

Potential disadvantages in the use of crop-resistance bred for a particular herbicide are:

- increased use of that one herbicide with possible environmental and crop residue implications;
- increased selection pressure for weeds to develop resistance;
- the costs/opportunity costs involved in developing the herbicide-resistant crop that could have been directed at other aspects of cultivar improvement for which there is no acceptable alternative;

- societal resistance to release of genetically modified organisms in the environment;
- costs of registration of particular herbicides for specific cultivars.

As seed companies continue their steady progress in increasing the yield and quality factors in new crop cultivars, the presence or absence of herbicide resistance will have to affect significantly the yield and production cost aspect if it is to become a significant crop cultivar selection criterion. At least, it will give farmers the opportunity for more choice in their weed management systems. However, it seems unlikely to become a major factor during the time scale of this review, i.e. until the year 2000.

Biological Control Products

The control of weeds by their naturally occurring enemies such as fungal pathogens, or insect pests has long received attention and is a subject of research at many universities and research institutes around the world.

The two basic approaches are to:

- introduce an exotic pest or pathogen into the environment where the weed is a problem, and allow the biological control agent to build-up and damage the weed population within a particular ecosystem or region over a long period;
- apply the pest or pathogen directly in large quantities to control the particular weed directly and rapidly, with no long-term control expected or intended.

The first approach has been successfully employed for the control of single weed species over large areas, e.g. the prickly pear control in Australia with the moth *Cactoblastis cactorum*. Today, targets for such an approach are bracken in the UK and New Zealand, gorse in New Zealand, and the control of water weeds in lakes and slow-moving waterways in many parts of the world (Anon, 1990). The second approach is well exemplified by the use of fungal pathogens or mycoherbicides, applied to a crop to kill-off or inhibit a specific weed.

Both approaches are, in theory, technically feasible, and in these days of heightened concern over the wide use of synthetic chemicals, appear to offer an environmentally friendly natural solution to weed problems. In practice, however, there are many problems associated with the development and effective use of biological control agents that will probably prevent them becoming a significant method of weed control in arable crop production at least during the time scale covered by this review.

The particular pest or pathogen would need to be produced in large quantities, be formulated and packed to be kept viable for transport and storage, be storable without special conditions on the farm, be easily usable, and give reliable cost-effective activity, and should not cause problems to any non-target organism . The movement of such pathogens across national boundaries could cause difficulties with plant quarantine authorities, and release of genetically modified organisms may also encounter regulatory barriers.

The expertise and resource needed to develop, formulate, package and sucessfully market biological control products lies with the large multinational agricultural chemical companies, and yet the nature of the products (highly specific for single weed species in mixed weed populations, environmentally demanding for effective activity, more suited to range-land than annual crops) mitigates against these companies devoting the substantial resources necessary for developing, registering and supporting what would generally be "niche" products. Weeds that are hardest to control with biological control agents will be sexually reproducing weeds that are either annuals, or perennials with long growing periods, with large reserves, high powers of re-growth and large seed banks.

For problems of single noxious weeds invading conservation areas, e.g. gorse in parts of New Zealand, or bracken over large watershed areas of Wales and Scotland, the national agriculture research service institutes and co-operating universities are the organisations most likely to have the interest and the funds to proceed with funding and testing of potential biocontrol agents. It is also probable that in these situations partial or slow control effects will be better accepted than in intensive arable agriculture.

Concerns, Challenges and Influencing Factors

Humans: Their Numbers, Needs and Fears

The food need aspect of the population has been mentioned in the introduction to this paper. There will be ca. 800 million more people in the world by the year 2000 than there were at the start of the 1990s. They will need to be fed, clothed, housed and productively employed. This has to mean:

- more crop land;
- more efficient/productive use of current crop land, and not just short term but on a sustainable basis.

Soil erosion has to be minimised and soil fertility maintained. This will require intelligent vegetation management, and the development and use of appropriate tillage systems. In the tropics and sub-tropics there will be a pressing need for rapid increase of food-grain production, as the greater proportion of population increases will occur in these regions.

In the current developed temperate countries, agricultural production is already highly advanced and productive, and populations are either only very slowly growing or stable. Hence, these countries have been able to generate surplus food grains and dairy products, and these populations are generally well fed and even over-fed. These populations, from the benefit of their comfortable existence, are challenging the processes and the technology that have brought them to this comfortable situation, and have become much concerned with the effects of modern technology, including modern agricultural technology, on the environment and on the quality of foods. For agriculture specifically, there is a growing concern about the effects of the chemicals used in agriculture, i.e. fertilisers, and plant protection products including herbicides, on food quality and longer-term environmental consequences. It is of course natural and desirable that societies should be concerned with environmental sustainability, but we should try to avoid an over-reaction and to address the issues based on an assessment of risks and benefits, and not on an emotional level in which "natural" is good, and "man-made" is bad. An analogy with the motor-car may help put over the point: the motor-car is a potentially lethal machine, which when misused can cause death and destruction, yet we recognise the benefits of the motor-car and would not advocate its abandonment to more natural means of transport (walking, horse and cart). What we do is to improve the design of cars to make them intrinsically safer, we educate users on the safe ways to use them, and we introduce rules and laws to enforce the safety of their use. This is surely the sensible approach also for plant protection products. The process of improving design and educating users is well underway, and needs to continue. In many countries there is increasing legislative control on the use of plant protection products. This concern and pressure will generate more research on the costs, risks and benefits of modern agricultural interventions and lead to a more open debate on the issues involved. The studies required on the potential risks of new agricultural plant protection products become even more complex and expensive, which prolongs the time from discovery to introduction and increases development costs. The consequences for the agricultural indusrry are, therefore, a slower rate of availability of new technology, and a higher cost of new technology when it is introduced. Such new technology, because of the costs involved, may not be generally available to the farmers in the developing world.

Environmental Protection

Two major concerns in intensive agriculture in Europe and North America are the finding of agricultural chemicals, including herbicides, in groundwater and run-off water, and the potential carry-over risks of the new low-rate products to subsequent rotational crops. To reduce these concerns there will be considerable attention to the way in which current herbicides are used, and to the design of new products. An axiom that would apply to the use situation might be "as little as possible, and as much as necessary". This will require clear information on the components of the weed flora, the effects of certain levels of infestation on crop performance, and the efficacy and cost of potential means of control. The goal will be to reduce use rates and frequencies of herbicide applications, in combination with other practices, to achieve the degree of weed control needed. The generation of such information is ongoing, and will require the continued co-operation of industry, officials and farmers in order to generate the information necessary to allow such integrated use of weed control practices.

Disposal of contaminated primary packagings becomes a significant issue and this will pressure industry to develop alternatives, such as wettable powders packed in water-soluble bags, packed in easily burned or disposed of secondary packaging. The design of new products has already been discussed under "Technology" above with attention to such factors, as low rates, soil breakdown and mobility behaviour, and formulation and application modifications. Care for the environment is vital if we are to achieve sustainable productive agriculture. In the weed management context we should take care to focus efforts and resources on the high priority issues that include:

- prevention of soil erosion;
- maintenance of soil fertility;
- conservation of limited moisture.

all of which are influenced by vegetation management practices and tillage systems.

The agricultural chemical industry needs to work closely with national programme researchers on the best ways of weed management in the context of alley cropping, cover cropping or mixed cropping, e.g. a low-growing inhibited weed cover will be preferable to bare ground in some situations; exactly what, how and where needs to be worked out. There is a need for open discussions on the risks and benefits of current technology and practice, with attention to the facts. Industry will be able to provide the needed products and use systems, but the costs will have to be borne by the agricultural community and subsequently by the consumer.

Herbicide Resistance in Weeds

The development of very selective and yet effective herbicides has been a major contributory factor in the development of highly productive modern intensive agriculture. However, weed populations are dynamic, and as genetic variability exists within as well as between species, the selection pressure exerted by selective herbicides can select for:

- species that never were susceptible to the herbicide;
- naturally occurring resistant biotypes within species that are generally highly sensitive to the herbicide.

Thus, two potential consequences of the widespread use of a particular selective product or family of products (i.e. with the same basic mode of action) are:

- a shift in species to harder-to-control weeds; and/or
- the selection and spread of herbicide-resistant biotypes of once susceptible species.

At a recent count there were 111 species of weeds that have been found to have developed resistance to various herbicides that originally were effective (Le Baron, 1990).

Much has been learned in the last 20 years about the prevention, delay and management of resistance to agricultural chemicals, and there is no reason why herbicide-resistant weeds should become a major problem, provided that industry, officials and farmers work together in the application of some basic principles to counter the selection, survival and dissemination of herbicide-resistant weeds. The Herbicide Resistance Action Committee (HRAC) of the International Group of National Associations of Agrochemical Manufacturers (GIFAP) is committed to the sensible management of herbicide resistance, and seeks to work with relevant officials and the farming community to develop and implement effective anti-resistance strategies. The key component of such strategies is to avoid continuous selection pressure with the same mode of action products. In practice this means integration of herbicide use with other relevant and cost-effective weed management practices (e.g. mechanical tillage) and the rotation of herbicides (with different modes of action) and where relevant, the rotation of the crop.

The Problem of *Striga*

Striga is a parasitic weed that severely damages the food grain crops in the tropical and sub-tropical regions of Africa and Asia. In the African Sahel, *Striga hermonthica* is a major cause of loss to the sorghum and pearl millet crops, and as maize has been moved further north into the savannah regions of West Africa, it has been widely attacked by this devastating parasitic weed. Other species of *Striga* also cause serious problems — *S. asiatica* in graminaceous crops, and *S. gesneroides* in food legumes such as cowpea. During the last 25 years, considerable resources have been directed at the *Striga* problem: studies of the biology of the parasite, studies on chemicals or trap crops to germinate the dormant seeds, studies on genetic resistance in host plants. While much has been learned, the *Striga* problem is not at all under control.

It is possible to manage *Striga* selectively in cereal crops with spot-applied post-emergence herbicides, but by the time the plant has emerged above ground it has already done substantial damage, and the sorghum and millet farmers in the semi-arid tropics generally do not have access to the selective herbicides nor the equipment to apply them. Trap crops, e.g. cotton, stimulate *Striga* seed germination, but achieve inconsistent results. They provide no desirable economical solution since:

- a food crop during each season is necessary to meet the needs of an increasing population in the concerned countries;
- in major areas of *Striga* occurrence the water requirements for growing a trap crop such as cotton are too high.

Soil sterilisation or chemical stimulation of seed germination (Strigol) is too expensive. Furthermore, another specific stimulation agent (Sorgoleone) acts as a contact allergen and is too fast degraded in soil to achieve sufficient-lasting control effect. Breeding of crops with genetical resistance to *Striga* attack up to date has not been successful; no viable resistance has been found in crop lines suitable for developing countries.

The parasite sets a profuse amount of seed each season, which then can lie dormant in the soil for many years to infect subsequent crops. If the plant can be prevented from replenishing its seedbank, then the reservoir in the soil will diminish and run-out. One method that could be funded and tried on a village level basis would be to offer a bounty on pre-flowering *Striga* plants collected and brought to the village headman for destruction, for payment in valuable agricultural kind (e.g. seeds, fertiliser, seed-treatment materials). This could well be the most cost-effective and productive method of control. It would take a longer time to achieve, but as the million of dollars spent on more sophisticated methods over the past 25 years have not been successful, a chance should be given to such a "cultural" control programme. Various other options to be pursued, include:

- seed treatment with chemical agents which either kill *Striga haustoria* when invading the host's roots or which inhibit *Striga* seed germination;
- blocking of signals needed for *Striga* development, i.e. growth hormones sent from the host plant, by application of a chemical agent;
- continuation of resistance breeding, e.g. by cultivars not excreting stimulation agents or those not offering host conditions after germination;
- biological control possibilities.

There is an urgent need for innovation in the management of *Striga*, for the benefit of food production in Sub-Saharan Africa (Parker, 1991).

Key Crops

Maize

Three major concerns in maize production in temperate regions that will have an impact on weed control technology in the crop are: (a) run-off and erosion from maize production on sloping land; (b) the increase and spread of triazine-resistant weeds (e.g. *Chenopodium, Amaranthus* and *Solanum* spp.); (c) the detection in groundwater of herbicides widely used in the maize crop.

The increased use of cover-crops and/or extension of no-till systems are the likely consequences of concern (a). Concern (b) is being addressed through the development of new classes of herbicides (e.g. sulfonyl-ureas) for use in maize, and the extension of other widely used herbicides to maize either using "safeners" or resistance to the herbicide built into the maize cultivars to allow farmers to avoid continuous selection pressure and yet still enable them adequately to manage their weed problems. Concern (c) is receiving much attention in Western Europe at this time with open questions on the identity and sources of the chemicals detected, and on the significance or risk to humans of the small amounts that are found. Industry is working closely with officials to find answers to these questions. Consequences are likely to include some restriction of use of the currently most widely used products (rate and/or location restrictions, e.g. in key watershed areas), the use of alternative products, the growing of alternative crops, and the development of reduced leaching formulations. There needs to be a serious and unemotional debate, based on the known facts, on the relative risks and benefits of the continued use of key beneficial products in the maize crop.

Rice

One of the major factors affecting rice cultivation practices in many developing countries, is the steady movement of labour from the farms to urban areas, as industrialisation creates more attractive employment opportunities. This will continue throughout the 1990s and become a greater problem for existing farmers. As it is generally the younger, more active segment of the population that moves off the land to the cities, the effects on the farm will be considerable. For rice particularly, there will be increasingly a move from hand transplanting to direct seeding, a move that is already well underway in several countries in SE Asia, facilitated by reliable irrigation systems, short-duration modern rice cultivars and the availability of cost-efficient herbicides (De Datta, 1989; De Datta & Nantasomaran, 1990). This in turn causes a shift in weed populations, e.g. annual grasses such as *Echinochloa* spp. and *Leptochloa chinensis* become increasingly important and seriously affect yield quantity and quality. The reduction of the labour forces and the move to direct sowing makes difficult hand-weeding and mechanical weeding, and thus leads to a greater reliance on herbicides to control the grasses which are favoured by the new cultivation system. There is also some evidence that where the direct seeding plus herbicide system of rice cultivation has been in practice for some time, a further shift of the weed flora to the harder-to-manage perennials such as *Scirpus maritimus* and *Paspalum distichum* can occur in certain situations. Research is underway to learn how best to manage the shifting weed problems in the new cultivation systems. A combination of various factors in an integrated weed management system will probably include aspects of land preparation, the use of clean seed, careful water management and the use of cost-effective herbicides (De Datta, 1989). The continuous use of one type of herbicide should not be employed, or this will encourage development of resistance or shifts to hard-to-control weeds. The move to direct seeding, precipitated primarily by increased real labour costs and enabled by available technology, has shown farmers the distinct economical advantages of such a system. The move could accelerate, so long as the necessary technology is locally available, and even extend to areas where, as yet, labour is still plentiful. The final choice of weed management techniques will depend upon the severity of weed infestation, and the available technology, knowledge and experience (with economic resources of the farmer being a major component influencing availability).

Small-grain Cereals

The small-grain cereals, particularly wheat and barley, will remain with rice as the major food providers for mankind, throughout the next decade and beyond. The experience in Western Europe in the past 20 years, where average yields/ha have virtually doubled to 4.5 tonnes, indicates what can be achieved when the price received for the grain makes cost-effective the use of high yielding varieties and the inputs they need, including herbicides. New developments in herbicide technology over the next ten years will be influenced primarily by the economics of production, and the political/regulatory pressures on crop inputs. Use of herbicides according to threshold values of weed infestation will increase, made possible by new post-emergence herbicides. Application rates will decrease, with newer more active products being registered and introduced. Sensitivity of rotational crops will also be a major determinant of products and dose rates used. The set-aside schemes will continue to take land out of cereal production in Western Europe, which will reduce areas treated with herbicides. Farmers will, however, farm the remaining crop land intensively in order to maintain income levels.

Potentials for weed resistance development will be greatest in cereal monoculture situations. The emergence of weeds with multiple resistance to herbicides of different modes of action will force growers to use novel mixtures or alternations of different herbicide families, and to introduce appropriate rotational crops. In cereal-fallow systems, selective herbicides labelled for use in the cereal crops may not be allowed to be used during the fallow. The development of herbicide-resistant cultivars is unlikely to have a major impact over the next ten years.

Plantation Crops

The diversity of plantation crops around the world, including rubber, oil palm, coffee, cocoa, tea, citrus, grapes, pome-fruits and various nuts and stone-fruit, and the countries in which they are grown, make it difficult in a short time to cover all the important aspects of weed control developments. If we look closer at the tropical plantation crops of rubber and oil palm, which are the most important in this southeast Asian region, then the factors that will exert most influence on the weed control technology over the next ten years will be cost and availability of labour, cost and availability of herbicides and the cost, availability and maintenance capacity of the machines to apply the products. (Sungha, 1990)

As already mentioned for rice, the reduction of availability of labour, and productivity benefits from moving to alternative means of crop management, will increase the need for more mechanisation in the SE-Asian plantation industry. Increased mechanisation of herbicide applications using tractor mounted spraying equipment, and increases in ULV applications will require investment in machinery, and could lead the way to providing bulk-filling systems and returnable re-fillable containers, reducing the problem of disposal of contaminated herbicide packaging. For the smallholder, who will come under pressure due to the technological and economy-of-scale advantages of the commercial estates, more user- and environment-friendly formulation and packaging will be available, but is likely to cost more than the traditional products and therefore may not actually be within the reach of such users.

Concern will grow over the effects of the applied products on the environment. Possible water contamination is likely to be the most important area for analysis with potential increase in regulations on "what", "when" and "how much". The need for longer persistence, coupled with groundwater concerns requires novel non-leaching formulations. The agricultural chemical industry will have to work in close patnership with the plantation managers to provide acceptable products.

Conclusions

The cost-efficient control of weeds will continue to be an essential part of productive agriculture, and agriculture will need to continue to be steadily more productive over the next 30 years in order to meet the food needs of the world's increasing human population. The technology needed and the way in which it is best used, will continue to be provided by industry working in close cooperation with officials and farmers. No new types of weed control technology are envisaged in the next 10 years, but there will be much adaptation and development for the improvement of currently used technology.

Public concern about environment and food safety will have a significant effect on the types of herbicides available and the way they are used, with increased emphasis on low-rate, shorter-persistence post-emergence products packed in easily used and disposed of packagings. Registration requirements will lengthen the time to develop new herbicides and make them more expensive. This will accelerate the move to "need to spray" decisions based on infestation levels, weed flora composition, and economic threshold information. Herbicide resistance risks will guide strategies in the use of the new selective products, but herbicide resistance built especially into crop cultivars will not have a major impact before the year 2000. Weed populations, farmers, officials and industry are all responsive to the pressures upon them. There will continue to be a dynamic equilibrium among these components to provide an effective and sustainable weed control through the year 2000.

Acknowledgements

I am grateful to the following people for providing source material and ideas for inclusion in this paper: Dr J.L. Allard, Professor A.H. Bunting, Dr K.M. Chin, Dr L. Ebner, C.J. Hare, R. Istead, Dr K.F. Kon, K. Maag, Dr B. Miflin, Dr K. Moody, Dr N. Nelgen, Dr D. Nevill, Dr M. Quadranti, Dr J. Rufener, Dr J. Scheuring, Prof. F.J. Schwinn, Dr D. Wissler, Dr A. Zoschke.

References

Anon (1990) Alternatives to the chemical control of weeds: *Proceedings of an International Conference held at the Forest Research Insitute, Rotorua, New Zealand, 25–27 July 1989* (Eds. Bosset, C., Whitehouse, L.J. and Zobkiewiez, J.A.). *FRI Bulletin* 155, Ministry of Forestry, Forest Res. Inst., New Zealand.

Barbier, Edward, B. (1989) Sustaining Agriculture on Marginal Land: A Policy Framework. *Environment* 31 (9), 13–39.

Brown, Lester R. (1988) The Changing World Food Prospect: the Nineties and Beyond. Worldwatch Paper 85. Worldwatch Institute.

Brown, Lester R. *et al.* (1990). *State of the World 1990.* A Worldwatch Institute Report on Progress Toward a Sustainable Society. W.W. Norten & Company, New York & London.

Borlaug, Norman E. (1990) The challenge of feeding 8 billion people. *Farm Chemicals International,* Special Issue Summer 1990, No. 5, pp. 9–12.

Bunting, A.H. (1991) Notes on Food and Agriculture 1991. Position paper prepared for Royal College of Defence Studies.

De Datta, S.K. (1989) Technology and Economics of Weed Control in Small Rice Farms in the Asian Tropics. Symposium “Weed Problems and Their Economic Management” in the 12th Conference Asian-Pacific Weed Science Society, pp. 23–37, Seoul, Korea.

De Datta, S.K. and Nantasomaran, P. (1990) Present status and future prospects of direct seeded flooded rice in tropical Asia. The International Rice Research Conference 1990. Focus on Irrigated Rice, 27–31 August 1990, Seoul, Korea.

Le Baron, Homer M. (1990) Weed science in the 1990s: Will it be forward or reverse? *Weed Technology 4,* 671–689.

Parker C. (1991) Protection of crops against parasitic weeds. *Crop Protection* 10, 6–22.

Sungha, Bonaventure P. (1990) The plantation sector in the 1990s. *The Planter* 66 (772), 348–364.

Chapter nineteen: **Management of Weeds in Plantation Crops in 2000**

Y.K. Chee[1], Ahmad Faiz Alif[1] and G.F. Chung[2]
[1]*Rubber Research Institute of Malaysia*
[2]*Sime Darby Plantations Bhd, Kuala Lumpur, Malaysia*

Abstract

In the year 2000, the major plantation crops in Malaysia will be oil palm, rubber and cocoa. These crops will contribute significantly in export earnings to the national economy. In order to enchance Malaysia's competitiveness in the international primary commodity market, high yields and lower production costs must be achieved. In the cultivation of these crops, weed management in the immature phase is one of the more important and expensive inputs. The weeds need to be managed effectively to avoid suppression of growth and yield of the main cultivated crops. Efficient crop and weed management will also minimise soil erosion.

With the constraints in labour availability and high cost of labour in the plantation sector, weeds need to be managed economically. The future trend is to use cost-effective herbicides with low toxicity to man, fish, beneficial insects and wildlife. These herbicides should have minimal pollution effects on soil, water and harvested products. The practical problem of herbicides on weed succession and weed resistance are discussed.

In order to lower the cost of weed management, there is a need to reduce the rates of herbicides used and the number of spraying rounds. Mechanised spraying on flat terrain and ultra low volume and controlled droplet application of herbicides on undulating and hilly areas will receive more attention in alleviating labour shortage, therefore enhancing higher productivity.

The weed species in the plantation crops are classified to serve as a guide for efficient weed management and selective use of herbicides.

An integrated system of weed management involving cultural methods, herbicides, sheep grazing and biological agents (insects) to reduce weed infestation and minimise crop loss of the three main plantation crops will be discussed. This balanced approach is in line with the international thrust towards the protection of the environment and sustainable agriculture.

Introduction

In the year 2000, the major plantation crops in Malaysia will be oil palm, rubber and cocoa. The World Bank forecasts that in the year 2000, Malaysia will produce 7.6 million, 1.6 million and 0.76 million tonnes of palm oil, rubber and cocoa respectively (Duncan, 1986). This constitutes 54.4%, 26.8% and 28.8% of the world's total production. As a result of the high production, these crops will contribute significantly to the export earnings of the Malaysian national economy. In order to enhance Malaysia's competitiveness in the international primary commodity market, high yields and lower production cost are important factors.

In the cultivation of these crops, weed management in the immature phase is one of the important and expensive inputs. A large proportion of the total expenditure is incurred in weed management. It constituted 27.7%, 26.0% and 16.8% of the total planting cost as in Table 1 (Wood and Edward, 1987). This high cost in weed management is due to 13 to 25 rounds of spraying (Table 2) during the immature period which is three years for oil palm and cocoa and five years for rubber (Chee *et al.*, 1990).

Table 1. Planting cost (M$/ha) on large estate for Malaysia's three main plantation crops (Wood and Edward, 1987)

	Oil palm*	Rubber*	Cocoa*
Land preparation	1300	1300	1300
Plants and planting	700	100	1000
Covers/shade	200	200	250
Upkeep (weed control)	1250	1325	1930**
Manuring	1025	1275	660
Total	4475	4200	5140
Weed control (%)	27.9	31.6	18.77

* Immaturity period: Oil palm and cocoa 3 years, rubber 6 years.
** In cocoa, weed control is 50% of the total upkeep cost.

Under Malaysian conditions of high rainfall of 2100 to 2500 mm, temperature of 21 to 35°C and humidity of 52 to 96%, weeds need to be managed well so that they do not interfere with agricultural operations and suppress the growth and yield of the main crops. Since weed management is an important component of effective crop and vegetation management systems, this paper will discuss and highlight crop loss by weeds and weed management

Table 2. Number of spraying rounds/year for weed management of plantation crops during immature period (Chee *et al.*, 1990)

Year	Rubber*	Oil Palm**	Cocoa**
1	6–8	6–8	6–8
2	4–6	4–6	4–6
3	3–4	3–4	3–4
4	3–4	–	–
5	2–3	–	–
Total	18–25	13–18	13–18

* Immature period rubber = 5 years
** Immature period oil palm and cocoa = 3 years

options in relation to sustainable agriculture and safety to the operator and environment.

Crop Loss by Weeds

Weeds affect the main crops directly by competing for nutrients, water, light and allelopathy. They also affect the crops indirectly by their thick coverage which poses problems of accessibility and leads to poor standards of agronomic practice in manuring, pest and disease control and pruning. This results in irregular and poor harvest of oil palm, cocoa and rubber. The poor harvest is more acute in hilly areas. The direct effects of weeds are quantified by the reduction in growth of 51 to 77% in two-month-old young rubber seedlings (Chee *et al.*, 1990) and 7 to 13% in the girth of rubber (Table 3, Watson *et al.*, 1964a). There was a reduction of 6 to 20% in yield of palm (Table 4, Gray & Hew, 1968) and 46.5 to 51.2% for cocoa (Table 5, Chung & Lam, 1990). As a result of a large crop loss due to weed competition, many management options can be adapted in the management of weeds in plantation crops in the year 2000.

Weed Management Options

There will be many constraints in the weed mananagement of plantation crops in the year 2000. Weed management will be costly because herbicides and labour will be expensive due to labour shortages. Efficient and cost-effective

Table 3. Effects of ground cover on girth of rubber trees (Watson *et al.*, 1964)

Treatments	Mean girth (mm)	% Girth compared to control	% Reduction
Legumes (control)	439	100.0	-
Grasses	407	92.7	7.3
Naturals	386	87.9	12.1
Mikania	380	86.6	13.4

Table 4. Effect of different cover crops on total yield of oil palm for 4 3/4 years on Selangor series clay (Gray & Hew, 1968)

Treatments	Yield (t/ha)	% Yield compared to control	% Reduction
Legumes (control)	118	100.0	-
Paspalum conjugatum (grass)	111	94.1	5.9
Nephrolepis biserrata (ferns)	104	88.1	11.9
Mikania micrantha (Broadleaved)	94	79.7	20.3

Table 5. Effect of different ground cover on yield of hybrid cocoa after 36 months of competition from weeds (Chung & Lam, 1990)

Treatments	Yields (kg/ha)	% Yield compared to control	% Reduction
Control (blanket spraying)	1300	100.0	-
Strip spraying* + 100% *Asystasia* (inter-row)	696	53.5	46.5
Asystasia - 100% unweeded	626	48.2	51.8

* 2.8 L Paraquat + 2.8 L Diesel + 2.8 L Lissapol in 450 L/ha

weed management options will be very important in the success of plantation crops in Malaysia.

Herbicide and Application Technology

Weed management needs to be economical. The future trend is to use cost-effective herbicides which have low toxicity to man, fish, beneficial insects and worms (bees and earth worms) and wild life. These herbicides should minimally pollute the soil, water and harvested products. Another challenge is the concern for the effect of herbicides on the quality of groundwater. It is clear some herbicides are mobile in the soil, and others leave behind residues. Herbicides should not be used near water catchment areas, rivers and ponds. The time intervals after spraying and entry of grazing animals is important, depending on the herbicides applied.

The future trend in application technology is towards low volume and ultra low volume of spray for weed management in plantation crops. The volume of spray at 450–1100 litres was reduced to 200–550 litres by 1984 and today to 20–450 litres (Lee, 1991). Weed management in the future will be achieved by mechanised application of herbicides. In oil palm, the cost of spraying circle and harvesting paths was M$4.20/ha and M$6.00 for tractor/pump compared to M$6.00/ha for conventional knapsack. The area covered was 1.5 ha for conventional knapsack sprayer compared to 4.8 ha on slightly hilly terrain and 12 ha for flat land with good accessibility and without stumps for tractor mounted twin cones. (Lee & Yaakob Hitam, 1989).

In the adoption of very low volume and ultra low volume, there should be more attention paid to appropriate personal protective clothing and equipment. There should be greater awareness of the spray operator. In 1960, many workers wore shorts, short-sleeved shirts, slippers or simple canvas shoes, though a towel was sometimes wrapped around the face. Today long pants, long sleeved shirts, and rubber boots are worn by the sprayers. In the year 2000, it is envisaged that the majority of spray operators will wear aprons, masks, face-shields, goggles and respirator for the application of herbicides.

Classification of Ground Vegetation

The over-reliance and repeated usage of the same herbicide mixtures for weed management and eradication of a certain weed species have resulted in the colonisation of tolerant or resistant weed species. The increasing use of glyphosate mixtures (dicamba, fluroxypyr, metsulfuron methyl) in rubber and oil palm have generally provided a longer period of control of soft grasses and broad leaves than paraquat plus diuron mixtures. The weeds that succeed in the glyphosate treated area are the regenerated original grasses plus broad leaved weeds. This has resulted in the rapid upsurge of the highly adaptable *Asystasia gangetica* in oil palm and rubber areas which have posed greater management problems than the original soft grasses like *Paspalum conjugatum, Digitaria ciliaris* and *Ottochlea nodosa.*

In the light of the above scenario, it is clear that the weeds in the plantation need to be classifield for efficient weed management and selective use of herbicides. The class A plants are *Calopogonium caeruleum, Calopogonium mucunoides, Centrosema pubescens* and *Desmodium ovalifolium* and *Pueraria phaseoloides.* They are beneficial plants and should be exploited further to suppress noxious weeds and to cut down weeding cost. Class B plants are *Axonopus compressus, Borreria latifolia* and *Paspalum conjugatum.* These plant species are acceptable but need control when necessary. Class C plants are noxious weeds: *Asystasia gangetica, Imperata cylindrica, Mikania micrantha.* In medium- to long-term weed management strategies, weed suppression may prove to be more useful than total eradication in undulating and hilly terrain which will cause erosion. To control *Imperata cylindrica* (Teoh and Pang, 1989) it was recommended that after spraying with herbicides, plant legume covers *(C. caeruleum)* be planted.

In the management of weeds in plantation crops, legume cover crops are planted to control weeds and erosion. Legume cover crops, *C. caeruleum* and *D. ovalifolium,* are planted because they are shade tolerant. There are many benefical effects of legume covers. The organic matter from the legume covers enhances soil fertility and crop growth. The amount of litter returned by legume covers was 6038 kg/ha compared to 4096 kg/ha for *M. micrantha* at 24 months after planting as in Table 6 (Watson *et al.*, 1964a). The amount of nutrient return over a five-year period for legumes was nitrogen 226–353, phosphorus 18–27, potassium 85–131 and magnesium 15–27 kg per ha compared to a low return of nutrients from *Mikania micrantha* of nitrogen 74–119, phosphorus 9–14, potassium 63–99 and magnesium 9–24 as in Table 7 (Watson *et al.,* 1964b).

Table 6. Amount of litter (kg/ha) of different covers at 24 months after planting

Covers	Litter (kg/ha)
Legumes	6038
Grasses	6140
Naturals	5383
Mikania	4096

Table 7. Total amount of nutrients (kg/ha) returned and to the soil over a five-year period under cover crops

Cover plants	N	P	K	Mg
Legumes	226–353	18–27	85–131	15–27
Grasses	24–65	8–16	31–86	9–15
Naturals	13–117	3–10	46–140	3–18
Mikania	74–119	9–14	63–99	9–24

Biological Control

Sheep have proven to be good biological agents for weed management in rubber. Investigations (Wan Mohamad, 1978) have showed that species growing in the inter-rows of rubber were palatable to ruminants and were highly nutritious. The weed species that are frequently found in immature and mature rubber are given in Table 8 (Chee & Ahmad, 1990). In immature rubber of one to five years, most of the species are grazed by sheep except *Imperata cylindrica* and woody shrubs, such as *Chromolaena odoratum, Melastoma malabathricum, Lantana camara* and *Clidemia hirta.* Among the legume species, the sheep graze *P. phaseoloides* and *C. pubescens* but will only feed on the young shoots and leaves of *C. caeruleum.* In mature areas, the sheep feed on most species except woody shrubs and ferns. Young shoots of the ferns *Nephrolepis biserrata* and *Dicranopteris linearis* are also eaten. Tan & Abraham (1981) found that there was a reduction in undergrowth cover in grazed treatments over an experimental period of one year. The decline in percentage of undergrowth cover in grazing treatments either rotationally or free range, was twice as much as in the fields where no sheep were kept.

In the maintenance of rubber plantations, purification of legume cover crop is carried out by spraying glyphosate, or paraquat based herbicides. Sheep grazing can be used to advantage for this purpose and also where *C. caeruleum* is infested with various mixed weeds. The selective grazing habit of sheep mainly for *M. micrantha, P. conjugalium* and *O. nodosa* helps to purify the legume cover and increase the percentage of *C. caeruleum* in the stand.

The effect of sheep grazing on the growth of rubber was recorded by Tan & Abraham (1981). They found the girth increment of trees in the fields where undergrowth was grazed by sheep, either rotational or free range, was generally higher by 7.5 cm and 8.00 cm respectively than those in the fields where undergrowth was not grazed (6.0 cm). The difference in the growth of the trees may be attributed to a reduction in weed competition and the manure from sheep. Results showed that using sheep to graze the weeds in immature and mature rubber could save 17 to 30% of the total weeding cost (Ani Arope *et al.*, 1985; Abdul Karim, 1990). These varations in the level of savings in

Table 8. Plant species present in the inter-row of rubber

	Age of rubber		
	1–2	3–5	6–10
Legumes	*Pueraria phaseoloides** *Centrosema pubescens* *Calopogonium caeruleum*	*Pueraria phaseoloides** *Calopogonium caeruleum*	*Calopogonium caeruleum* *Centrosema pubescens**
Grasses	*Paspalum conjugatum** *Axonopus compressus** *Digitaria spp.** *Ottochloa nodosa* *Imperata cylindrica*	*Paspalum conjugatum** *Ottochloa nodosa** *Axonopus compressus**	*Cyrtococcum* spp. *Centotheca lappacea** *Ottochloa nodosa**
Broad leafed	*Asystasia intrusa** *Mikania micrantha**	*Asystasia intrusa** *Mikania micrantha** *Borreria latifolia**	*Asystasia intrusa** *Mikania micrantha**
Ferns			*Nephrolepis biserrata* *Dicranopteris linearis*
Woody shrubs		*Chromolaena odoratum* *Melastoma malabathricum* *Lantana camara* *Clidemia hirta*	*Chromolaena odoratum* *Lantana camara* *Clidemia hirta* *rubber seedlings*

* Plant species grazed by sheep.

weed control costs depend on grazing management, stocking rate and age of rubber. The integration of sheep with rubber will reduce both the amount of imported herbicides for weed control and the chemical pollution to the environment. Another beneficial effect of grazing sheep as compared to herbicides is that the sheep do not completely kill the weeds and expose the land to soil erosion.

The search for natural bioagents will continue with the hope that *mycoherbicides* and *bioherbicides* can be developed for weed management. The introduction of insects like *Liothrips mikaniae* for the control of *M. micrantha* and other bioagents will be the future challenge to weed management of plantation crops.

Cultural Practices

Good cultural practices like minimal tillage and its correct timing, optimum manuring, pest and disease control and use of advanced planting material will reduce the immaturity period. As a result of these good cultural practices, the number of spraying rounds will be less because of early maturity.

An optimum planting density and distance will encourage early closure of the crop canopy resulting in less light transmission in plantation crops and poor weed growth. Yoon (1971) showed that at two and a half years after planting, the percentage light transmission for rubber clone RRIM 600, at 296, 398 and 556 plants/ha was 80%, 57% and 47% respectively.

In rubber and oil palm, the use of vigorous and high yielding planting material will reduce the immaturity and maintenance period. S.K. Leong, (pers. comm., 1991) found that planting "core" stump as advanced planting material can reduce the immature period to between 4 and 4.5 years compared to 5 to 6 years for conventional two whorl budded plants. In oil palm, G.F. Chung (pers. comm., 1991) found that the immaturity period for planting 18-month-old polythene bag *tenera* palms was 24 months compared to 28 to 36 months for planting material of 8 to 12 months old. The mulching of the palm with empty fruit bunches also reduced the number of spraying rounds from six to four during the first year after planting.

Conclusion

Weed management for plantation crops in the year 2000 will face many constraints in terms of shortage of workers, high cost of labour and herbicides, and pollution to the environment and operator. Integrated weed management involving cultural methods, herbicides, sheep grazing and biological agents (insects) to reduce weed infestation and minimise crop loss of the three main plantation crops is recommended.

Weed management should be a balanced approach that involves the plantation crops and vegetation management. Safety to the environment and ecology should be taken into consideration. This balanced approach is in line with the international thrust towards protection of the environment. It must be low input and cost effective to sustain agricultural production.

The low input weed management system will involve the planting of vigorous advanced planting material, legume cover, good cultural practices and maintenance of natural covers. Mulching with empty oil palm fruit bunches, pruned leaves and rice straw will enhance crop growth and reduce herbicide spraying. The biological method of using sheep will be ideal for maximising the use of weeds as forage species. The inter-rows can be planted with intercrops to minimise weed infestation and also will give a cash return. In the year 2000, fewer spraying rounds and herbicides should be used because of their high cost. Judicious use of herbicides should be practised. Selective spraying of noxious weeds should be recommended. Herbicides should have low toxicity to operators and beneficial insects combined with minimal pollution to the environment and wild life. Weed suppression is more useful than total eradication in undulating and hilly terrain in our plantation crop system because of soil erosion.

Acknowedgements

The authors wish to thank Dr Abdul Aziz S.A. Kadir (Director, Rubber Research Institute of Malaysia) and Dr Ho Chai Yee (Research Director, Ebor Research, Sime Darby Plantations Berhad) and Dr Ismail Hashim (Head, Crop Protection and Microbiology Division, Rubber Research Institute) for their help and encouragement.

References

Abdul Karim, K. (1990) Sheep rearing - FELCRA progress and experience. *Proceedings 13th Annual Conference*. Malaysian Society of Annual Production, pp. 256-259.

Ani Arope, Tajuddin, Ismail and Chong, D.T. (1985) Sheep rearing under rubber plantations (Eds. Sivarajasingam, S., Hutogalung, R.I. and Kassim Hamid) *Proceedings 9th Annual Conference Malaysian Society of Animal Production*. Serdang, Malaysia, pp. 117–123.

Chee, Y.K. and Ahmad, F. (1990) Sheep and chemical weed control in rubber. *Proceedings of Australian Centre for International Agriculture Research Workshop on Forages for Plantation Crops*. Bali, Indonesia, 27–29 June 1990.

Chee, Y.K., Lee, S.A., Ahmad, A.I., Teo, L., Chung, G.F. and Khairuddin, H. (in press) Crop and loss by weeds in Malaysia. *Proceedings 3rd Tropical Weed Science conference, 4–6 December, 1990*. Kuala Lumpur. Malaysian Plant Protection Society.

Chung, G.F. and Lam. K.S. (1990) Effect of weed control on cocoa yield. *Proceedings*

3rd Tropical Weed Science Conference, 4–6 December, 1990. Kuala Lumpur. Malaysian Plant Protection Society.

Duncan, R.C. (1986) Statistical Handout. World Bank Focus for Commodities. *Proceedings of the International Seminar on Commodities 21–27 July 1986, Kuala Lumpur.* Ministry of Primary Industries, Malaysia.

Gray, B.G. and Hew, C.K. (1968) Cover crop management on oil palm on the West Coast of Malaysia. *Proceedings of Conference on Oil Palm Development in Malaysia.* pp. 56–65 (Ed. Turner, P.D.) Incorporated Society of Planters, Kuala Lumpur.

Lee, S.A. (1991) Changing trends and challenges in weed management and safe application of herbicides. In Incorporated Research Centre Colloquium, Batu Anam, Johore, Malaysia. 29 April 1991.

Lee, S.A. and Yaakob Hitam (1989) New developments in very low volume and ultra low volume application with emphasis on efficiency and spray operator exposure. *Proceedings 5th course on Pesticide Application Technology, 20–26th November, 1989.* University Agriculture, Serdang, Malaysia.

Tan, K.H and Abraham, P.D. (1981) Sheep rearing in rubber plantations. pp. 163–173. *Proceedings of Rubber Research Institute Malaysia Planters' Conference,* Kuala Lumpur.

Teoh, K.S. and Pang, K.F. (1989) Preliminary results using Assault herbicide in association with minimal tillage techniques of legume establishment. *Proceedings 2nd Tropical Weed Science Conference.* Phuket, Thailand.

Wan Mohamed B.W.E. (1978) Utilisation of ground vegetation in rubber plantation for animal rearing. *Proceedings Rubber Research Institute Malaysia. Planters Conference.* Kuala Lumpur. pp. 265–281.

Watson, G.A., Wong, P.W. and Narayanan, R. (1964a) Effect of cover plants on soil nutrient status and on growth of Hevea. III: Comparison of legume creepers with grasses and *Mikania cordata, Journal Rubber Research Institute Malaysia* 18(3), 80–95.

Watson, G.A., Wong, P.W. and Narayanan, R. (1964b) Effect of cover plants on soil nutrient status and growth of Hevea. IV: Legumes crops compared with grasses, *Mikania cordata* and mixed indigenous covers on four soil types. *Journal Rubber Research Institute, Malaysia* 19(3), 123–145

Wood, B.J. and Edward, N.C. (1987) The role of research and development in plantation profitability. Part 1 – Elements of profitability. *The Planter,* 63, 236–248. Incoporated Society of planters, Kuala Lumpur.

Yoon, P.K. (1971) Assimilation and Productivity in Tropical plants. Doctor of philosophy thesis, University of Singapore.

Chapter twenty:
Quarantine in 2000 with Reference to Weeds

S.S. Sastroutomo
ASEAN-PLANTI Quarantine Centre and Training Institute (PLANTI)
43400 Serdang, Selangor Darul Ehsan
Malaysia

Abstract

Many serious weeds in southeast Asian agriculture are of foreign origin and introduced to the region accidentally through the importation of planting materials or unintentionally as ornamental plants (e.g. the water hyacinth, *Eichhornia crassipes*), potential forage crops (e.g. *Panicum maximum, Pennisetum purpureum, Brachiaria mutica*), cover crops (e.g. *Mikania micrantha*), or used to prevent or control soil erosion of river banks (e.g. *Mimosa pigra*).

Recently, weedy species have been introduced as contaminants in internationally traded seed during both small- and large-scale movements of seeds of improved cultivars and new crops. They pose a grave threat to ASEAN agriculture and eventually the economy. With increasing trade, germplasm exchange and travel by the year 2000, it is necessary to develop preventive measures against this introduction. Appropriate regulations and inspection procedures similar to those used to guard against insect pests and diseases should be developed for weeds in the ASEAN region.

The most effective means of dealing with weed seed contaminants in crop seeds for planting are prevention and decontamination. Seed certification and related seed control measures are also proven means for preventing the contamination of planting seeds with seeds of objectionable weeds.

Germplasm materials originating from research institutions sometimes accidentally carry noxious weed seeds besides insect pests or seedborne diseases despite treatment and certification. Measures that have been agreed upon for the ASEAN countries for import restriction on seed lots are described.

Introduction

Risks of introducing or spreading insects, plants, or other biological pests are related to the movement of plant or animal products and objects capable of harbouring the source of biological hazard. These risks are dependent upon the commodity involved, origin of the commodity and tolerance of the pest to transportation conditions at destination areas which favour establishment and proliferation of hazardous biological activity (Dollar & Loharanu, 1985).

Animal and plant quarantine authorities in most southeast Asian countries are authorised to take measures to prevent the introduction and dissemination of new or not widely distributed hazardous biological agents into importing areas or regions.

Some economically important pests, such as insects and diseases have been the prime targets of quarantine inspection for a long time; however, little attention if any has been given to weeds by plant quarantine authorities during their regular inspection at ports or entry points. The negative effects of weeds on crop production have been reported by various researchers. Crop losses to weeds ranged from 15 to 42% in lowland rice, 47 to 87% in upland rice, 16 to 82% in maize, 18 to 68% in soybean, 10 to 50% in groundnut, and 6 to 62% in cassava.

Most noxious weeds in southeast Asian agriculture are of foreign origin and introduced into the region accidentally through the importation of planting materials, especially seeds, or unintentionally as ornamental plants. Historically, the wide dissemination of weedy species followed the course of exploration and colonisation. In modern times, weedy species have been and still are disseminated among continents and countries, and within countries, as contaminants in internationally traded seeds, grain, straw, other packing materials, bird seed, whole spices, and many other types of agricultural products. Recently, weedy species have been introduced as contaminants in both small- and large-scale movements of seeds of imported cultivars and new crops.

Noxious weeds are responsible for causing high losses in agricultural production. With increasing trade, germplasm exchange and travel by the year 2000 it is necessary to develop preventive measures against the introduction of noxious weeds into the ASEAN region. One inexpensive means of weed management is through stringent application of plant quarantine regulations.

Weed Seed Interceptions

A large number of introductions of grass or cover crop seeds and other planting materials have been made during the past ten years in the ASEAN region. The introduction has originated from a number of countries, and international research institutions. Generally, grasses have been introduced in large quantities for the livestock industry, and landscaping and cover crops for plantations (Sastroutomo, 1990).

A study on the interception of weed seeds from cover crop seed consigments imported to Malaysia from Indonesia, the Philippines and Thailand showed that out of 15 samples of four cover crop seeds inspected, i.e. *Calopogonium caeruleum, C. mucunoides, Centrosema pubescens* and *Pueraria javanica*, ten samples were contaminated with weed seeds. Thirty-three species of weed seeds were identified from these samples, with *Mimosa pudica, Cassia tora, Euphorbia esula, Crotalaria mucronata* and *Urena lobata* as predominant ones. Other weed seeds included *Mimosa pigra, Sida rhombifolia, Mikania micrantha* and *Echinochloa colona*. These weeds are considered noxious in Malaysia and the introduction of *M. pigra* is prohibited by legislation (Sastroutomo, 1988).

All intercepted weed seeds from cover crops were already established in the country. However, a previous report recorded the interception of two species not found in the country, i.e. *Cannabis sativa* and *Convolvulus sepium* (Zakaria & Nair, 1985).

Sastroutomo (1989) reported a total of 41 species of weed seeds from nine samples of grasses imported from USA, Australia and India with *Paspalum dilatatum, Avena fatua, Lepidium virginicum, Panicum miliaceum* and *Vicia hirsuta* as predominant ones.

A recent study on the interception of weed seeds from grass seed consignments imported to Malaysia from Australia showed that out of 16 samples of 10 grass seeds inspected, i.e. *Axonopus affinis, Brachiaria decumbens, B. humidicola, Cynodon dactylon, Panicum maximum, P. trichogluma, Paspalum notatum, P. plicatulum, Setaria anceps* and *S. kazungula*, 14 samples were contaminated with weed seeds. Seventy-nine species of weed seeds were intercepted with *Fumbristylis dichotoma, Malva silvestris, Verbena officinalis, Solanum rostratum* and *Chloris* sp. as predominant ones (Tasrif, 1990). Almost all the intercepted weed seeds have not been recorded in Malaysia. Their ability to proliferate under tropical conditions is questionable, but if by coincidence the grass seeds containing these seeds were planted in areas suitable for their growth, then there would be a chance that the weeds would establish in this country. Furthermore, if the grass seeds can grow in the country from where they arrived, the weeds can possibly establish as well.

Based on these results it is emphasised that clean seeds for planting materials form the basis for sound production and quarantine activity. Thus, plant quarantine officers should be aware of their important role in preventing the entry of noxious weeds.

Tolerance Level for Weed Seed

ASEAN PLANTI, through its technical meetings in 1987 in the Philippines and Malaysia, established the ASEAN tolerance level of weed seeds in seed consignments (Anon, 1987a,b). For A1 weeds, i.e. noxious weeds which are not yet introduced in ASEAN countries, the tolerance level should be zero. However, for A2 weeds, i.e. weeds already introduced into some countries in ASEAN but not others, the tolerance level should be zero if the particular weed is not present in a country. The tolerance level could be above zero in a situation where the weed is present. The tolerance levels for some crop seeds are given in Tables 1 and 2.

Table 1. Tolerance level of various components of purity analysis including seeds in various standard crop seeds

Crop	Percentage by weight*			
	Pure seed seeds	Other crop seeds	Inert matter	Weed seeds
1. *Oryza sativa*	98.0	2.0	0.5	0.2
2. *Arachis hypogaea*	97.0	3.0	0.5	None
3. *Zea mays*	98.0	2.0	0.2	None
4. *Glycine max*	97.0	3.0	0.7	0.5
5. Vegetables (*Brassica* spp., tomatoes, lettuce, etc.)	98.0	2.0	0.1	0.2
6. Cover crops (*Calopogonium mucunoides, C. caeruleum, Centrosema pubescens, Pueraria javanica, Desmodium* spp.)	98.0	1.0	0.5	0.5
7. Grasses (*Axonopus* spp., *Brachiaria decumbens, Cynodon* spp.)	98.0	1.0	10.5	0.5

* Except for pure seed, all values are maximum tolerable values in each component.

Prevention and Control

The most effective means of dealing with weed seed contaminants in crop seeds for planting are prevention and decontamination. Seed certification and related seed control measures are also proven systems for accomplishing the crucial steps and important actions for preventing the contamination of planting seeds with seeds of objectionable weeds (Delouche, 1988; Sastroutomo, 1988).

Planting of germplasm material originating from research institutions sometimes also carries noxious weed seeds besides insect pests or seedborne diseases despite treatment and certification. The following measures for

Table 2. Tolerance levels of weed seeds in various crops

Crops	Found (%)	Stands (%)	Objectionable weeds
Rice	0.1	0.2	Wild rice or red rice, *Echinochloa crusgalli, Striga asiatica, Scirpus maritimus*
Groundnut	None	None	
Maize	None	None	
Soybean	0.2	0.5	*Echinochloa colonum, Rottboellia exaltata*
Vegetables	0.05	0.2	*Argemone mexicana, Spergula arvensis, C. bursa-pastoris, Galinsoga parviflora, Polygonum nepalense, Rumex acetocella, Althernanthera philoxeroides*
Cover crops	0.2	0.5	*Mimosa pigra, Rottboellia exaltata, Pennisetum setosum, P. polystachyon, P. purpureum, Asystasia intrusa*

imported seed lots have been agreed upon for the ASEAN countries (Anon, 1985):

- All importation of economic seeds for planting should be made through the main entry points.
- Seeds for planting, including weeds in laboratories, should be examined for pests with the necessary equipment, manpower and expertise.
- Each ASEAN country should establish a seed health testing unit adequately equipped to carry out pest detection and certification activities on samples intended for import, export and local distributions.
- The unit should work in close co-operation with the PQ service. The unit should also develop treatment schedules and sensitive techniques for detecting important pathogens.
- When international or regional centres are available, the unit should work closely with them. The unit should be given the regulatory functions of inspection, treatment and destruction of seed consignments.

Weed Identification Training for Plant Quarantine Officers

The plant quarantine system is the only solution to prevent the introduction of noxious weeds. To achieve this goal it is necessary to have plant quarantine (PQ) officers with a strong knowledge of weed taxonomy and methodology for interceptions. Due to lack of skilled PQ inspectors in the region, most seed consignments have been examined on a visual basis for insect pests and diseases. As such little attention has been paid to noxious weed seeds found in these consignments, a special training programme for PQ inspectors in ASEAN was launched by PLANTI in 1983. So far eight short courses on identification of weeds have been conducted with a total of 124 participants (Table 3). Besides this, PLANTI also holds a Cerificate Course in PQ (five months) where weed science in general is taught.

With increasing trade, germplasm exchange and travel by the year 2000, PQ in the region will play an essential role in safeguarding agricultural crops from invading new and noxious weeds. PQ inspectors form the first line of defence against the introduction of noxious weeds. In view of their heavy responsibilities, more well-trained personnel are required to carry out inspection, treatments, and other techniques effectively and efficiently. Therefore, training of plant quarantine staff should be conducted on a regular basis to keep abreast of biological changes and weed distribution patterns which are likely to change from time to time.

Conclusion

Noxious weeds are responsible for causing high losses in agricultural production. With increasing trade, germplasm exchange and travel by the year 2000, more and more noxious weeds will be introduced to the ASEAN region if preventive control measures are not taken seriously.

Application of stringent plant quarantine regulations is proven to be the most inexpensive solution to this problem. Seed certification and related seed control measures are also proven systems for preventing the contamination of crop seeds from objectionable noxious weed seeds.

The inspection method and tolerance level of weed seeds in seed consignments established by ASEAN PLANTI should be implemented in the ASEAN region and should be reviewed from time to time.

Training on identification of weed seeds of PQ importance should be given regularly for plant quarantine inspectors, in the ASEAN region.

Acknowledgements

The author thanks the Director of ASEAN PLANTI for the encouragement and permission to present this paper. Thanks are also due to Ms Azura A. Mahyudin and Ramidah Rosdi for their assistance.

Table 3. Short courses in weed identification conducted by PLANTI for PQ officers in the ASEAN region

Title of the course	No. of participants						
	Brunei	Indonesia	Malaysia	Philippines	Singapore	Thailand	Total
1. Weeds of Plant Quarantine Importance (1983)	-	3	5	-	2	3	13
2. Weeds of Plant Quarantine Importance (1985)	-	4	4	3	-	4	15
3. Interception of Weeds and Sampling Techniques (1986)	1	4	7	2	-	4	18
4. Weed Contaminants in Seed Consignments (1987)	-	4	7	4	-	4	19
5. Advanced Course on Weed Identification (1988)	-	3	4	2	-	3	12
6. Collection, Preservation and Identification of Weeds (1984)	-	4	5	3	-	4	16
7. Interception and Identification of PQ Pests in Seed Consignments (1990)	1	3	3	3	1	4	15
8. Advanced Course on Weed Seed Identification (1991)	-	4	5	3	-	4	16
TOTAL	2	29	40	20	3	30	124

References

Anon (1985) Report of the ASEAN PLANTI Meeting on Seedborne Diseases in the ASEAN and Seed Health Testing. 15–17 October 1985. Chiengmai, Thailand. 32pp. + Annexes.

Anon (1987a) Report of the ASEAN PLANTI Meeting on ASEAN Planting Materials and Certification Scheme. 1–3 April 1987. ASEAN PLANTI, Malaysia. 27pp. + Annexes.

Anon (1987b) Report of the ASEAN PLANTI Meeting on Standardization for Weed Interception. 30 September–2 October 1987. Manila, Philippines. 30pp. + Annexes.

Delouche, J.C. (1988) Weeds as rice contaminants. In: *Rice Seed Health. Proc. Intern. Work. Rice Seed Health.* IRRI, Manila, Philippines. pp. 179–187.

Dollar, A.M. and Loaharanu, P. (1985) Irradiation and other alternative treatments for the export of fruits and vegetables. *Proceedings of Regional Conference on Plant Quarantine Support for Agricultural Development.* ASEAN PLANTI pp. 105-112

Sastroutomo, S.S. (1988) Germination and viability of weeds intercepted from cover crop seeds. *PLANTI Proc.* No. 3. pp. 169–178.

Sastroutomo, S.S. (1989) Weed seeds intercepted from imported grass seeds. *PLANTI Proc.* No. 4. pp. 273–278.

Sastroutomo, S.S. (1990) *Preventive Control of Weeds Through Seed Regulations in the ASEAN Region.* BIOTROP Spec. Publ. No. 38. pp. 243–256.

Tasrif, A. (1990) *Weed Seeds Intercepted from Grass Seeds and Germination Ability.* BIOTROP Spec. Publ. No. 38. pp. 237–242.

Zakaria, N. and Nair, J. (1985) Weed seeds and other contaminants in cover crop seeds imported into P. Malaysia and their quarantine implication. *J. Plant Prot. Tropics,* 2(1), 53–59.

Chapter twenty-one:
Management of Weeds Discussion

Chairman: Dr T.C. P'ng

Vincent Ng (Jabatan Pertanian) Can you please indicate what type of mycoherbicides are available, for what crops, their costs, and how widespread is their usage? In costs, how do they compare with chemical herbicides?
Dr R.J. Williams In the USA, two mycoherbicides based on *Colletotrichum* sp. are in the market (Collego and Devine). They are specific for weeds (vetch in rice), but are effectively restricted niche products. I am unable to advise on costs.
Dr Lakchai Meenakanit We have some problems on the leaching of herbicides which are extensively used in fruit plantations in Thailand. In many areas we found the leaching of pesticides into the underground water. When farmers pump up water to their crops, they cause damage to the crops. Do you have any comments or ideas on how to prevent this?
Dr R.J. Williams Several measures are available, including the correct use according to label instructions, the prevention of point source spills during spray-tank filling and correct disposal of spray-tank washings. Closed-filling systems have been developed to prevent spills during tank filling; and research is underway to develop new, rapidly degradable products, and controlled-release formulations of leachable products.
Dr Lakchai Meenakanit Can you please comment on the comparison between cover crops and herbicides? Which is preferable, economically and environmentally?

Secondly, are your proposals for protective clothing for spray applicators really practicable in the hot, humid SE Asian climate?
Dr Y.K. Chee Legume cover crops are planted in the inter-rows of rubber and oil palms for weed control, to minimise soil erosion and to return litter and nutrients to the soil. When these legume covers spread and strangle the rubber and oil palm, herbicides are used to control them. In terms of economic and environmental impact, legume covers are better than using chemicals for weed control.

In the development of protective clothing, the problems of suitability are always present. Cotton with long sleeves and pants have been found to be the most suitable.

Definitely by the year 2000, some modification of the protective clothing discussed will be practical.

Dr S.A. Lee (MARDI) Dr Sastroutomo's paper highlighted the large number of noxious weed seeds that are imported into Malaysia. Have you followed some of these weed seeds from the point of import to field planting? Do some of these imported noxious weed seeds establish themselves in Malaysia? If a seed consignment fails your seed purity test, what do you do with that consignment?

Dr S.S. Sastroutomo We did a germination test on the intercepted weed seeds from the consignment and 60–70% of the seeds germinated under laboratory conditions. We also did a survey of grasses that we intercepted from grass consignments from Australia and New Zealand. About 70% of them can germinate and grow under laboratory conditions.

We also did a survey in the field where the imported grass seeds are planted. In some cases, we found there were new species of grasses which could not be identified by using books on the local flora.

Regarding the seed or the consignment that does not pass the purity test, we submit the report of the analysis to the Department of Agriculture. It is up to them to take action. ASEAN PLANTI has no legal status to enforce plant quarantine regulations.

PART 6
Biodiversity and Pest Management

Chapter twenty-two: **Potential for Loss of Biodiversity in Malaysia, Illustrated by the Moth Fauna**

J.D. Holloway[1] and H.S. Barlow[2]
[1]*International Institute of Entomology, 56 Queen's Gate, London SW7 5JR, UK*
[2]*P.O. Box 10139, 50704 Kuala Lumpur, Malaysia*

Summary

A profile of biological diversity in Malaysia is presented, using moths as an exemplar insect group. Because of the phenomenon of host-plant specificity, moths may indicate potential for loss of diversity general to both plants and animals.

Diversity within Malaysian ecosystems is very much higher than elsewhere in Malesia, with the exception of the island of New Guinea. Indications are that diversity is greatest at the intergrade between lowland forest types and lower montane forests. At this point an intrinsically diverse lower montane association of species has its diversity augmented by overlap with both lowland and upper montane elements.

The effects of forestry practices and conversion to managed systems such as plantation or field crops are discussed. Loss of diversity is inevitable in such circumstances but the effects vary; in some instances (e.g. logging cycles in natural forest) recovery of diversity may be possible. Examples are given from Borneo and Peninsular Malaysia, with some reference to Indonesia.

Such effects differ depending on taxonomic group: some higher taxa (families, sub-families) are much more vulnerable than others. This is demonstrated with moth data for Borneo, tabulated group by group with respect to biogeographic and ecological categories. A classification of groups on the basis of these tables indicated there is a spectrum of vulnerability. At one extreme there are groups such as Lasiocampidae (perhaps paralleled by the Dipterocarpaceae) with endemism and high representation in lowland forest and therefore most vulnerable to human disturbance, with high potential for loss of diversity, including extinction. At the other extreme are groups such as some sections of the Noctuidae (the Gramineae may be comparable) with many geographically widespread taxa with specialism to open and disturbed habitats.

These include many taxa that respond positively to human activities and are regarded as pests (or weeds).

The family Limacodidae is an exception to this, combining high endemism with high pest status. The implications of this for conservation of biodiversity within forestry and cash crop plantations are considered.

Introduction

Before any predictions of loss of biodiversity can be made for a tropical country such as Malaysia, it is essential to have an accurate profile of that diversity, an inventory of its quality and ecological distribution. This can then be set beside projections of habitat loss due to human activity to provide estimates of potential loss of species.

In tropical countries there is often available a fairly comprehensive inventory of higher plants and vertebrates with habitat data. The impact of human activity on vegetation is readily apparent, and some qualitative assessment of the potential effect on floristic diversity can usually be obtained. But it is rare to find quantitative data to any degree of detail.

The greater size of vertebrates renders them far more apparent than insects and other invertebrates, but it is also the basis of their considerably lower richness in species. Contrast the lymantriid moth genus *Redoa*, with over 70 Bornean species, with the recorded fauna of 92 bat species, and the family Lymantriidae itself, with over 300 Bornean species, with the total land mammal fauna of 221 species (Payne *et al.*, 1985). The total butterfly and moth fauna is probably almost two orders of magnitude greater than the mammal fauna in Borneo.

However, vertebrate diversity is thought to correlate more with vegetation structure than with floristic diversity (MacArthur (1972: 178), for birds). It would follow from this that attempts to assess loss of animal diversity due to human disturbance with reference to vertebrates alone is likely to give a very incomplete picture that will primarily reflect the effects of gross structural modification of vegetation only. The actual situation appears to be more complex, for example in the effects of shifting cultivation in New Guinea on various bird foraging guilds in comparison with butterfly and reptile diversity (Bowman *et al.*, 1990). More comparative studies of this nature are needed.

Plant-feeding insect groups that exhibit specificity along floristic lines will be likely to provide a much more precise profile of diversity, reflecting general patterns for both flora and fauna, though the correlation is not without exceptions (Holloway, 1984). If these insect groups can be sampled readily by standard techniques, then these observations can be placed on a quantitative basis. Moths, sampled with light-traps, are such a group, and have the additional advantage that they have been extensively collected and studied in the past so they provide a biogeographically comprehensive database that is founded on reliable systematics. They will therefore be used here as an exemplar group in profiling Malaysian biodiversity and considering the threats to it.

Before proceeding with this aim, it is worth noting that insects can match vertebrates in trophic versatility, including herbivores, predators and carrion feeders, but also include diverse guilds, largely lacking in vertebrates, such as detritivores, fungivores (including lichen browsers) and parasitoids, and form, with the Arachnida, a major component of the soil fauna. Plant-feeding insects make up at least one-quarter of global biodiversity, and may support a similar number of parasitoid insects. These also exhibit a degree of specialism and probably help maintain high insect herbivore diversity through suppression of dominance (LaSalle & Gauld, 1991). On this combined diversity depend insect predators and vertebrate insectivores. So phytophagous insects not only may reflect diversity in the plant kingdom but also provide the major resource base for most of the rest of the animal kingdom. They therefore have the potential to provide indicators of most aspects of biodiversity and the effects of our actions on it. This is explored in more detail by Holloway & Stork (1991).

A Profile of Moth Diversity in Malaysia

Moth diversity within Malaysia ecosystems is very much higher than elsewhere in Malesia except for the island of New Guinea. Figure 1 shows profiles of diversity (the alpha-statistic of Fisher *et al.*, 1943) revealed by quantitative moth samples over altitude transects in Malaysian Borneo, the islands of Sulawesi and Seram in Indonesia, Papua New Guinea (high altitudes only) and islands in the SW Pacific (Vanuatu, New Caledonia, Fiji).

A point for samples pooled over a year at 610 m in the Genting Sempah of Peninsular Malaysia indicates that diversity there is of the same order as in Borneo, but this site will be discussed in more detail below. Data from the lowlands of New Guinea are needed to confirm whether the high levels at altitude are paralleled lower down. Also, no data are available for the NE Himalayan region, identified as a centre of species richness for many oriental butterfly genera (Holloway, 1969, 1974) and therefore likely to harbour high moth diversity as well.

In Borneo, Sulawesi and Seram the diversity profile has a typically humped form, the highest diversity being recorded between 500 m and 1000 m. Studies of zonation in all three localities suggest this altitude range covers the transition zone between lowland forest types and lower montane forest. Here an intrinsically diverse lower montane association of species has its diversity augmented by overlap with lowland and upper montane elements.

Quantitative analyses of data from Borneo have enabled these species associations to be delineated (Holloway, 1970, 1989), and their contribution to diversity at different altitudes is illustrated diagrammatically in Fig. 2.

The somewhat counter-intuitive shape of the curve, with maximum

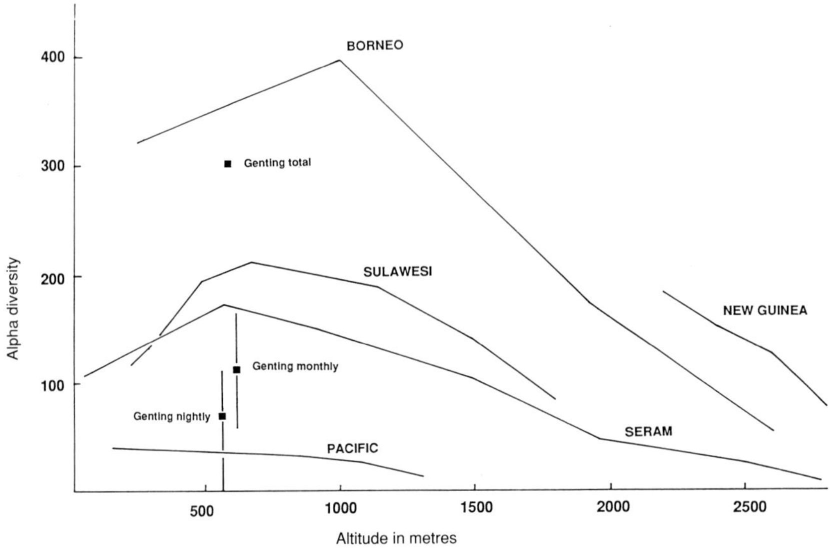

Fig. 1. Profiles of moth diversity with altitude in natural vegetation from various localities in the Indo-Australian tropics. The average diversity and range of diversity recorded at the Genting Sempah locality for nightly, monthly pooled and total pooled samples is also indicated

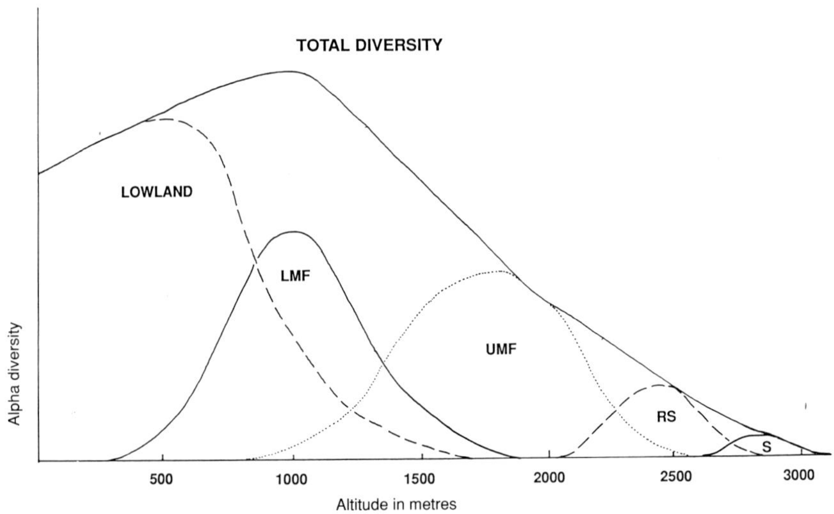

Fig. 2. Diagram of the contribution to total diversity with altitude in Borneo made by associations of species in five overlapping zones: lowland forest, lower montane forest, upper montane forest, Radio Sabah (named after a relay station in the zone on G. Kinabalu) and summit. The last two are only developed to any great extent on G. Kinabalu

diversity occurring at other than the lowest altitudes, is enigmatic. Possible explanations of this lower than expected lowland diversity include: the relative unavailability to moth larvae of dipterocarp foliage biomass, a major component of the lowland forest (Holloway, 1989); elements of beta-diversity or habitat diversity, e.g. specialism to forest types, being greater in the lowlands but not being assessed in individual samples (Holloway, 1989); specialist lowland taxa being more prone to extinction during depressions of vegetation zones in the tropics at glacial maxima (Holloway *et al.*, 1990). The third explanation has the virtue of generality to all three areas where the humped curve has been detected, but the other explanations may also be valid in Malaysia.

Effects on Moth Diversity of Forestry, Shifting Cultivation and Conversion to Managed Systems

Though our assessments of natural profiles of moth diversity for the Indo-Australian tropics are increasing in number, similar measures from logged or otherwise disturbed forest, plantations and field crop systems are much fewer. A scattering of pilot samples from Malaysia and Indonesia provide pointers to the effects of these human activities on diversity, but a more systematic programme of sampling is needed urgently. Diversity values from the pilot studies are plotted in Fig. 3.

The effects of shifting cultivation have been studied in Seram, Indonesia (Holloway, in press), with samples indicated by open circles in Fig. 3. Those three with lowest diversity (coincidentally at the lowest altitude) were from freshly cleared and planted garden areas. The three samples with higher diversity were made in areas abandoned a decade or so previously, with bamboo and secondary forest well developed. Samples from cultivated areas abandoned several decades back and now well forested gave diversity values comparable with that of forest at similar altitudes undisturbed by man. It is therefore possible that a sustainable cycle of forest exploitation or farming that permits a full regeneration cycle will only temporarily depress biodiversity, though this may be dependent on the proportion of climax forest present throughout the progression of such cycles (see below).

Two samples from Bornean alluvial forest, illustrated by stars in Fig. 3, also illustrate the capacity of forest diversity to recover, albeit a forest type that is no doubt adapted to the instability of terrain inherent in a mature river course. The lower value is for a sample made in forest regenerating in an area that had been clear and farmed up to ten years previously. The upper value is for an area where farming is thought to have ceased 100 years previously (Holloway, 1984, 1987; J. Proctor, pers. comm.), with diversity approaching the maximum so far recorded for that altitude.

Conversion of lowland forest to field crop or cash crop will depress diversity to a much lower level as seen in samples from coconut plantations

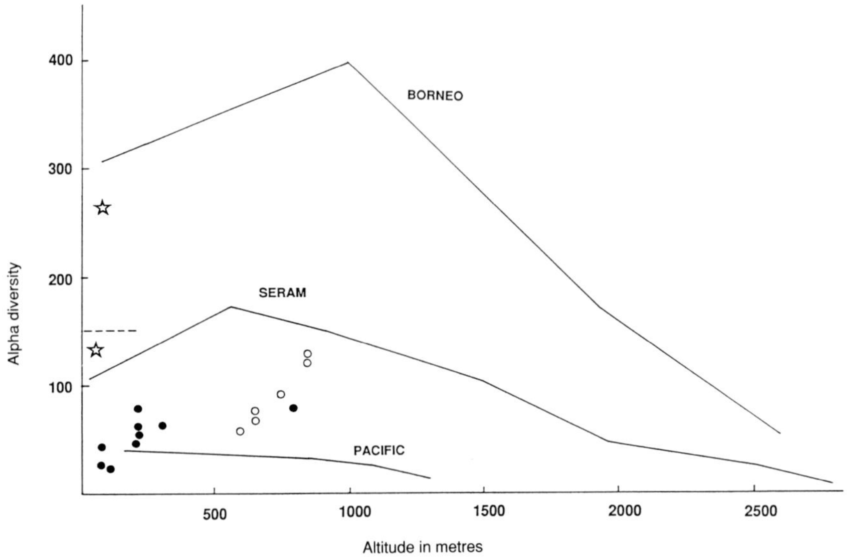

Fig. 3. Three of the natural diversity profiles from Fig. 1 are contrasted with samples from habitats disturbed or managed by man. Solid circles: field crop systems and coconut plantations in Sulawesi. Open circles: phases of shifting cultivation in Seram. Stars; 10 year (lower) and 100 year (upper) regenerating alluvial forest in Borneo. Horizontal dashed line: maximum diversity recorded in softwood plantations in lowland Borneo.

and a zone of rice/soya/maize cultivation in Sulawesi (solid circles). Diversity in these systems is not significantly greater than that in samples from moderately remote archipelagos of the southwest Pacific. The quality of the fauna is also markedly changed (Holloway, 1987, and see below). It is possible that, were such areas of permanent cultivation to be abandoned, forest cover of some sort would come back and moth diversity would increase. The extent to which this would progress would be dependent on the degree to which plants and animals could recolonise the area from localities where undisturbed habitats persist.

An example of where this appears to have occurred is at the Genting Sempah locality mentioned above: a small clearing beside a stream, with advanced secondary forest on the surrounding hills and a small patch of primary forest on the top of one of them. A moth sampling programme (Barlow & Woiwod, 1989) was carried out in the area during the late 1970s using a Rothamsted light-trap with a source of illumination that is much less powerful than the mercury vapour lamps used to sample in Borneo and elsewhere. This appears to give a much more localised sample with relatively low diversity recorded on a daily basis or even for pooled monthly samples. However, when the samples made over 13 months were pooled, alpha diversity values of about 300 were obtained for macrolepidoptera and 386 for macrolepidoptera and Pyraloidea (Fig. 1). This may be an effect of habitat patchiness, as the area includes a mosaic of forest types and open and disturbed habitats. Nevertheless, this pooled value is comparable with those obtained at similar altitudes in Borneo and significantly higher than that obtained (319 for macrolepidoptera and Pyraloidea) using Rothamsted traps in primary lowland forest at about 200 m in northern Sulawesi (Barlow & Woiwod, 1990).

This is all the more remarkable as it has come to our notice that the Genting site, as recently as the early 1950s, was apparently clean cleared for the growing of peanuts and other field crops. The lower parts at least of the surrounding slopes were planted with tea bushes that were allowed to grow up to encourage seed production. Figures 4a & b show the area as it was in the early 1950s. Figures 5a & b show the same site today (little changed from the period of the Rothamsted samples). Differences in perspective are due to a wider angle of lens for the 1950s photographs.

These photographs illustrate the speed with which rainforest can reassert itself, not necessarily in a commercially productive manner, for there are few trees in these forests of any timber value. The moth sampling programme indicated the extent to which faunal diversity can recover. Important factors over the 20–25 years of recovery were that the land was untouched during that period and that the area was surrounded by forest that, though logged over, had never been cleared.

In the ten years since the Rothamsted trapping exercise there has been little change to the slopes immediately surrounding the area, but encroachments of shifting cultivators have almost cleared the patch of primary forest on top of one of the hills and have generally reduced the extent of forest cover in the neighbourhood. The habitat mosaic therefore now consists of a higher proportion of low diversity patches of vegetation. It has not yet proved possible to repeat the Rothamsted trapping exercise. However, general observations nearby from the 1970s to the present day of moths coming to a whitewashed wall illuminated by a tungsten lamp suggest that moth diversity has declined substantially.

All these observations indicate that a degree of forest disturbance or clearance is not necessarily detrimental to the long-term maintenance of biological diversity provided the seeds of recovery, in the form of refuges of undisturbed forest, are maintained with the system. Considerable work needs to be done to assess factors of both time (cycle frequency) and space (reserve size and frequency within the mosaic) necessary to maintain optimum diversity within zones of multiple land use.

Extensive areas of lowland Malaysia are being converted to plantations of various types. A well maintained oil palm plantation has moderate structural diversity (canopy, trunks, ground layer) and low floristic diversity (palm, cover crop, epiphytes such as ferns on the trunk and lichens and moss on the foliage).

Fig. 4a & b. Genting site, early 1950s

Fig. 5 a & b. Genting site, 1991

No quantitative assessment of diversity in oil palm is known to us, but coconut plantation equivalents in Sulawesi support moth diversity of the same order as field crop systems (Fig. 3).

Another option in the forestry sector is for softwood plantations of various types, but particularly of the N Australian legume, *Acacia mangium*. Pilot studies from *Acacia mangium* and *Pinus* plantations in Sabah (Chey, 1990) have produced samples with maximum diversity of 150 (alpha). This level is indicated on Fig. 3, and represents half the maximum lowland rainforest level for Borneo. Most plantations have a floristically poor understorey of weeds, vines and shrubs that will contribute to the level of insect diversity. The pilot samples may have included moths from nearby secondary forest which would have boosted diversity. Now Mr Chey is currently engaged in a more intensive survey where these problems will be addressed.

A particular problem with legume monocultures is that the family Leguminosae is rated most highly by many groups of plant-feeding insects, including defoliating moth larvae (e.g. Holloway, 1989). It is therefore possible that a diverse fauna of arboreal legume feeders will develop over time in these monocultures. This contribution to maintenance of biodiversity is unlikely to be welcomed by foresters, a point that will be considered further below.

Maintaining the Quality of Diversity: the Spectrum of Vulnerability among Higher Taxa

In the conversion by man of natural tropical forest systems to very much simpler, managed systems, the very considerable loss of diversity observed affects some taxonomic groups far more seriously than others. This can be illustrated by contrasting the proportions of various moth families in a forest system with those in field and plantation systems.

In Fig. 6 three Bornean forest types are compared with managed systems in Sulawesi. This comparison is feasible because the family mix in the two island faunas is comparable, even though diversity overall is generally lower in Sulawesi.

The three forest types support a diverse mixture of families and are approximately comparable with each other. The regenerating and mature alluvial forest mentioned earlier are included with hill dipterocarp forest from Gunung Mulu. The regenerating alluvial forest, though of much lower diversity than its montane counterpart, shows an increase in Sphingidae (mostly macroglossine species) and a reduction of Noctuidae but otherwise a very similar representation of families. Similarly, in the shifting cultivations cycle in Seram mentioned earlier, the fall in diversity caused by clearing was not accompanied by changes in family proportions (Holloway, in press). In both instances the response seems to have been comparable to that in natural cycles of disturbance and regeneration within the forest system. Moth taxa abundant in the early stages of the swidden cycle are probably gap phase specialists in the natural forest.

In the two managed systems, the family proportions are distinctly different, Noctuidae, Sphingidae and Arctiidae becoming more predominant relative to Geometridae, with the virtual eclipse of some of the smaller families such as the Cossidae, Drepanidae, Uraniidae and the bombycoid families (excluding the Sphingidae).

These samples are also marked by the presence of a high proportion of open habitat specialist taxa, particularly in the Noctuidae and Arctiidae. The specialist arboreal feeding groups are those that show the greatest decline.

Vulnerable higher taxa can also be identified by segregating their component species among biogeographic and ecological categories and considering the threats posed by particular types of development.

Malaysian moth species (here represented by the Bornean fauna), and probably those in most other groups of organisms, can be divided among a number of general biogeographic categories: endemics; those found in most areas of Sundaland (Peninsular Malaysia, Sumatra, Java, Borneo); those found in Sundaland plus the Philippines and/or Sulawesi (Wallacea); those extending from the NE Himalaya to Sundaland; those more widespread in the oriental tropics (India to tropical Australasia). It might be expected that the more widespread a species is, the less vulnerable it is to extinction.

The same species can also be assigned to habitat categories. The analyses of associations in relation to altitude in Borneo has facilitated this (see Fig. 2), but in addition to these restricted altitude associations must be added categories that span a wider altitude range (lowland to montane, general montane) as well as an open habitat specialist category.

Data have been taken from the checklists in published parts of *The Moths of Borneo* series (Holloway, 1983–89) and from a draft checklist for the Ennominae, to be covered in the part currently in preparation. A total of approximately 1200 species is involved, perhaps one-third of the Bornean fauna of macrolepidoptera. For most of these some indications of habitat preference are available, albeit often based on only a few specimens.

For each of 15 higher taxa, species were allotted to boxes in a two-way table — biogeographic v. habitat categories. The 15 higher taxa were as follows:

1. Cossoidea (Cossidae, Metarbelidae, Ratardidae) (43 species).
2. Limacodidae (97 species).
3. Noctuidae: "lower trifines" (Acronictinae, Agaristinae, and Amphipyrinae from *Sasunaga* onwards in the published checklist) (104 species).
4. Noctuidae: "higher trifines" (Noctuinae, Heliothinae, Hadeninae and Amphipyrinae up to *Sasunaga*) (75 species).
5. Noctuidae: Euteliinae (71 species).
6. Noctuidae: Stictopterinae (90 species).
7. Arctiidae: Arctiinae (39 species).

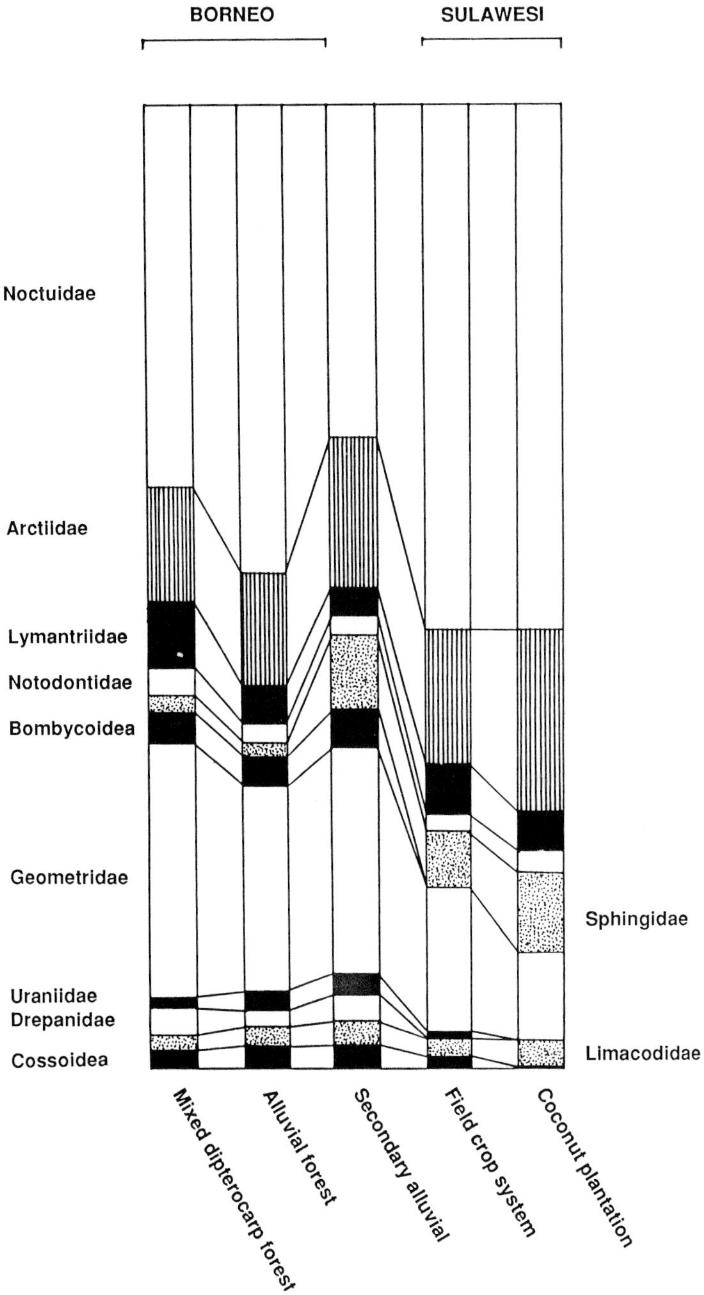

Fig. 6. Proportions of different moth families represented in some of the samples illustrated in Fig. 3: mature and regenerating lowland forest in Borneo; field crop and plantation agroecosystems in Sulawesi

8. Notodontidae (122 species).
9. Lasiocampidae (62 species).
10. Sphingidae (93 species).
11. Geometridae: Ennominae tribe Hypochrosini (60 species).
12. Geometridae: Ennominae tribe Baptini (including *Synegia* and allied genera) (48 species).
13. Geometridae: Ennominae, *Cassyma/Zamarada* group of genera (39 species).
14. Geometridae: Ennominae tribe Boarmiini (in a broad sense) (154 species).
15. Geometridae: Ennominae, residue (106 species).

The tables for these 15 groups and for the total sample, with numbers of species in each category box expressed as a percentage of the total, are presented in Fig. 7–9.

Given that lowland forest is the habitat type most at risk from development, then it follows that higher taxa with a high proportion of endemism restricted to lowland forest are those where the risk of significant erosion of diversity is greatest. The Cossoidea, Limacodidae, and the Ennominae groups Hypochrosini, Baptini and the *Cassyma/Zamarada* group all have over 15% of their species in this category.

Many groups also have numerous Sundaland species restricted to lowland forest. The risk here is lower but will be dependent on joint Malaysian/Indonesian policies for conserving biodiversity. The Cossoidea, Limacodidae, Lasiocampidae and the *Cassyma/Zamarada* group each have about half their species in this or the previous category, with the Notodontidae, Hypochrosini and Baptini following at around 40%. Just under one-third of species in the total sample fall within this category. Conversely, those higher taxa with low representation in these categories, but higher numbers of geographically widespread taxa, particularly those specialist in open habitats, are generally at low risk. Both groups of trifine Noctuidae are examples of this as are the Sphingidae and Arctiidae. The noctuid groups Euteliinae and Stictopterinae are intermediate, having many widespread species though no open habitat specialism.

Some groups have high representation in montane categories such as the higher trifine Noctuidae, the Arctiinae and some of the sub-groups of the Ennominae. Diversity in montane habitats is at lower risk than that in lowland ones, though it is not immune from threat from leisure developments or projects to grow temperate vegetables.

In general, vulnerability for a higher taxon is high if its species are concentrated in the top left of the table and low if they are concentrated to the bottom right but these are just the extremes of a spectrum of vulnerability.

It is possible to produce a classification of these higher taxa from the tables, summing the extent of overlap of percentages in each category box in pairwise comparisons. This gives a similarity coefficient that can range from zero for no overlap to 100 for total similarity. The array of pairwise comparisons can be subjected to standard phenetic classificatory methods. In

1	Low	LMF	UMF	RS + S	LMF + UMF	Low + Mont.	Open
Endemic	20.93		4.65				
Borneo + Wallacea							
Sundaland	27.90				4.65		
Sundaland + Wallacea	13.95				2.33	2.33	
Himalaya + Sundaland	4.65						
Wide Oriental	16.28						2.33
Indo-Australian							

Cossoidea

2	Low	LMF	UMF	RS + S	LMF + UMF	Low + Mont.	Open
Endemic	22.58	4.30	6.45		1.08	1.08	
Borneo + Wallacea	1.08						
Sundaland	40.86	2.69	2.69		1.08	2.15	
Sundaland + Wallacea							
Himalaya + Sundaland	4.30				1.08	1.08	
Wide Oriental	6.45						
Indo-Australian	1.08						

Limacodidae

3	Low	LMF	UMF	RS + S	LMF + UMF	Low + Mont.	Open
Endemic	8.25		5.15		7.22		
Borneo + Wallacea					1.03		
Sundaland	11.34	1.03	4.12		1.03	8.25	2.06
Sundaland + Wallacea	4.12					1.03	
Himalaya + Sundaland	3.09	1.03	1.03			4.12	
Wide Oriental	7.22	1.03	2.06			5.15	1.03
Indo-Australian	2.06	2.06	2.06			3.09	10.31

Lower trifine Noctuidae

4	Low	LMF	UMF	RS + S	LMF + UMF	Low + Mont.	Open
Endemic			21.13	11.27	0.70		0.70
Borneo + Wallacea			1.41	1.41			
Sundaland	2.82		2.82			1.41	
Sundaland + Wallacea			4.23				
Himalaya + Sundaland							
Wide Oriental	2.82		2.82	4.23	0.70	1.41	3.52
Indo-Australian	1.41		4.23			1.41	28.16

Higher trifine Noctuidae

5	Low	LMF	UMF	RS + S	LMF + UMF	Low + Mont.	Open
Endemic	7.46		1.49		1.49	1.49	
Borneo + Wallacea	2.99						
Sundaland	22.39	1.49	1.49		2.99	5.97	
Sundaland + Wallacea	4.48				1.49		
Himalaya + Sundaland		1.49					
Wide Oriental	4.48		1.49		1.49	1.49	
Indo-Australian	16.42		1.49		4.48	11.94	

Euteliinae

6	Low	LMF	UMF	RS + S	LMF + UMF	Low + Mont.	Open
Endemic	3.53		8.24			2.35	
Borneo + Wallacea			1.18				
Sundaland	8.24		5.88		3.53	12.94	
Sundaland + Wallacea	2.35	1.18	1.18		2.35	3.53	
Himalaya + Sundaland		1.18	1.18		1.18	3.53	
Wide Oriental	2.35		4.71			4.71	
Indo-Australian	3.53	1.18	2.35		3.53	14.12	

Stictopterinae

Fig. 7. Percentage representation of species in higher taxonomic groups among biogeographic and habitat categories as discussed in the text. Groups 1–6

7	Low	LMF	UMF	RS + S	LMF + UMF	Low + Mont.	Open
Endemic	5.26	5.26	15.79		10.53	2.63	2.63
Borneo + Wallacea	2.63						
Sundaland	10.53	2.63			2.63	2.63	2.63
Sundaland + Wallacea						2.63	
Himalaya + Sundaland					2.63		
Wide Oriental					5.26		10.53
Indo-Australian						5.26	7.89

Arctiinae

8	Low	LMF	UMF	RS + S	LMF + UMF	Low + Mont.	Open
Endemic	8.33		5.83			1.67	
Borneo + Wallacea							
Sundaland	30.00		8.33		1.67	5.83	1.67
Sundaland + Wallacea	6.67					4.17	1.67
Himalaya + Sundaland	3.33					1.67	
Wide Oriental	8.75	1.25	3.33		0.83	3.33	
Indo-Australian			0.83			0.83	

Notodontidae

9	Low	LMF	UMF	RS + S	LMF + UMF	Low + Mont.	Open
Endemic	9.68		6.45		8.06	1.61	
Borneo + Wallacea							
Sundaland	41.93		6.45		1.61	8.06	
Sundaland + Wallacea							
Himalaya + Sundaland					1.61	3.23	
Wide Oriental	8.06				1.61	1.61	
Indo-Australian							

Lasiocampidae

10	Low	LMF	UMF	RS + S	LMF + UMF	Low + Mont.	Open
Endemic						1.10	
Borneo + Wallacea							
Sundaland	15.38	1.10			2.18	6.59	1.10
Sundaland + Wallacea						1.10	
Himalaya + Sundaland	6.59		1.10				
Wide Oriental	16.48				3.30	10.99	4.40
Indo-Australian	4.40		2.18			12.09	9.89

Sphingidae

11	Low	LMF	UMF	RS + S	LMF + UMF	Low + Mont.	Open
Endemic	15.83	0.83	9.17	2.50	1.67	1.67	
Borneo + Wallacea					1.67		
Sundaland	25.83	1.67	1.67		1.67	6.67	0.83
Sundaland + Wallacea							
Himalaya + Sundaland	6.67		1.67		3.33	1.67	
Wide Oriental	5.00		1.67		3.33	1.67	
Indo-Australian		1.67	1.67				

Hypochrosini

12	Low	LMF	UMF	RS + S	LMF + UMF	Low + Mont.	Open
Endemic	18.75	2.08	12.50	2.08		4.17	
Borneo + Wallacea		2.08					
Sundaland	17.70	11.46	4.17		2.08		
Sundaland + Wallacea	4.17		2.08				
Himalaya + Sundaland						2.08	
Wide Oriental	2.08	2.08	2.08			4.17	
Indo-Australian	2.08				2.08		

Baptini

Fig. 8. Percentage representation of species in higher taxonomic groups among biogeographic and habitat categories as discussed in the text. Groups 7–12

13	Low	LMF	UMF	RS + S	LMF + UMF	Low + Mont.	Open
Endemic	29.73	1.35	4.05			5.41	
Borneo + Wallacea	2.70						
Sundaland	21.62	2.70	2.70		2.70	8.11	
Sundaland + Wallacea	2.70						
Himalaya + Sundaland		2.70			2.70		
Wide Oriental	5.41					5.41	
Indo-Australian							

***Cassyma* group**

14	Low	LMF	UMF	RS + S	LMF + UMF	Low + Mont.	Open
Endemic	8.78	4.05	11.49	2.03	2.70	2.03	
Borneo + Wallacea	0.68		0.68				
Sundaland	16.22	4.05	5.41		0.68	6.08	0.68
Sundaland + Wallacea	2.70				1.35	0.68	
Himalaya + Sundaland	2.70	2.70	1.35		1.35	0.68	
Wide Oriental	4.05	2.03	4.05		1.35	2.03	0.68
Indo-Australian	4.73					2.03	

Boarmiini

15	Low	LMF	UMF	RS + S	LMF + UMF	Low + Mont.	Open
Endemic	3.96	3.96	7.92		1.98	1.98	
Borneo + Wallacea							
Sundaland	15.35	2.48	1.98		2.97	6.93	
Sundaland + Wallacea	4.95		1.98		0.99	2.97	0.99
Himalaya + Sundaland	0.99	1.98	3.96			2.97	
Wide Oriental	10.89	0.99	2.97		2.97	1.98	
Indo-Australian	1.48	0.50				5.94	0.99

Other Ennominae

TOTAL	Low	LMF	UMF	RS + S	LMF + UMF	Low + Mont.	Open
Endemic	9.43	1.54	7.76	1.15	1.58	2.30	0.13
Borneo + Wallacea	0.51	0.09	0.43	0.09	0.09	0.09	0.09
Sundaland	20.14	1.92	3.63		1.71	6.06	0.73
Sundaland + Wallacea	2.99	0.09	0.60		0.60	1.37	0.26
Himalaya + Sundaland	2.39	0.85	0.94		0.77	1.54	
Wide Oriental	6.19	0.64	2.05	0.26	1.32	3.07	1.07
Indo-Australian	3.28	0.38	1.02		0.60	4.01	4.27

TOTAL SAMPLE

Fig. 9. Percentage representation of species in higher taxonomic groups among biogeographic and habitat categories as discussed in the text. Groups 13–15, and figures for the total sample (groups 1–15 and several additional groups too small to be included in their own right)

Fig. 10 a single-link cluster analysis dendrogram is supported by a linkage diagram. The latter illustrates all linkages (coefficients) down to a certain level connecting points representing, in this instance, the higher taxa.

Both figures indicate strong similarities between those groups with a high lowland forest representation of localised species, with groups with montane or open habitat specialism placed peripherally, as are those with numerous widespread taxa. The first probably represent the norm for tropical plant and animal groups: most of these will be very vulnerable if loss of lowland forest habitats is extensive.

This analysis shows clear correlation with the change of family proportions on conversion to managed systems indicated in Fig. 6. The vulnerable groups

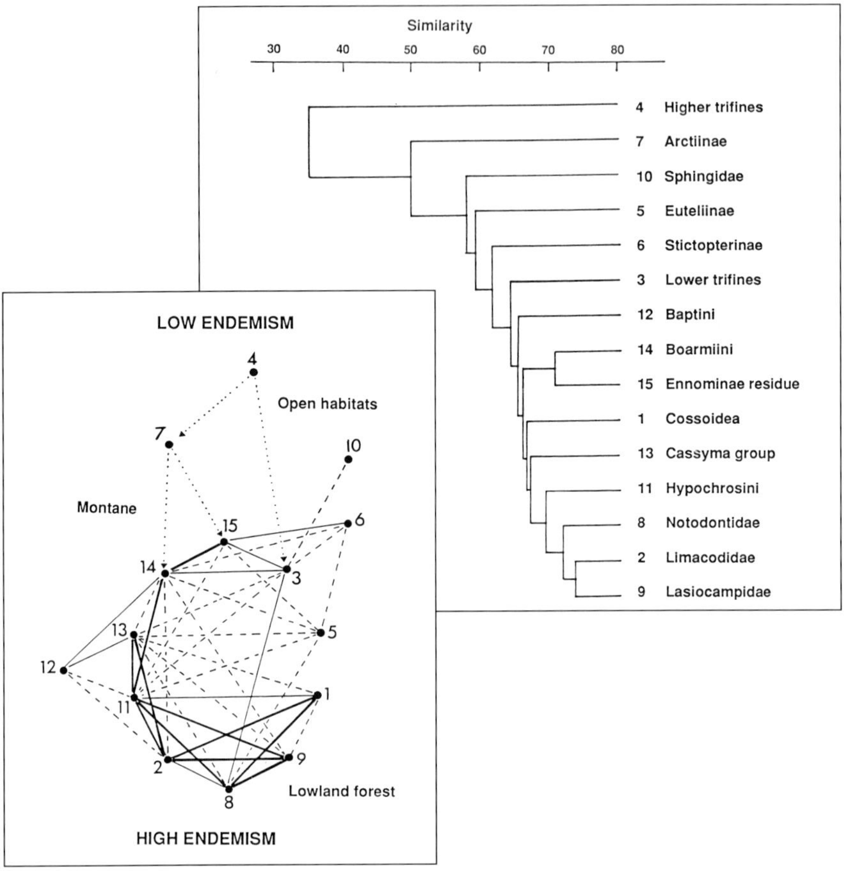

Fig. 10. Single-link dendrogram and linkage diagram classifying higher taxonomic groups (1–15 in Figs 7–9) according to pairwise similarities of percentage representation among biogeographic and habitat categories as discussed in the text. In the linkage diagram heavy solid lines indicate pairwise similarities of 66–100%, light solid lines indicate similarities ranging over 55 to 59%. Dotted lines with arrows indicate the pairwise similarities of percentages lower than 55 that associate the remaining two groups (4 and 7) with the main cluster of groups.

show decline or disappear altogether and the least vulnerable ones come to predominate or, where pest open habitat specialists are involved, to thrive.

Therefore, to maintain the quality of biodiversity in Malaysia, there is no alternative to the preservation of significant representative tracts of natural lowland forest. This need not necessarily preclude the harvesting of forest products, including timber, provided this is done in a sustainable manner with due regard to natural cycles of regeneration. At present, there is a lack of information on the effects of methods of timber extraction on the diversity of animal groups except, perhaps, for vertebrates. There are a number of proposals on the table aimed at obtaining such information for Malaysia, and both of us are collaborating with forest entomologists here to establish the pattern for the moth fauna.

The Limacodidae: a 'Vulnerable' Pest Group

This symposium is focused on pest management in relation to the environment, so the concluding section examines a group of moths that, though highlighted as potentially vulnerable in the analysis just described, nevertheless contains a significant number of serious pest species.

The family Limacodidae is well known for its impact on oil palm, coconut and other cash crop plantations in Southeast Asia. Yet, while most open habitat pest species and many of those on arboreal crops are geographically widespread, most Limacodidae are very localised geographically, a localisation that has become more apparent following the elucidation of previously unsuspected pest complexes (Cock *et al.*, 1987). In the Bornean fauna, one-third of the species are endemic and over three-quarters are restricted to Sundaland. Three-quarters are restricted to lowland forest types.

Despite their pest status, the Limacodidae are an unusual group with, in their larvae at least, undoubted aesthetic qualities. They manifest themselves as pests in sporadic outbreaks with the population apparently synchronised, sometimes even multispecific. The trigger for such outbreaks is unclear, but a catastrophic event such as climatic extreme or an insecticide application is a current hypothesis (Cock *et al.*, 1987). It is apparent that natural enemies and pathogens suppress limacodid populations for much of the time and may rapidly limit the outbreaks when they occur, reducing the need for application of chemicals (preferably by trunk injection for systemic control). Indeed, parasitoid natural enemies may resolve the paradox of "vulnerable" pest groups by maintaining high pest species diversity through suppression of tendencies in any one species to become extremely abundant (LaSalle & Gauld, 1991). They prevent potential pests from becoming actual pests.

An integrated pest management strategy would appear to recommend itself as a means of ensuring that diversity in a fascinating group of insects may be preserved while minimising its impact on man's economic activities. We must hope that similar solutions can be found for the Lepidoptera diversity that may well develop in legume softwood monocultures. The comparative study of the dynamics of associations of Lepidoptera in plantation and natural forest systems is likely to be a prerequisite for attaining such solutions.

Acknowledgements

We are grateful to Dick Vane-Wright and Gaden Robinson for their comments on an earlier draft of this paper. Our particular thanks are due to Phillipa Holloway for preparing the typescript and assisting with the illustrations. Slides for the verbal presentation were photographed by Mike Chittenden and André Berly.

References

Barlow, H.S. and Woiwod, I.P. (1989) Moth diversity of a tropical forest in Peninsular Malaysia. *Journal of Tropical Ecology* 5, 37–50.

Barlow, H.S. and Woiwod, I.P. (1990) Seasonality and diversity of Macrolepidoptera in two lowland sites in the Dumoga-Bone National Park, Sulawesi Utara. In: *Insects and the Rain Forests of South East Asia (Wallacea)* (Eds. Knight, W.J., Holloway, J.D.), pp. 167–172. Royal Entomological Society, London.

Bowman, D.J.M.S., Woinarski, J.C.Z., Sands, D.P.A., Wells, A. and MacShane, V.J. (1990) Slash-and-burn agriculture in the wet coastal lowland of Papua New Guinea: response of birds, butterflies and reptiles. *Journal of Biogeography* 17, 227–239.

Chey, V.K. (1990) *A Study of the Moth Species found in Plantation Forests in Sabah.* Unpublished Dissertation, Postgraduate Diploma in Applied Insect Taxonomy, University of Wales, Cardiff, 87 pp.

Cock, M.J.W., Godfray, H.C.J. and Holloway, J.D. (eds) (1987) *Slug and Nettle Caterpillars, the Biology, Taxonomy and Control of the Limacodidae of Economic Importance on Palms in South-east Asia*, CAB International, Wallingford.

Fisher, R.A., Corbet, A.S. and Williams, C.B. (1943) The relation between the number of species and the number of individuals in a random sample of an animal population. *Journal of Animal Ecology* 12(1), 42–58.

Holloway, J.D. (1969) A numerical investigation of the biogeography of the butterfly fauna of India, and its relation to continental drift. *Biological Journal of the Linnean Society* 1, 373–385.

Holloway, J.D. (1970) The biogeographical analysis of a transect sample of the moth fauna of Mt Kinabalu, Sabah, using numerical methods. *Biological Journal of the Linnean Society* 2, 259–286.

Holloway, J.D. (1974) The biogeography of Indian butterflies. In: *Ecology and Biogeography in India* (Ed. Mani, M.S.), pp. 473–499. W. Junk, The Hague.

Holloway, J.D. (1983–1989) *The Moths of Borneo.* Part 4 (1983), Notodontidae, *Malayan Nature Journal* 37, 1–107. Part 14 (1985) Noctuidae (part), *Malayan Nature Journal* 37, 157–317. Part 1 (1986), Key to families, Cossidae, Limacodidae, etc., *Malayan Nature Journal* 40, 1–166. Part 3 (1987), *Bombycoidea*, Southdene, Kuala Lumpur. Part 6 (1988), *Arctiidae* (part), Southdene, Kuala Lumpur. Part 12 (1989). Noctuidae (part), *Malayan Nature Journal* 42, 57–226.

Holloway, J.D. (1984) The larger moths of Gunung Mulu National Park. In: *Gunung Mulu National Park, Sarawak, Part 2* (Eds. Jermy, A.C., Kavanagh, K.P.), pp. 149–190. *Sarawak Museum Journal*, 30: Special Issue 2.

Holloway, J.D. (1987) Macrolepidoptera diversity in the Indo-Australian tropics: geographic, biotopic and taxonomic variations. *Biological Journal of the Linnean Society* 30, 325–341.

Holloway, J.D. (1989) Moths. In: *Tropical Forest Ecosystems of the World*, 14B (Eds. Lieth, H., Werger, M.J.A.), pp. 437–453. Elsevier, Amsterdam.

Holloway, J.D. (in press). Aspects of the biogeography and ecology of the Seram moth fauna. In: *The Natural History of Seram* (Eds. Proctor, J., Edwards, I.).

Holloway, J.D. and Stork, N.E. (1991) The dimensions of biodiversity: the use of invertebrates as indicators of human impact. In: *The Biodiversity of Microorganisms and Invertebrates: Its role in sustainable agriculture* (Ed. Hawksworth, D.L.). CAB International, Wallingford, UK.

Holloway, J.D., Robinson, G.S. & Tuck, K.R. (1990) Zonation in the Lepidoptera of northern Sulawesi. In: *Insects and the Rain Forests of South-East Asia (Wallacea)* (Eds. Knight, W.J., Holloway, J.D.), pp. 153–166. Royal Entomological Society, London.

LaSalle, J. and Gauld, I.D. (in press) Parasitic Hymenoptera and the biodiversity crisis. *Redia*.

MacArthur, R.H. (1972) *Geographical Ecology: Patterns in the distribution of species*. Harper & Row, New York.

Payne, J., Francis, C.M. and Phillips, K. (1985) *A Field Guide to the Mammals of Borneo*. The Sabah Society, Kota Kinabalu.

Chapter twenty-three:
Botanical Diversity in Malaysia

Salleh Mohd Nor and S.K. Yap
Forest Research Institute Malaysia
Kepong

Introduction

Biological diversity or biodiversity in short refers to the number of species existing in and adapted to an area. The greatest manifestation of biodiversity is seen in the tropical rainforests where about half of the entire world biota is found, making up about 45% of all vascular plants (Wilson, 1988). In the Asian region, the tropical forests still cover about 300 million ha amounting to about a third of the total land area (Spears, 1988). There is now global concern on the depletion of these forests as seen by the number of international fora, conventions and publications on the issue. Within each region there is also an increasing awareness among the public of the value of biodiversity and the need for conservation.

Diversity of Malaysian Flora

Malaysia, being located in the tropics has a wide range of ecohabitats within the extensive forest that still covers about 61% of its land area (Anon, 1988). There are about 12,000 species of seed plants documented (Zakri *et al.,* 1989) and new species are still being found every year. In Peninsular Malaysia alone, 8000 species are recorded with 2580 species of herbaceous plants (Kiew, 1988) and the recently completed four volumes of *Flora of Malaya* (Forest Department and Forest Research Institute Malaysia, 1972-89) document 2830 tree species (Ng *et al.,* 1990). There are about 199 species of palm indigenous to Peninsular Malaysia representing 7.7% of the global palm flora (Dransfield, 1988a). For the non-flowering plants, Holttum (1988) estimated over 500 fern species and for higher fungi alone 302 species have been recorded (Corner, 1988), while many still remain undiscovered or undescribed. In Sarawak, with a richer flora than Peninsular Malaysia, there are 231 species of Dipterocarpaceae (Ashton, 1982) and for the family Moraceae, 125 species were documented by Primack (1983).

This diversity is most evident in the lowland forests which also have the greatest number of habitat types. Enumeration of trees greater than 1 cm in diameter in a 50 ha plot in Pasoh FR gives a figure of 820 species, with an average of 210 species per ha (Kochummen *et al.*, 1990). A similar study in Sg. Menyala FR by Wyatt-Smith (1949) recorded 150 species per ha.

There is also a high degree of endemism in our flora. Ng *et al.*, (1990) recorded 26.4% of the tree flora in the Malay Peninsula to be endemic. Some of these species (18%) are so limited in distribution, that they are endangered with the gradual reduction of the habitat types. Trees of *Shorea gratissima* are now confined to pockets of coastal hills in Jugra FR, Selangor and southeastern tip of Johore while *Canarium perlisanum* has only been collected from Kaki Bukit, Perlis. The diminishing peat swamp forests are home to *S. platycarpa* and *Crytostachys renda*. The limestone hill palm flora is very restricted having only two species of *Maxburretia* that are strict calcicole elements (Dransfield, 1988a). One of the species is confined to Batu Caves and Bukit Takun while the other is found only in the Langkawi Islands. In the Malay Peninsula, 19 of the 28 bamboo species are endemic (Dransfield, 1988b). An extreme degree of endemism is observed in the families of herbs, Begoniaceae and Gesneriaceae, with 90% and 80% respectively of their species confined to the Peninsula. Some of these species are rare and narrow in distribution: for example, the whole population of *Didymocarpus primulina* is confined on a rock face, while *Loxocarpus semitorta* is endemic to Mt Ophir (Kiew, 1988).

The Value of Botanical Diversity

Only a small proportion of the diverse flora in the forests has been exploited, with about 300 of the 12,000 species harvested for utilisation (Zakri, *et al.*, 1989). The most important is the production of timber, with a gradual increase in demand for minor forest products like rattan and bamboos. The forestry sector has become the second highest revenue earner for the country. Many of the other species have potential economic importance especially pharmaceutical products which have just begun to be tapped. This means that there are still many more forest plants in the country with uses waiting to be discovered.

As an example, there are numerous wild fruit tree species in our forests bearing fruits suitable for human consumption. Studies show that 76 of the recorded 820 species of trees over 1 cm in diameter in Pasoh bear edible fruits (Saw *et al.*, 1991). Out of these, 51 species are related to the fruit trees commonly cultivated in orchards. These wild relatives make up the genetic pool from which selection for improved traits of the domesticated trees can be made.

Assessment of the value of plants in pharmaceutical production has been limited (Latif, 1985; Lowry, 1971, 1973) although a large number of plants have been recorded to have medicinal properties and have been used traditionally (Burkill, 1935). Best known will be the roots of *Eurycoma longifolia* (with the dubious function of aphrodisiac) and the bark of *Chukrasia*, *Lansium*, *Sandoricum* and *Xylocarpus* reputed to be good for treating dysentery (Burkill, 1935). Systematic analysis and testing of these species have just begun (Zakaria & Ibrahim, 1986). An alkaloid, sebiferine, isolated from the lauraceous tree *Alseodaphne perakensis* has been found to reduce sensibility to pain and also lower blood pressure in addition to being a uterine stimulant (Nordin Lajis, 1990). The same author also reported the extraction of securinine from *Breynia coronata* (Euphorbiaceae), a compound used for the treatment of paralysis.

Of greater interest to this meeting will be the role of plants in the control of pests and diseases. The complex ecosystem of the forest with its large number of species has resulted in a delicate balance between pests and hosts. Through the evolutionary process, plants have developed a series of secondary compounds to protect themselves against insects or fungi, commonly referred to as host plant resistance. Leaves of the leguminous plant, *Hymenaea courbaril*, a tree introduced into Malaysia from Central America, have been shown to contain caryophillene epoxide which is a broad spectrum fungicide (Hubbell *et al.*, 1983). Antifungal activities of *Avicennia*, *Excoecaria*, *Derris* and *Acanthus*, common mangrove plants of our coastline have also been recorded (So, 1990).

Roots from two climbers of the leguminous family (*Derris elliptica* and *D. malaccensis*), common in the Malaysian forest, are important sources of fish poison commonly known as tuba. In addition they are also used as a cheap source of insecticide. Rotenoids are insecticidal compounds extracted from *D. elliptica*. The roots of this plant can be ground into powder or the active components extracted by chemical solvents (Cremlyn, 1980).

The resistance of plants with latex to insect predation has been documented in a number of unrelated families of tropical plants, *viz.* Apocynaceae, Euphorbiaceae, Guttiferae, Moraceae and Sapotaceae (Janzen, 1985). The absence of insect borers in the living tree trunk of Jelutong (*Dyera costulata*) when compared to the large number of these predators in a log of the same species is a clear example of the effect of the latex.

Another range of plant products is the allelochemicals, which are receiving increasing attention in insect pest control. Azadirachtin, isolated from fruits and leaves of the commonly planted avenue trees, *Azadirachta indica*, has been used as an antifeedant in insect control (Gill & Lewis, 1971). Many species of plants have been screened for the presence of substances mimicking juvenile hormone or JH in short. From a common weed in Malaysia, *Cyperus aromaticus*, Toong (1981) has isolated and identified JH III which affects the metamorphosis of insects.

In addition to higher plants, there are a series of fungi which are able to control pests. Antagonistic micro-organisms which can control various plant diseases have been well documented in the Third International Conference on Plant Protection 1990. It will suffice here to re-emphasise the role of the micro-organisms. Parasitic nematodes have been successfully controlled by nematophagous fungi such as *Arthrobotys* sp. and *Dactyllela* sp. (Davide, 1990).

Many oil palm plantations have root disease caused by *Ganoderma* and studies into its possible control by using another fungus, *Trichoderma*, have been initiated. There are also several reports of the antagonism of ectotrophic mycorrhizal fungi on pine trees to root pathogenic fungi and soil bacteria (Marz, 1969; Zak, 1964).

Tan & Loke (1985) argued that the complicated chemical structure of these plant secondary products is difficult to synthesise in economic quantities. For example, organotin compounds which have properties close to the plant antifeedants are more toxic than the plant extract. This puts further emphasis on the importance of utilising the wide spectrum of plant materials available in the forest.

The Conservation of the Floristic Diversity

Realising the strategic importance of maintaining the diversity of the forest ecosystem, the Malaysian Government has, through various acts established protected areas for conservation of these resources and plans for the sustainable management of the forestry industry in the future. Despite these efforts there is the need for greater R & D input to supplement and support conservation efforts in the country.

FRIM has initiated programmes to make inventories of the resources available in the forest. As mentioned earlier, there is the documentation of all trees above 1 cm diameter in Pasoh and we will be starting to assess the floristic composition in undisturbed areas and areas that have been commercially logged in the Greater Templer Park. These studies will assist in the understanding of different ecological habitats, which may have their own specialised flora which would have to be included in the present system of protected areas. They will also provide indicators to the impact of forestry practices on the level of diversity. In addition to *in situ* conservation, many of the highly vulnerable plant species wili have to be protected through *ex situ* means. The campus of FRIM has a collection of 722 species of trees established and this list contains both indigenous as well as exotic species (Yap *et al.*, 1990).

Attempts are also being made to store propagules, especially seeds, organs or tissues. Initial success has been obtained through cryogenic storage of forest tree seeds (Krishnapillay *et al.*, 1990) but more work will be needed to develop suitable protocols for the wide spectrum of species.

It is, however, important that conservation should not be considered as keeping the resources under lock and key. It involves the sustainable use of genetic resources through proper management so as to benefit Man. In the process of land conversion or commercial exploitation, sufficient remnant patches of forests or refuges can be the source of genetic materials for the regeneration of many of the forest plants. These will also provide essential corridors between large conservation areas so as to allow occasional genetic transfers between populations. Research into the strategies of establishment and management of these refuges is needed.

Conclusion

With the wide range of species diversity present in the country it is critical that concerted efforts be made to capture the economic benefits from this biodiversity. Besides the range of pharmaceuticals and pesticides that can be extracted from the plants, there is a whole wealth of potential plant products that have to be investigated, isolated, identified and commercialised. This requires the R & D effort of both the public and private sectors. However, the current investment into R & D in this area is still limited. Much more needs to be done urgently if we are to maximise the potential benefit that can be obtained from our botanical diversity.

References

Anon (1988) *Forestry in Malaysia. An exercise on balanced landuse.* Ministry of Primary Industries, Malaysia.

Ashton, P.S. (1982) *Flora Malesiana* 1, 9 (2): 237–552.

Burkill, I.H. (1935) *A Dictionary of the Economic Products of the Malay Peninsula.* 2 volumes. Crown Agents for the Colonies, London.

Corner, E.J.H. (1988) Higher fungi. In: *Key Environments Malaysia* (Ed. Earl of Cranbrook), pp. 88–101. Pergamon Press, Oxford.

Cremlyn, R. (1980) *Pesticides Preparation and Mode of Action.* John Wiley & Sons Ltd.

Davide, R.G. (1990) Biological control of plant parasitic nematodes in the Philippines. Abstracts Third International Conference on Plant Protection in the Tropics.

Dransfield, J. (1988a) Forest palms. In: *Key Environments Malaysia* (Ed. Earl of Cranbrook), pp. 37–48. Pergamon Press, Oxford.

Dransfield, S. (1988b) Forest bamboos. In: *Key Environments Malaysia.* (Ed. Earl of Cranbrook), pp. 49–55. Pergamon Press, Oxford.

Forest Department and Forest Research Institute Malaysia (1972-89) *Tree Flora of Malaya,* 3 vols. (Eds. Whitmore, T.C. and Ng, F.S.P.) Published by Forest Research Institute, Malaysia.

Gill, G.S. and Lewis, G.T. (1971) Systemic action of an insect feeding deterrant. *Nature* 232, 402–403.

Holttum (1988) Ferns. In: *Key Environments Malaysia* (Ed. Earl of Cranbrook), pp. 77–87. Pergamon Press, Oxford.

Hubbell, S.P., Wiemer, D.F. and Adejare A. (1983) An antifungal terpenoid defends a neotropical tree (*Hymenaea*) against attack by fungus-growing ants (*Atta*). *Oecologia* 60, 321–327.

Janzen, D.H. (1985) Plant defences against animals in the Amazonian rainforest. In: *Key Environments Amazonia* (Eds. Prance, G.T and Lovejoy, T.E), pp. 207–217. Pergamon Press, Oxford.

Kiew, R. (1988) Herbaceous flowering plants. In: *Key Environments Malaysia* (Ed. Earl of Cranbrook), pp. 56–76. Pergamon Press, Oxford.

Kochummen, K.M., LaFrankie, J.V. and Manokaran, N. (1990) Floristic composition of Pasoh Forest Reserve, a lowland rain forest in Peninsular Malaysia. Abstracts of the International Conference of Tropical Biodiversity.

Krishnapillay, B., Marzalina Mansor and Yap S.K. (1990) Is long-term storage of seeds possible? Pre-proceedings of the Conference on Forestry and Forest Products Research CFFPR 1990 pp. 210–216.

Latif, A. (1985) Ethnobotanical and phytochemical studies of Malaysian Menispermaceae. *Malays. Applied Biolo.* 14, 68–75.

Lowry, J.B. (1971) Conserving the forest, a phytochemical view. *Malay. Nat. J.* 24, 225–230.

Lowry, J.B. (1973) Phytochemical prospecting and conservation of lowland tropical rain forests. In Symposium on Planned Utilization of Lowland Tropical Forests, Cipayung, 1971. Bogor.

Marz, D.H. (1969) Antagonism of mycorrhizal fungi to root pathogenic fungi and soil bacteria. *Phytopathology* 59, 153–163.

Ng, F.S.P., Low, C.M. and Mat Asri Ngah Sanah (1990) Endemic trees of the Malay Peninsula. Research Pamphlet No. 106. Forest Research Institute of Malaysia.

Nordin Lajis (1990) Chemical products from the forest. Pre-proceedings of the Conference of Forestry and Forest Products Research, CFFPR 1990 pp. 236–240.

Primack, R.B. (1983) *Forester's guide to the Moraceae of Sarawak.* Forest Department, Sarawak.

Saw, L.G., LaFrankie, J.V., Kochummen, K.M. and Yap, S.K. (1991) Fruit trees in a Malaysian rain forest. *Economic Botany* 45(1), 120–136.

So, M.L. (1990) Antifungal activities of mangrove plants. Abstracts of the Third International Conference on Plant Protection in the Tropics, p. 165.

Spears, J. (1988) Preserving biological diversity in the tropical forests of the Asian region. In: *Biodiversity* (Ed. Wilson, E.O), pp. 393–402. National Academy Press, Washington, DC.

Tan, K.H. and Loke, W.H. (1985) Insect growth regulators and semiochemicals: novel chemicals for insect pest management. In: *Integrated Pest Management in Malaysia* (Eds. Lee, B.S., Loke, W.H. and Heong, K.L.), pp. 105–128. The Malaysian Plant Protection Society.

Toong, Y.C. (1981) Isolation and identification of JH III from *Cyperus aromaticus*. First Asian Chemistry Conference. Singapore. 1981.

Wilson, E.O. (1988) The current state of biodiversity. In: *Biodiversity* (Ed. Wilson, E.O), pp. 3–18. National Academy Press, Washington, DC.

Wyatt-Smith, J. (1949) A note on tropical lowland evergreen rainforest in Malaya. *Malay. Forester* 12, 58–64.

Yap, S.K., Ratnam, L., Chua, S.L., Lesmi Tipot, Krisnapillay, B. and Marzalina Mansor (1990) Conservation of Malaysian forest genetic resources: current developments and future approaches. Pre-proceedings of the Conference on Forestry and Forest Products Research CFFPR 1990, pp. 201–216.

Zak, B. (1964) Role of mycorrhiza in root disease. *Ann. Rev. Phytopath.* 2, 377–392.

Zakaria, M.B. and Ibrahim, H. (1986) Phytochemical screening of some Malaysian species of Zingiberaceae. *Malaysian J. Sci.* 8, 125–128.

Zakri, A.H., Saw, L.G. and Rajanaidu, N. (1989) National conservation strategies of plant genetic resources in Malaysia. In: *Genetic Resources of Under-utilised plants in Malaysia* (Ed. Zakri, A.H.), pp. 161–179. Malaysian National Committee on Plant Genetic Resources.

Chapter twenty-four:
Zoological Biodiversity in Malaysia

B.H. Kiew
Department of Zoology
University of Malaya
59100 Kuala Lumpur
Malaysia

Summary

Malaysia has a rich and diverse fauna. The richness and diversity of the fauna is due to the prevailing climatic conditions present in the area over a long period of evolution. The measure of richness and diversity is not fully known owing to lack of appropriate expertise, lack of interest in the non-economic species and lack of a workable methodology which would provide meaningful results with limited manpower and time. The fauna of Malaysia is under threat: from erosion of overall numbers and species. The approach taken by local conservation groups is to try to preserve sufficient viable and representative areas of habitat types present in the country. Agriculturalists can do their part in maximising the use of altered habitats already under human management and reducing their demands on natural habitats.

Introduction

Malaysia is fortunate in that its vertebrate fauna has been relatively well documented. The mammals have been published by Davies & Payne (1982), Harrison (1966, 1973), Medway (1969, 1977), and Payne *et al.*, (1985). The birds have been documented by Madoc (1956), Medway & Wells (1976), and Smythies (1981). The amphibians have been recorded by Berry (1975), Boulenger (1912), Smith (1930), Inger (1966) and Inger & Stuebing (1989). The reptiles have been reported in Tweedie (1983). The fishes have been published in Scotts (1957), Mohsin & Azmi (1983) and Inger & Chin (1962).

The invertebrates are less well documented. The butterflies have been documented by Corbet & Pendlebury (in press), and Fleming (1974). Corals have been reported in Betterton (1981), Searle (1956) and Wood & Tan (1987). Most of the animals which are of no economic or aesthetic importance have yet to be compiled into a single volume for easy reference. Numerous specialised articles are available in the scientific literature scattered in scientific journals published all over the world. It will take a dedicated individual considerable time and resources to gather all of them for a complete compilation.

No attempt has been made to calculate the index of diversity of the Malaysian fauna. The richness and diversity of the fauna is best reflected in the checklists of species of the various groups of animals compiled for baseline evaluation for conservation purposes in Kiew (1982, 1984a,b) and Kiew & Davison (1982).

Zoological Diversity in Malaysia

There are 331 known species of mammals in Malaysia. There is a total of over 600 species of birds recorded within the political boundaries of the country. The number of species of snakes known is 168. The lizards consist of more than 140 species and 27 species of tortoises and turtles can be found in the sea, rivers and on the land in the country. There are three species of crocodiles recorded in the country. The freshwater has over 300 species of fishes while the number of species of fish in the sea easily exceeds 2000.

It is difficult to provide numbers for the various groups of invertebrates. The best studied, the butterflies, have over 1000 species with the moths estimated as over 8000 (Barlow, pers. comm.). There are recorded 146 species of corals (Betterton, 1981). There have been few attempts to provide an index of diversity for studies of specific habitat of specific groups of animals. Lloyd *et al.* (1968) found the rainforest of Nanga Tekalit in the Third Division of Sarawak to have the species diversity measured by the Shannon function H' as 3.301 for frogs, 3.096 for lizards and 2.856 for snakes producing a total of 4.265. Barlow & Woiwod (1989) have attempted to estimate the index of diversity for moths in Malaysia.

Reasons for Richness and Diversity of Fauna

There are many basic reasons for the richness of the Malaysian fauna. Situated on two separate land masses, Peninsular Malaysia is linked with Thailand. East Malaysia is on the west coast of Borneo. Thus Malaysia has the advantage of having two separate units for independent biological evolution on land, surrounded by a vast expanse of warm shallow sea. West Malaysia has its own endemics resulting from evolution *in situ* as well as faunistic intrusion from the Indochinese sub-region and the island of Sumatra. East Malaysia, being on the second largest island in the world has the advantage of space and habitat diversity to provide the spatial isolation required in the evolution of animal population and species.

The warm and gentle climate which does not select species with any severity also contributes to the production of a rich fauna generated by the inherent biological potential for change and adaptation. The energy from the sun which provides the warmth also provides for higher productivity in the biological systems in operation. This means more growth and greater reproduction which ultimately leads to faster evolution. The differences in the local physical environment in terms of topographical terrain which provides light and energy availability gradients, water availability and stress gradients, contribute to the creation of more niches. This enhances selection and adaptation. The varying mosaic patterns of distribution of minerals and rainfall also create more axes for differentiation causing increases in the permutation of numbers of niches. The different characteristics of the three main ecosystems, land, water and air, set the basic three spheres for evolution and specialisation among animals. The existence of osmotic gradients in water also calls for further adaptation and the evolution of new life forms in answer to these environmental stresses.

With the successful profusion in the evolution of plant life in the tropics, the animal life which is dependent on it has also evolved equally extensively and intensively to survive and overcome the defence mechanisms developed by the plants. This has provided an increase in the adaptation and development of zoological biodiversity. Host specificity is not an unusual phenomenon in the tropics among parasites, saprophytes as well as herbivores. The location of Malaysia in an archipelago further enhances evolution, through isolation of animal populations providing the mechanism for the creation of races and sub-species, the precursors for the formation of valid biological species. Such a phenomenon is evident from the large number of races and sub-species, with sibling and cryptic species found among animals in Malaysia. Of the 600 species of birds in Malaysia some 499 have recognised sub-species. Of the 331 species of mammals, 155 have sub-species.

In summary, the richness of zoological diversity is a factor of time, space, habitats, food plants, interacting biological entities, light gradients, water availability, mineral factors, osmotic variation, topographical variation and inherent biological potential for genetic and phenotypic changes of all living things in a favourable environment.

Problems in the Study of Zoological Diversity

What is the actual zoological biodiversity in Malaysia? It is hard to tell. The problem is the lack of a sound, practical, effective method for accessing the diversity and the limited number of proficient taxonomists available in the country and the world who are interested in such tasks.

A number of indices of diversity are available for use in the literature. The most popular index is associated with the name Shannon such as the Shannon–Weaver index. The other commonly used index is the Simpson's index. Both these indices require the listing of the species present with their number. The full data required for the calculation of these indices are difficult to obtain in short surveys with limited manpower and expertise. In Malaysia, the zoological diversity of an area is still commonly indicated by just the species list which is found to grow with the size of an area surveyed and the manpower and time put into the work.

Threats to the Zoological Diversity in Malaysia

With the current rate of development, the zoological biodiversity in Malaysia is under threat and in need of protection and preservation. This is evident from the rate of loss of the lowland rainforest and coastal habitats which house more than 70% of the Malaysian fauna. Many areas of lowland forest under some form of protection are observed to be eroded. Many quickly become unviable to sustain populations of larger animals and birds.

The threat to the fauna is chiefly due to the loss of viable habitats, where viability is a function of size. The demand for land for conversion to agriculture and for mining and urban development has contributed considerably to the loss of many natural areas. This erosion of natural areas housing the Malaysian fauna is unacceptable, if vast areas already altered and put to use by man are left idle and unproductive while claims to excise more natural areas are said to be for increasing productivity. Logging, although disruptive and destructive at the time, is less of a threat as the natural forest can be allowed to regenerate, if managed, for sustained yield production and wildlife protection.

The conservation of Malaysia's rich diverse fauna could be best achieved through the preservation of an adequate network of viable units of its diverse ecological habitats. Attempts have been made in the preparation of a Blueprint for Conservation by the Malayan Nature Society and a National Conservation Strategy by World Wide Fund for Nature Malaysia to be presented to the government for the formulation of a workable master plan towards this end.

We can only hope that our government will receive the ideas favourably and put into effective action the proposals needed for the conservation of Malaysia's rich, diverse fauna in the immediate future.

References

Barlow, H.S. and Woiwod, I.P. (1989) Moth diversity of a tropical forest in Peninsular Malaysia. *Journal of Tropical Ecology* 5, 37–50.

Berry, P.Y. (1975) *The Amphibian Fauna of Peninsular Malaysia.* Tropical Press Kuala Lumpur, 130 pp.

Betterton, C. (1981) A guide to the hard corals of Peninsular Malaysia. *Malay. Nat. J.* 34, 171–336.

Boulenger, G.A. (1912) *A Vertebrate Fauna of the Malay Peninsula from the Isthmus of Kra to Singapore Including the Adjacent Islands. Reptilia and Batrachia.* Taylor and Francis, London. 294 pp.

Corbet, A.S. and Pendlebury, H.M. (in press) *The Butterflies of the Malaya Peninsula.* 4th edn. J.N. Eliot. Malayan Nature Society, Kuala Lumpur. 578 pp.

Davies, G. and Payne, J. (1982) *A Faunal Survey of Sabah.* World Wildlife Fund, Malaysia, Kuala Lumpur. 279 pp.

Fleming, W.A. (1974) *Butterflies of West Malaysia and Singapore.* Longman, Kuala Lumpur. 2 vols.

Harrison, J. (1966) *An Introduction to Mammals of Singapore and Malaya.* Tien Wah Press, Singapore. 340 pp.

Harrison, J. (1973) *An Introduction to the Mammals of Sabah.* Sabah Society, Kota Kinabalu. 244 pp.

Inger, R.F. (1966) The systematics and zoogeography of the amphibia of Borneo. *Fieldiana: Zool.* 52, 1–402.

Inger, R.F. and Chin, P.K. (1962) The freshwater fishes of North Borneo. *Fieldiana: Zool.* 45, 1–268.

Inger, R.F. and Stuebing, R.B. (1989) *Frogs of Sabah.* Sabah Parks, Kota Kinabalu Publ. No. 10, 129 pp.

Kiew, B.H. (1982) Conservation status of the Malaysian fauna: I Mammalia. *Malay. Naturalist.* 35(3), 3–19.

Kiew, B.H. (1984a) Conservation status of the Malaysian amphibians. *Malay. Naturalist.* 37(4), 6–10.

Kiew, B.H. (1984b) Conservation status of turtles, terrapins and tortoises. *Malay. Naturalist.* 38(2), 2–3.

Kiew, B.H. and Davison, G.W.H. (1982) Conservation status of the Malaysian fauna: II Birds. *Malay. Naturalist.* 36(2), 2–34.

Lloyd, M., Inger, R.F. and King, F.W. (1968) On the diversity of reptile and amphibian species in a Bornean rain forest. *Am. Nat.* 102, 497–515.

Madoc, G.C. (1956) *An Introduction to Malayan Birds.* Malayan Nature Society, Kuala Lumpur. 234 pp.

Medway, Lord (1969) *The Wild Mammals of Malaya and Offshore Islands including Singapore.* Oxford University Press, Kuala Lumpur. 127 pp.

Medway, Lord (1977) *Mammals of Borneo.* Malaysian Branch of the Royal Asiatic Society, Kuala Lumpur. Special Publication. 172 pp.

Medway, Lord and Wells, D.R. (1976) *The Birds of the Malay Peninsula. Vol. 5. Conclusion and Survey of Every Species.* Witherby, London.

Mohsin, A.K.M. and Ambak, Mohd. Azmi (1983) *Freshwater Fishes of Peninsular Malaysia.* Universiti Pertanian Malaysia Press. 284 pp.

Payne, J., Francis, C.M. and Phillips, K. (1985) *A Field Guide to the Mammals of Borneo.* Sabah Society and World Wildlife Fund (Malaysia).

Scotts, J.S. (1957) *An Introduction to the Sea Fishes of Malaya.* Government Press, Kuala Lumpur. 180 pp.

Searle, A.G. (1956) An illustrated key to Malayan Hard Corals. *Malay. Nat. J.* 11, 1–26.

Smith, M.A. (1930) The reptilia and amphibia of the Malay Peninsula. *Bull. Raffles Mus.* 3, 1–149.

Smythies, B.E. (1981) *The Birds of Borneo.* Sabah Society and Malayan Nature Society. 473 pp.

Tweedie, M.W.F. (1983) *The Snakes of Malaya.* Singapore National Printers. 167 pp.

Wood, E.M. and Tan, B.S. (1987) Hard coral—The coral reefs of the Bodaya Islands (Sabah: Malaysia) and Pulau Sipadan. *Malay. Nat. J.* 40 (3&4), 189–224.

Chapter twenty-five: Biodiversity and pest management *Discussion*

Chairman: Dr J.K. Waage

Dr Maryati Mohd. (UKM) It is impossible to work with SE Asian fauna without constant reference to overseas centres of taxonomic expertise. I would like to propose the setting up of a biosystematic centre in this region. May I ask what international assistance could be available?

Dr Holloway First, I would like to thank Dr Maryati for advance notice of her question. It is a very important question so I appreciate being given time to prepare a response. I would then like to pass it on to Professor Greenland as it has an international perspective in its relevance to efforts in hand to bring more cohesion to work in biosystematics at a global level.

It is a question that I know strikes a chord with many Malaysian scientists, as it does with me and my co-author, Henry Barlow. Indeed we once published a paper exploring various aspects of helping biosystematics institutions to develop in tropical countries (Holloway & Barlow, 1983).

The answer lies first and last within Malaysia. Most ingredients are here, perhaps to a greater extent than in most tropical nations, though it should be noted that Indonesia already has such an institution. There is a highly sophisticated scientific community that includes internationally recognised biosystematists. There is a strong and growing tradition for natural history, epitomised by the Malayan Nature Society. Its activities range over scientific exploration, publication and education. Its president and another of its most active members are sitting on the platform with me. Education is another key, and this country's system is producing numbers of highly motivated young biologists. I am delighted to find I may be helping two of them with doctoral projects that involve a major input of biosystematics. A further ingredient is demand: integrated pest management requires sophisticated systematics.

Outsiders like myself, and the organisations to which we belong, can help in two main ways: in training and in transfer of information. In training, Malaysia is again a world leader, thanks to Dr Maryati, the Universiti Kebangsaan, and many other collaborating institutions who have joined with CAB International to mount two highly successful courses in insect systematics for Asia and the Pacific. On Friday I will be meeting them to plan for the third in 1993. By the year 2000 we may well have done five with a total throughput of over 100 trainees from as far afield as Germany and Kiribati.

By transfer of information I mean the collating of biosystematic data in existing major centres of biosystematic activity, such as the Natural History Museum in London, and making them available out here in the form of user-friendly guides to the flora and fauna, keys and databases, both as hard copy and in electronic form. We thus help to provide both skills and tools. I find it particularly gratifying that Malaysians who have participated in our training courses use my guides to the Bornean moth fauna, published by the Malayan Nature Society, to set up their own institutional reference collections: at MARDI, the UKM campuses and at the Forest Research Institutes here and in Sabah. It is the more satifying because many of the ingredients I mentioned earlier are coming together.

I see two important ingredients as missing at present. First, a consensus must evolve among Malaysian biologists on the form an institute for biosystematics might take. Certainly it must seek to co-ordinate the many growing points of biosystematics in the country without stifling them. Secondly, there must develop firm political support for the concept so that the emerging institution is adequately staffed and equipped. This political ingredient is the most important of all, but it may rest with Malaysian scientists to ensure a biosystematic institute is on the political agenda as a goal for the year 2000. I would also hope that its strengths would be shared more widely in southeast Asia, building on the spirit of cooperation and desire to share information and experience that is always evident among trainees on the UKM/CABI systematics courses.

Dr Salleh I entirely support this proposal. One of our problems has been the lack of local taxonomists.

Dr Greenland In response to the question by Dr Maryati I would like to confirm that CABI fully recognises the need to collaborate with national programmes to help improve their natural collections, and capabilities on biosystematics. We believe that the importance of biosystematics and taxonomic research, which underpins all of biological science, has received too little recognition. We are endeavouring to draw international attention to this problem, and to contribute ourselves through training and in other ways. We recognise that Malaysia has a special interest in the biodiversity existing in its richly endowed forests, and already has special strengths in taxonomic work. We have been pleased to collaborate with Malaysian organisations in providing training courses for the region, and hope that we may be able to assist in further strengthening of Malaysia's potential to create a regional centre of excellence on biosystematics.

Reference

Holloway, J.D. and Barlow, H.S. (1983) The role of taxonomy reference works and insect collections in tropical enology. *Antenna* 7(2), 50–53.

PART 7
New Technologies

Chapter twenty-six: **Biological Control in the Year 2000**

J.K. Waage
International Institute of Biological Control
Silwood Park, Buckhurst Road
Ascot, Berks SL5 7TA, UK

Introduction

Biological control is the use of living organisms as pest control agents. The year 2000 will see its eighteenth century of application in Asia. In a Chinese document dated AD 304, a reference exists to the sale of ant nests in markets for the control of pests on citrus (Yang, 1984). These ants were the red tree ant, *Oecophylla smaragdina*, which make arboreal nests and feed on insects in the tree canopy. From the seventeeth century, published records indicate that citrus growers in what is now Guangdong Province had a well-established method for manipulating these nests in orchards so as to increase pest control.

The method is still in use today, in a somewhat more sophisticated form: concrete "moats" are constructed around trees to confine ants to the tree canopy and bamboo sticks strung between trees expand the ants' foraging areas and with them, the area of pest control. In the past few years, scientists in Guangzhou have developed mathematical simulation models for citrus pests in which the effects of such predation can be evaluated, and more importantly, predicted for the first time (Anon, 1989). Implicit in this story are the two elements which will characterise the development of biological control into the next century: *technological innovation, and a strengthening of traditional appreciation for the value of natural enemies in crops.*

To understand the present position of biological control, we must look to its recent, rather than its ancient history. The popularity of biological control in the early part of this century was dealt a severe blow by the advent of synthetic organic pesticides in the 1950s. When compared with these new products, natural enemies were seen as inefficient and unreliable. Mounting problems with pesticide resistance and environmental pollution led scientists in the 1970s to develop the concept of integrated pest management, or IPM. This was an attempt to free pest management from the domination of chemical control by the addition of other technologies, among them biological control.

Today, many scientists acknowledge the potential value of biological control in IPM, but relatively little thought has been given to what kinds of biological control we need to develop in this context, or how we will truly integrate them with other control methods. More importantly, almost 40 years of preoccupation with chemical pesticides have left most scientists, industrialists, policy makers and farmers poorly informed about the role of natural enemies and unconvinced of their power in pest management. Before the technology of biological control can be advanced and implemented, it will be necessary to re-establish this understanding.

Revealing the Power of Natural Enemies

An understanding of the potential value of natural enemies can be gained from a study of population dynamics in predator–prey or parasite–host systems. Simple mathematical models of these systems reveal how a pest's rapid rate of population growth can be transformed into a force for its own suppression, through the action of a natural enemy which not only kills pests, but reproduces itself in the process of doing so. Theory shows that natural enemies which are efficient at killing pests over a range of pest densities (functional response), and show a high reproductive rate over those densities (numerical response) can cause substantial depression of pest population numbers (Hassell, 1978). A degree of variation in the pattern of this mortality, such as might be caused by non-random search or the ability of some pest individuals to avoid discovery, will cause this interaction to be stable, that is, to persist with little fluctuation in pest numbers over time (Pacala *et al.*, 1990).

Unfortunately, population dynamic theory is not easily appreciated by the non-professional. Indeed, the reason that we use models in biological control is to understand outcomes of population interactions which are too complex to be intuitively obvious. To reveal the power of biological control to a sceptical agricultural community, hard evidence is required. Today, this evidence comes from two primary sources: *classical biological control* and the phenomenon of *pest resurgence*.

Classical Biological Control

The introduction of exotic natural enemies for control of pests, or classical biological control, provides a striking demonstration of the potential impact of natural enemies.

To give a recent example, in the early 1980s, the Asian mealybug became a devastating pest of citrus and mango throughout West Africa (Agounké *et al.*, 1988). A classical biological control programme was initiated, which involved exploration by IIBC in India and southeast Asia, for the natural enemies of this pest which, incidentally, was rare throughout the Asian region. A large complex of natural enemies were found—quite possibly the cause of the pest's scarcity in its area of origin.

A host-specific parasitic wasp, *Gyranusoidea tebygi*, was introduced in 1988, with rapid and dramatic results. Spreading at up to 100 km per year, this tiny parasitoid caused dramatic declines in mealybug populations from outbreak levels, to stable non-damaging numbers (Agricola *et al.*, 1989). This change in pest abundance is a measure of the impact of this agent, and it is a permanent effect. The classical biological control of *Rastrococcus invadens* is the latest in a series of successful programmes to control exotic insects with their natural enemies. Over the past century, over 4000 introductions have been made of exotic insect natural enemies against insect pests, and about 30% of these have contributed substantially to the pest's control. Characteristically, successful exotic agents can reduce insect pest population levels by two to three orders of magnitude (Beddington *et al.*, 1978). In Asia, perhaps the most significant recent success has been the control of *Plutella xylostella* by exotic parasitoids in highland vegetable growing areas (see Ooi, this volume).

Classical biological control has also been successfully undertaken against weeds. In Asia, the most notable recent success has been the control of water fern in Sri Lanka and Papua New Guinea by the introduction of the South American weevil, *Cyrtobagus salviniae* (see Whitten, this volume). Overall, about 1000 introductions have been made of plant feeding insects and pathogens, with about a 40% rate of success.

For the most part, classical biological control has been targeted against exotic pests, as in the examples I have just given. By the next century, I believe we will see a substantial increase in classical biological control programmes against indigenous pests. This involves finding natural enemies from the same pest elsewhere, or from other pest species, which can fit into its natural enemy complex and augment existing mortalities (Carl, 1982).

Nowhere is the potential for this new initiative more apparent than in Asia, where the many geographical barriers between agricultural regions have left pest species in certain areas without effective natural enemies present in others. Thus, a study of the Asian Corn Borer, *Ostrinia furnicalis*, in the Philippines revealed a low level of egg parasitism, in contrast to the European, *O. nubilalis*. The egg parasitoid from this species, *Trichogramma evanescens*, was introduced into the Philippines and is now established (Tran *et al.*, 1988). It now contributes to the control of this indigenous Asian pest.

Other indigenous species which are the target of recent or current introductions of exotic agents include:

- the rhinoceros beetle, *Oryctes rhinoceros*, against which a Malaysian baculovirus has been released on Pacific and Indian Ocean islands (e.g. Zelasny *et al.*, 1990);
- the vegetable pest, *Helicoverpa armigera*: parasitoids from Europe have been established in New Zealand (Cameron, 1989), and parasitoids from Pakistan are being introduced into Myanmar (A.I. Mohyuddin, pers. comm.);

- stem borers of sugar cane (*Chilo* spp.): parasitoids from southeast Asia have been established in Pakistan (Mohyuddin, 1990).

Through these, and more traditional approaches to classical biological control, agricultural scientists are becoming increasingly aware of the power of exotic natural enemies to permanently suppress pest populations. It is tempting to see classical biological control as an unusual, one-off event, but this misses an important point: *there is no fundamental difference between these successes with exotic natural enemies and the action of indigenous species in our local crops.* The striking before and after picture of classical biological control simply isolates a process which is going on around us all the time, and largely undetected. What is revealed is an ecological phenomenon which is relevant to all pest management.

Pest Resurgence

The second phenomenon which reveals clearly the power of biological control is that of pest resurgence, in which pesticide application actually aggravates pest problems, or causes new ones, due to the elimination of important natural enemies. While classical biological control reveals the power of natural enemies by adding them to a pest population, resurgence does the same by the reverse process, by removing them with pesticides.

Resurgence is only one possible outcome of the interaction between pesticides, pests and natural enemies (Waage, 1989). Some of these may even be complementary, but in general, where natural enemies are important to pest regulation, pesticides which kill natural enemies exchange free natural pest control for expensive chemical control and, at worst, cause resurgence.

This is particularly true when pesticides are, for some reason, relatively ineffective against the pest itself. This may occur if the pest which resurges is not the target of the pesticide, which is inappropriate for that insect (e.g. organophosphate insecticides applied against fruitflies which cause resurgence of scale insects). Another major contributor to resurgence is pesticide resistance. Resistance, combined with the elimination of natural enemies, invites farmers on to the famous "pesticide treadmill" (Hansen, 1987). Growing resistance leads farmers to ever-increasing application, thereby causing more pest resistance and killing more natural enemies. In the end, the crop system collapses, as has occurred recently with cotton in some parts of the world.

In Asia, resurgence of insect pests is most frequently found in plantation crops, primarily with lepidopteran pests, on fruit crops with homopteran pests, and on several field crops with lepidopteran and homopteran pests. The most striking case of resurgence in Asia in recent years has been the outbreak of rice brown plant hopper, *Nilaparvata lugens*, on rice treated with broad spectrum insecticides. Research by national and international organisations has revealed that a major cause of this resurgence is the elimination of general predators, such as spiders, by these pesticides. Establishment of spray thresholds, and the use of selective pesticides restores this important, natural biological control. The value of conserving these poorly known natural enemies translates easily into economic terms. In Indonesia for instance, farmers who limited sprays and relied on these natural enemies fared far better than those using conventional methods of regular spraying. They sprayed their crops less often, saved money on pesticide purchase, and made less demands on government subsidies. Further, they got increased yields into the bargain (see Gallagher, this volume).

Looking into the next century, our growing understanding of the ecology of resurgence will aid scientists in predicting its occurrence and avoiding measures which encourage it. Pesticide resistance management, to which I will return at the end of this paper, will be an important element in this process.

Developing an Appreciation for Biological Control

The examples of classical biological control and resurgence provide us with the means to create an appreciation for the general role of natural enemies in pest management. The dramatic consequences of adding or removing natural enemies from a crop system reveal the unseen power of natural enemies in crops. Imparting this appreciation to a public raised with a pesticide mentality is an essential prerequisite to the acceptance of biological control's new technologies in the next century.

This process involves communication between scientist and non-scientist, on several levels. While only just begun, such communication will be a major feature of biological control in the next decade. Communication at the policy level is aimed at convincing decision makers in government of the wisdom of IPM, its economy and long-term environmental benefits. In Asia, this process is perhaps more advanced than in any other part of the world. National IPM policies exist in Philippines, Indonesia, China and India, and governments like Indonesia have taken major steps to reduce use of broad spectrum pesticides where they cause resurgence (see Gallagher, this volume).

Communication with the general public is a popular way of maintaining pressure on decision makers. Concern about pesticide effects is presently generating public interest in biological control. Beyond newspaper and magazines, introducing concepts of IPM and biological control into secondary and tertiary school curricula reaches an audience eager to find solutions to tomorrow's environmental problems. Notably, the rice IPM story, mentioned above has recently been used in the UK in educational material for secondary schools (Fullick and Fullick, 1990).

Communication with the farmer is, however, the immediate and the ultimate aim. Farmers are the true decision makers in pest management. The challenge to communicate biological control and IPM to farmers has been met in different ways. In Asia, field methods have been developed to show biological control to rice farmers, including small experiments on the exclusion of natural enemies, to generate a kind of pest resurgence. Recently, a manual on these methods has been published in Bahasa Malaysia (Ooi *et al.*, 1991). It is the aim of initiatives like these to ensure that the message of biological control comes across in a context appropriate to the farming community, which has the power both to understand and to implement it.

The Challenge to the Researcher

While examples from classical biological control and resurgence provide us with tools to change public opinion about the power of biological control, scientists are today challenged by the need to evaluate the action of natural enemies in situations where their effect is not so obvious. Adding or removing natural enemies from crops often reveals clear changes in pest numbers, but what about the action of the natural enemies already present in the crop? If they are important, how will we know, without the complex experiment of removing them?

An understanding of the impact of natural enemies in crops is essential to the development of IPM. If we know which natural enemies are contributing the most to pest depression at critical times, then we can select them for conservation measures or augmentation (see below). We can test pesticides to determine which are least harmful to these species and, generally, we can develop quantitative methods to evaluate the effect of IPM measures on the natural enemies which count. The methodology of natural enemy evaluation has hardly developed over the past 50 years, another legacy of the pesticide era. Fortunately, in the past few years, it has become one of the most rapidly growing areas of biological control research. The experience of this recent effort is that the more we study the problem, the more complex it becomes.

Nonetheless, much progress has been made in improving methods to assess parasitism by parasitoids (van Driesche *et al.*, in press) and to understand the action of natural enemies in relation to other mortalities through life table analysis (Bellows *et al.*, in press). With the aid of computers and expert systems, some quite powerful analytical methods are likely to be accessible to the IPM specialist over the next decade. Armed with improved methodology, the IPM specialist has several options for improving the contribution of important natural enemies in crops, be they native or exotic.

The Conservation of Natural Enemies

Conservation involves modifying cropping practices to improve the action of biological control agents. These practices may include the destruction of crop residues, cultivation, and the pattern and timing of planting. However, pesticide use is the cropping practice whose modification holds the greatest promise for improving the impact of natural enemies over the next decade. This is not only because of the strong antagonisms between many pesticides and biological control agents, but because pesticide technologies are relatively new to most farming systems: without a long tradition they are more easy to change.

Much attention has been paid in the last two decades to the value of increasing crop diversity to enhance natural enemy action. Traditional and experimental patterns of intercropping have been studied and compared to monocultures, to determine effects on pests and natural enemies. Effects on natural enemy numbers and diversity have been variable, and little effort has been made to determine the important natural enemies and to evaluate intercropping effects on these. However, there is evidence that, with some crops and pests, increasing crop diversity improves pest control through enhancing natural enemy action (Russell, 1989). For instance, in sugar cane in Pakistan, intercropping with clover gave equal yields to that without clover, but sampling and life table studies revealed higher natural enemies and mortality from them in intercropped fields, with a consequently lower cost to the farmer in pesticide application (Mohyuddin, 1990).

In future, efforts at conservation will be based increasingly on an understanding of the important natural enemies in a crop, and targeted at improving their action, rather than increasing natural enemy diversity and abundance generally. This will only improve the chances for successful conservation of natural enemies.

Augmentation of Natural Enemies

Sometimes the conservation of natural enemies is not sufficient to increase their useful contribution to pest management. This may be due to lack of resources in the modern crop monoculture, or the lack of continuity in such resources in seasonal crops. In such instances, methods exist for the *augmentation* of natural enemies, or the addition of natural enemies to crops for a short-term effect over, or within, a season. Adding insect natural enemies to crops has a long tradition, centred largely on the use of the egg parasitoids, *Trichogramma*. These insects are usually mass produced and mass released into crops at regular intervals.

Recent research in China, however, reveals that small, strategic releases of *Trichogramma* early in the crop season can give the same level of success as frequent mass releases: in one case, better control of sugar cane borers was achieved with early season releases of only one-twentieth of the number of wasps used conventionally in mass releases (Guo, 1988). This new approach, like future trends in conservation mentioned above, depends on an understanding of the action of natural enemies in crops, such that this can be predicted, and the right number introduced at the right time. It is another example of the opportunity for ecological research to substantially improve the cost efficiency and success of biological control in the decades to come.

Further, with this approach to augmentation, the need to produce fewer natural enemies may make it economical to augment a much wider range of species in the future. Improvement in insect mass production, and possibly in artificial diets for parasitoids and predators, may also allow new species of insect natural enemies to be augmented commercially in future.

Development of Biopesticides for Crop Protection

Biopesticides are the most rapidly growing technology of augmentation, indeed of biological control in general. They rely on the action of pathogens of insects, weeds and plant diseases which, although highly virulent, do not spread themselves in crops, and hence are appropriate for mass production and targeted release. The recent history of biopesticide development is one of promise and disappointment. Fuelled by the anticipation of IPM markets, and by the impressive virulence of pathogens in laboratory trials, a number of biopesticide products were developed in the 1970s which failed to survive in the marketplace. The problems were two-fold. First, the markets were not easy to penetrate, IPM had not taken off as anticipated, and few niches were available where a biopesticide would not face competition from an established, conventional product. More importantly, insufficient work had been done on ensuring the field efficacy of these products, and they proved variable in their effect, largely due to the influence of the environment on the survival of formulations.

Today, in the 1990s, we see a new surge of interest in biopesticides, and much more prospect of success. Exciting new markets have been created by political decisions to reduce use of broad spectrum pesticides in key agricultural areas: such as rice pests in Indonesia, soil dwelling insects in many countries, locusts in Africa, and forest pests in Canada. In this last example, the restriction of aerial application of chemical pesticides against spruce budworm, *Choristoneura fumiferana* in Ontario has led to considerable investment by ICI and Ciba-Geigy in developing *Bacillus thuringiensis* and *Trichogramma* spp., respectively, for this new and lucrative market.

Thus, where markets appear through government intervention in pest control, even multinational agrochemical companies may find it worth their while to invest in biological control. The recent programme of re-registration of pesticides in the USA will, in a similar manner, provide a boost to the development of biopesticides. Companies choosing not to register pesticides for certain uses, due to their low profitability relative to their high environmental exposure (e.g. malathion for fruitfly bait sprays), will lay open new markets for biopesticide development. To the extent that many developing countries follow US registration standards, this may have far-reaching effects in the tropics as well.

The other impetus to biopesticide development comes from new approaches to increasing field efficacy. For instance, the most important biopesticide today is *Bacillus thuringiensis* (*Bt*), a bacterium formulated for control of lepidopteran, coleopteran and dipteran pests. Its major shortcoming is its lack of persistence on plants, as the toxin which it contains breaks down rapidly on exposure.

To arrest breakdown of *Bt*, the company Mycogen has recently transferred the toxin-producing genes from *Bt* to another bacterium, *Pseudomonas fluorescens*, a common epiphytic micro-organism on plant surfaces. Inside the tough cell wall of its new host, the *Bt* toxins have much greater persistence in the field, and the resulting biopesticide is more effective and dependable (Soares and Quick, in press). Field efficacy may also be improved by improved formulation. Fungal pathogens hold, perhaps, the greatest promise for biopesticide development against insect, weed and plant disease targets. There are substantial constraints in their use against pests and diseases of aerial plant parts, because of the dry environment in which these surfaces exist.

The challenge to overcome this constraint is a key element of IIBC's new programme to develop a fungal insecticide for the desert locust, a denizen of particularly arid environments (Prior and Lomer, in press). To meet this challenge, novel work is underway at IIBC on oil formulation of fungal spores. Not only do oil formulations eliminate some problems of desiccation, but they may enhance the use of pathogens in ultra low volume (ULV) application systems. ULV systems are superior to conventional application methods in their targeting of droplets to the host, and may provide biopesticides with just the edge they need to achieve a field efficacy equivalent to chemical products.

Let me now bring us back down to earth from these exciting visions of the future. Biopesticides today command about 1% of the world pesticide market, and most of this is *Bt*. Optimistic biopesticide companies expect this percentage to treble by the year 2000. The large multinational agrochemical companies do not, but are nonetheless investing increasingly in biopesticide ventures, as we have seen. Thus, there is a long way to go before biopesticides rival, in any way, their chemical predecessors. However, these figures relate largely to the international commercial markets for pesticides. One advantage of the simple production technology for biopesticides is that they can be mass produced on a small scale for local markets, a technique which is already gaining popularity in tropical countries like Brazil, where both viruses and fungi are produced locally and commercially for application against insect pests on soybean and sugar cane, respectively (Moscardi, 1989; Prior, 1989).

The Future of Biological Control

Classical biological control, conservation, augmentation and biopesticides are the biological control technologies on which we will rely in the year 2000.

Not only can we expect new agents, methods and products over the next decade, but we can expect their use to have implications far beyond the pests against which they are targeted. Let me expand upon this last claim.

Introducing biological control into a crop pest management system usually permits a reduction in pesticide usage. This then leads to recovery of natural enemies of other pests in the crop. To protect this greater level of biological control, efforts are usually made to reduce use of the remaining pesticides in the crop system, and the entire pest management systems shift towards a biological strategy. This pest management revolution is illustrated in the history of biological control in temperate glasshouses (Hussey, 1985). In the 1960s problems with pesticide use led to the development of predatory mites for mite control and of parasitoids for whitefly control. Increasing dependence on these methods made it difficult to treat other glasshouse pests with insecticides, and biological control solutions were found: parasitoids for leafminers, predators for aphids, and biopesticides for these and a range of other pests. We might call this tendency of biological control to encourage its own development the "biocontrol treadmill", in the hope that it may replace a less desirable "treadmill" in years to come. I would suggest that over the next decade we may see the beginnings of a similar revolution in crops like cotton, vegetables or coffee in the Asia–Pacific region.

Lest we grow careless in our enthusiasm for biological control and its growing role in IPM, I should stress that the incorporation of biological control into existing pest management systems is not without its problems. A sad case in point is the recent development of resistance to insect growth regulators (IGRs) and *Bt* in Asia and the USA. In vegetable crops, where the moth *Plutella xylostella* is a worldwide pest, farmers entered the 1980s with few effective control measures. *Plutella* had demonstrated a capacity to develop resistance to all common pesticide groups, including carbamates, organophosphates and pyrethroids. In the 1980s, the impact of introduced parasitoids finally began to be noticed in parts of southeast Asia, and IPM packages were developed to rely on their important contribution. Where pesticides were still needed, products safe for natural enemies were sought, and found, in the form of new IGRs and new *Bt* products. However, before these could be introduced in an IPM package, farmers began to use them in much the same way as they had broad-spectrum compounds, in programmes of calendar application. Within a few years, resistance to IGRs was complete, and that to *Bt* growing. The demise of these products puts at risk an as yet-unrealized IPM solution to *Plutella* management.

Thus, IPM of *Plutella*, got off to a wrong start due to the persistence of traditional approaches to pesticide use. Had the contribution of natural enemies been used to reduce pesticide usage according to some threshold spray model (e.g. as devised in the Malaysian IPM programme, see Rahman, this volume), the life of these products would have been greatly extended. The example stands as a lesson in the importance of the word "integrated" in IPM.

Finally, this story illustrates the role of biological control in tackling one of the greatest challenges to pest management in the next decade: the management of pesticide resistance. Without the rapid development of methods for pesticide resistance management, we may lose some of the important products, chemical and biological, on which our future IPM programmes may depend, at least initially. How we manage resistance, and extend the life of these products, is a complex problem. Studies to date have revealed little more than our ignorance of the ecological genetics which underly the phenomenon. Nonetheless, most would agree that our most immediate and useful tool to extending pesticide life is to reduce pesticide use. To the substantial extent that this can be done by building biological control into pest management, natural enemies may soon become the surprising benefactors of a pesticide industry which had, only decades before, relegated them to obscurity.

References

Agounké, D., Agricola, U. and Bokonon-Ganta, H.A. (1988) *Rastrococcus invadens.* Williams (Hemiptera: Pseudococcidea), a serious exotic pest of fruit trees and other plants in West Africa. *Bulletin of Entomological Research*, 78, 695–702.

Agricola, U., Agounké, D., Fischer, H.U. and Moore, D. (1989) The control of *Rastrococcus invadens* (Hemiptera: Pseudococcidea) in Togo by the introduction of *Gyranusoidea tebygi* Noyes. *Bulletin of Entomological Research*, 79, 671–678.

Anon (1989) *Studies on the Integrated Management of Citrus Insect Pests.* Division of Citrus Pest Control, Guangdong Entomological Institute, Guangzhou, China. Academic Book and Periodical Press, Guangzhou. 226 pp.

Beddington, J.R., Free, C.A. and Lawton, J.H. (1978) Characteristics of successful natural enemies in models of biological control of insect pests. *Nature* 273, 513–519.

Bellows, T.S. Jr, van Driesche, R.G. and Elkinton, J.S. (in press). Life table construction and analysis in the evaluation of natural enemies. *Annual Review of Entomology.*

Cameron, P.J. (1989) *Helicoverpa armigera* (Huebner), tomato fruitworm (Lepidoptera: Noctuidae). In: *A Review of Biological Control of Invertebrate Pests and Weeds in New Zealand 1974 to 1987* (Eds. Cameron P.J., Hill R.L., Bain J. and Thomas W.P.), pp. 87–91. CAB International and DSIR, Wallingford, UK.

Carl, K.P. (1982) Biological control of native pests by introduced natural enemies. *Biocontrol News and Information* 3, 191–200.

Fullick, A. and Fullick, P. (1991) Biological pest control. Inside Science. *New Scientist*, 9 March 1991.

Guo Ming-Fang (1988) *New method of Trichogramma* utilization. In: *Trichogramma and Other Egg Parasites* (Eds. Voegele J., van Lenteren J.C. and Waage J.K.), pp. 469–476. INRA, Paris.

Hansen, M. (1987) *Escape from the Pesticide Treadmill.* Institute for Consumer Policy Research, Mount Vernon, New York. 185 pp.

Hassell, M.P. (1978) *The Dynamics of Arthropod Predator–Prey Systems.* Princeton University Press, Princeton, NJ, USA.

Hussey, N.W. (1985) History of biological control in protected culture. In: *Biological*

Pest Control: The glasshouse experience (Eds. Hussey N.W. and Scopes N.), pp. 11–29. Blandford Press, Poole, Dorset, UK.

Lomer, C. and Prior, C.J. (eds.) (1992) *Biological Control of Locusts and Grasshoppers.* Proceedings of a Workshop, Cotonou, Benin, 29 April–1 May 1991, CAB International, Wallingford.

Mohyuddin, I.A. (1990) Development of integrated pest management programme for sugarcane pests. Final Report, 1984–1989, International Institute of Biological Control, PARC-IIBC Station, Rawalpindi.

Moscardi, F. (1989) Use of viruses for pest control in Brazil: The case of the nuclear polyhedrosis virus of the soyabean caterpillar, *Anticarsia Gemmatalis. Memorias do Instituto Oswaldo Cruz* 84, 51–56.

Ooi, A.C.P., Shepard, B.M. and Kenmore, P.E. (1991) *Panduan Menunjukkan Kawalan Biologi kepada Penanam-Penanam Padi.* Jabatan Pertanian Semenanjung Malaysia.

Pacala, S.W., Hassell, M.P. and May, R.M. (1990) Host–parasitoid associations in a patchy environment. *Nature* 344, 150–153.

Prior, C. (1989) Biological pesticides for low external-input agriculture. *Biocontrol News and Information* 10, 17–22.

Russell, E.P. (1989) Enemies hypothesis: a review of the effect of vegetational diversity on predatory insects and parasitoids. *Environmental Entomology* 18, 590–599.

Soares, G.G. and Quick T.C. (in press) MVP, a novel bioinsecticide for the control of diamondback moth. In: *Management of Diamondback Moth and Other Crucifer Pests.* Second International Workshop, Tainan, Taiwan, 10–14 Dec. 1990, AVRDC, Tainan, Taiwan.

Tran, L.C., Bustamante, R. and Hassan, S.A. (1988) Release and recovery of *Trichogramma evanescens* Westw. in corn fields in the Philippines. In: *Trichogramma and Other Egg Parasites* (Eds. Voegele J., van Lenteren J.C. and Waage J.K.), pp. 469–476, INRA, Paris.

van Driesche, R.G., Bellows, T.S. Jr, Elkinton, J., Gould, J. and Ferro, D.N. (in press) The meaning of percent parasitism revisited: solutions to the problems of accurately measuring total losses from parasitism. *Environmental Entomology.*

Waage, J.K. (1989) The population ecology of pest–pesticide–natural enemy interactions. In: *Pesticides and Non-Target Organisms* (Ed. Jepson P.), pp. 81–93, Intercept Ltd, Andover, UK.

Yang Pei (1984) Development of the ancient techniques of using "Huang Jing Yi". *Agricultural History of China* 3 (10), 72–74.

Zelasny, B., Lolong, A. and Crawford, A.M. (1990) Introduction and field comparison of baculovirus strains against *Oryctes rhinoceros* (Coleoptera: Scarabaeidae) in the Maldives. *Environmental Entomology* 19, 1115–1121.

Chapter twenty-seven: **Trends in Biological Control in Malaysia**

P.A.C. Ooi
*International Institute of Biological Control (IIBC)
Malaysian Station, c/o CAB International Regional Office
Level 3, Block E, Kompleks Pejabat, Jln. Dungun,
Damansara Heights, 50490 Kuala Lumpur*

Abstract

Up to 1979, 23 biological control agents were introduced into Malaysia in an attempt to control 11 pests. However, successes were only realised after the 1970s for the control of the string bush, *Cordia curassavica* and the diamondback moth, *Plutella xylostella*. *Metrogaleruca obscura* and *Eurytoma attiva* were imported and successfully controlled the string bush, especially in coconut plantations.

Complete reliance on chemical insecticides led to the development of insecticide resistance in the diamondback moth. In the mid-1970s, *Diadegma semiclausem* and *Diadromus collaris* were established in the Cameron Highlands. Initially, they did not have much impact on the pest population because farmers used large amounts of insecticides. However, when farmers reduced the usage of insecticides the parasitoids controlled the pest effectively.

Both examples reflected the last resort attitude of existing pest management policy. However, biological control has proven to be both sustainable and cost effective. This has stimulated wide interest in biological control and there is a trend towards considering this approach on an *a priori* basis.

Conservation of natural enemies has been the mainstay of insect pest management in Malaysia since the 1950s. The role of natural enemies in controlling native pests was recognised in oil palm, cocoa, coconut and recently in rice. Continued outbreaks caused by indiscriminate use of insecticides reflected the need for more ecological studies to quantify and demonstrate the role of natural enemies. Besides training farmers to recognise their friends, biological control principles should be incorporated into the curricula of schools, institutions and universities.

Recent successes have stimulated greater interest in an ecological approach to pest management in Malaysia. The national co-ordinating committee on biological control will ensure that the present trend is maintained through the year 2000 and beyond.

Introduction

Biological control has gained popularity again in recent years because of its many ecological, sociological and economic benefits. However, many do not understand the true extent of biological control, especially in a developing country context. As it becomes increasingly popular, there are plant protection scientists who consider that any use of biological/botanical components is biological control. Hence, the use of neem or pyrethrum extract and host plant resistance occasionally appeared as biological control. This paper prefers the definition by DeBach (1964) which involves parasitoids, predators and pathogens to keep pest populations at low levels.

Often, biological control has been associated only with importation of natural enemies. This is a desirable approach for an exotic organism which becomes a pest in a new environment, in the absence of its natural enemies that keep it under control in its native home. Hence, it makes good sense to search for effective natural enemies in its native home and bring them to the environment where the organism is a pest. This approach is referred to as classical biological control.

A very neglected field of biological control is the one that exists naturally. This is characteristically ubiquitous in a tropical environment. It is ironical that while we recognise that few native organisms are serious agricultural pests, we are ignorant about the reasons for this. That this is so is revealed by the numerous cases of disruption of naturally occurring biological control through our careless technology. It has become increasingly clear, particularly in the tropics, that biological control should commence from an understanding of existing natural enemies.

Interest in biological control in Malaysia can be traced back to 1887. Teoh and Ooi (1986) observed that biological control featured prominently between 1940 and 1949 but declined in the next two decades (Fig. 1). Indications in the 1970s suggest that there is renewed interest in biological control. This paper examines whether this trend will persist through the year 2000. As there was more interest in the field of insect pests and weeds, this paper will examine case studies from both fields. Much of the information is gleaned from Ooi *et al.* (1981) and MAPPS (1986).

Classical Biological Control

Ooi *et al.* (1979) reported that between 1903 and 1979, a total of 23 biological control agents were imported to control 11 different pests. Despite the number of agents introduced, only a few resulted in successful control of the target organism. Reasons for the failures have been discussed by Ooi and Greathead (1986), Lim and Ooi (1989) and Khoo (1990). An important reason for failures is the lack of or inadequate ecological knowledge of target pests and natural enemies. For example, it was inappropriate that natural enemies were introduced in an attempt to control the rhinoceros beetle, *Oryctes rhinoceros* Linnaeus (Coleoptera: Scarabaeidae). The beetle was actually under natural biological control and this was illustrated by further studies which resulted in the discovery of a most effective *Baculovirus* that keeps the pest population in check. Similarly, attempts to control rice stem borers were unsuccessful as these native pests were already under natural biological control.

The successful biological control of the diamondback moth, *Plutella xylostella* (Linnaeus) (Lepidoptera: Yponeumeutidae) was preceded by extensive ecological control and in response to failures in chemical control of the pest in Cameron Highlands (Ooi, 1979; Lim, 1982). Recognition of an insufficient level of biological control resulted in the introduction and establishment of two effective parasitoids, *Diadegma semiclausem* Hellen (Hymenoptera: Ichneumonidae) and *Diadromus collaris* Gravenhorst (Hymenoptera: Ichneumonidae) (Ooi and Lim, 1989). Nevertheless, the pesticides treadmill

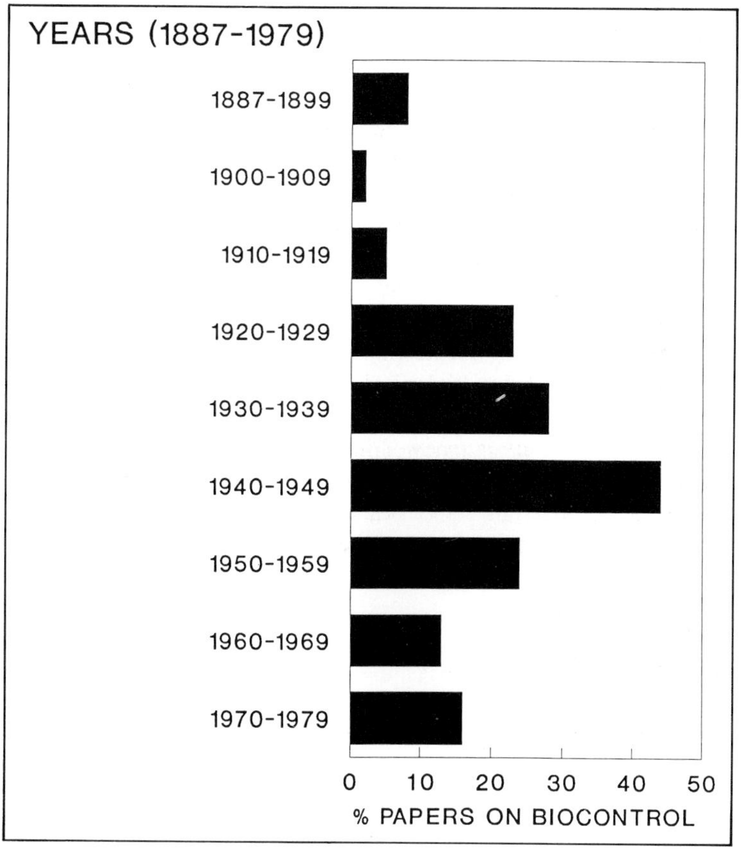

Fig. 1. Trend in the number of research papers covering biological control in Malaysia from 1887 to 1979 (adapted from Teoh and Ooi, 1986)

that cabbage farmers were trapped in delayed the benefits of biological control from reaching these farmers for more than ten years (Ooi, 1990).

Besides the introduction of entomophagous insects to control insects pests, biological control of weeds has been attempted in Malaysia since 1970 when insects were introduced to control Siam weed, *Chromolaena odorata* (L.) R.M. King and H. Robinson in Sabah (Syed, 1979). However, although the arctiid moth was established it did not contribute to effective biological control of the weed (Ooi *et al.*, 1988). The most successful biological control of a weed in Malaysia was the control of *Cordia curassavica* (Jacq.) R. & S., a serious weed problem in coconut plantations (Simmonds, 1980; Ung *et al.*, 1979). *Metrogaleruca obscura* DeGeer (Coleoptera: Galerucidae) and *Eurytoma attiva* Burks (Hymenoptera: Eurytomidae) were introduced from Trinidad and Mauritius respectively in 1977. Both insects established quickly and impressively controlled the weed throughout the Peninsula within a year. *M. obscura* was a most conspicuous defoliator while *E. attiva* destroyed seeds. The successful biological control of *C. curassavica* has provided the impetus for biological control of other weeds in Malaysia.

Conservation of Natural Enemies

Despite a rich array of crops and suitable conditions for growth of plants and phytophagous insects in Malaysia, there are very few insect pests of great importance. Often, major pests are those that occur occasionally in large numbers and decline soon after. It is believed that most indigenous phytophagous insects have a wide range of natural enemies that keep their populations in check. The recognition of natural enemies and their conservation has been translated to IPM programmes in oil palm (Wood, 1971) and cocoa (Conway, 1971). Very often, these integrated control approaches were based on cursory observations that conservation of natural enemies was important: the proof lies in the successful control of these pests based on these deductions. A recent successful integrated pest control programme was based on existing information about the effective natural biological control of the coconut leaf moth, *Artona catoxantha* Hampson (Lepidoptera: Zygaenidae). Ooi (1977) developed an integrated control of this moth employing trunk injection of a systemic insecticide which had minimal effect on the important tachinid parasitoid, *Bessa remota* Aldrich (Diptera: Tachinidae).

There are relatively few firm proofs of the importance of natural enemies in the management of pests. The study reported by Wood (1964) is among the few such demonstrating the importance of natural biological control in Malaysia. In an insecticide check experiment, Wood (1971) demonstrated that insecticide sprays resulted in peak population of bagworms in oil palms. A transition from purely qualitative assessment of natural control to that of quantifying the impact of such control is noted. Ooi (1986) demonstrated that a rice field which was treated with insecticides three times had higher *Nilaparvata lugens* (Stal) (Hemiptera: Delphacidae) populations as compared with a similar field without any spray. Follow-up studies using exclusion cage experiments showed that predators were important in keeping the *N. lugens* population low (Ooi, 1988).

Ecological studies combined with insecticide-check and exclusion cage experiments have shown that the indiscriminate use of insecticides can disrupt biological control (Wood, 1971; Kenmore, 1980; Ooi, 1988). An increasing role for biological control specialists is in the identification of biological control systems and preventing these from being destroyed by regular and unjustified use of insecticides. The conservation of biological control agents in Malaysia has been recognised in Malaysia since the 1950s but continued outbreaks arising from disruptive insecticide treatments remind us that more work is needed in this field of biological control.

Discussion

The successful biological control of *P. xylostella* and *C. curassavica* was the result of a last resort approach. In the case of *P. xylostella*, the pest was resistant to a wide range of insecticides and farmers almost gave up growing cabbages in the Cameron Highlands. As a result of ecological studies, an alternative pest management approach was made available to the farmers and the pest was brought under effective biological control in the cooler parts of the highlands.

Control of *C. curassavica* by herbicides was expensive and not very effective. The use of two biological control agents reduced the weed to a non-pest level. It is time to reflect whether we should approach biological control as a last resort measure. The successes achieved with the diamondback moth and *C. curassavica* have encouraged an *a priori* approach. This is reflected in the setting up of a natural co-ordinating committee on biological control to encourage this beneficial approach to pest management. Classical biological control is particularly suited to managing exotic pests and this is being pursued in the case of two recently introduced insect pests and several exotic weed species.

In the case of indigenous pests, field observations have suggested that natural enemies helped keep them in check. Conway (1972) suggested that in the past there was lack of proof of natural biological control because entomologists in the tropics were usually confronted with immediate pest problems. As a result they had neither time nor the resources to unravel relationships between prey and natural enemies. Recent studies have quantified the impact of natural enemies in both perennial and annual crops. However, more ecological studies are necessary to elucidate the value of natural enemies in all agroecosystems. Such impact studies have been demonstrated to rice farmers with some success and enabled many farmers to appreciate the role of natural enemies. Field demonstrations of natural enemies should be incorporated into curricula of schools, institutions and universities. This trend should be encouraged to avoid the disruption of existing biological controls in our agricultural systems. The prospects for biological control appear very bright for Malaysia especially in a society which cares for its environment.

Acknowledgements

I wish to thank the organising committee for the kind invitation to share my experience on biological control at this seminar. My gratitude to my institution for its support and to my friends G.S. Lim, P.E. Kenmore, B.M. Shepard and N.T. Talekar for sharing their views with me.

References

Conway, G.R. (1971) *Pests of cocoa in Sabah and their Control.* Kementerian Pertanian and Perikanan Sabah, Malaysia. 125 pp.

Conway, G.R. (1972) Ecological aspects of pest control in Malaysia. In: *The Careless Technology. Ecology and International Development* (Eds. Farvar, M.T. and Milton, J.P), Natural History Press, New York, pp. 467–488.

DeBach, P. (1964) The scope of biological control. In: *Biological Control of Insect Pests and Weeds* (Eds. DeBach, P. and Schlinger, E.I), pp. 3–20. Chapman and Hall, LA.

Kenmore, P.E. (1980) Ecology and outbreaks of a tropical insect pest of the green revolution, the rice brown planthopper, *Nilaparvata lugens* (Stal). Ph.D. thesis, University of California, Berkeley.

Khoo, S.G. (1990) Use of natural enemies to control agricultural pests in Malaysia. In: *The Use of Natural Enemies to Control Agricultural Pests* (Eds. Mochida, O. and Kiritani, K), FFTC/ASPAC, Taipei, Taiwan ROC, pp. 30–39.

Lim. G.S. (1982) The biology and effects of parasites on the diamondback moth, *Plutella xylostella* (L.). Ph.D. Thesis, University of London. 317 pp.

Lim G.S. and Ooi, P.A.C. (1989) Biological control of agricultural pests in Malaysia. In: *Biological Control of Pests.* BIOTROP Special Publication No. 36. pp. 15–37.

MAPPS (1986) First decade of MAPPS: 10 years of plant protection in Malaysia. Malaysian Plant Protection Society (MAPPS), Kuala Lumpur. 181 pp.

Ooi, P.A.C. (1977) The coconut leaf moth. Technical Leaflet No. 12. Ministry of Agriculture, Malaysia. 12 pp.

Ooi, P.A.C. (1979) Incidence of *Plutella xylostella* (L.) (Lepidoptera: Yponomeutidae) and its parasites, *Apanteles plutellae* Kurdj. (Hymenoptera: Braconidae) in Cameron Highlands, Malaysia. *Malaysian Applied Biology*, 8, 131–143.

Ooi, P.A.C. (1986) Insecticides disrupt natural control of *Nilaparvata lugens* in Sekinchan, Malaysia. In: *Biological Control in the Tropics* (Eds. Hussein, M.Y. and Ibrahim, A.G.), UPM, Serdang, Malaysia, pp. 109–120.

Ooi, P.A.C. (1988) Ecology and surveillance of *Nilaparvata lugens* (Stal) - implications for its management in Malaysia. Ph.D. Thesis, Universiti Malaya, Kuala Lumpur. 275 pp.

Ooi, P.A.C. (1990) Management of diamondback moth in Cameron Highlands. *MAPPS Newsletter* 13(4), 55-56.

Ooi, P.A.C. and Greathead, D.J. (1986) Biological control in Southeast Asia and the role of CIBC. In: *Proceedings Regional Conference on Plant Quarantine Support for Agricultural Development. December 1985* (Eds. Singh, K.G. and Manalo, P.L.), pp. 301–307.

Ooi, P.A.C. and Lim, G.S. (1989) Introduction of exotic parasitoids to control the diamondback moth in Malaysia. *Journal of Plant Protection in the Tropics* 6, 103–111.

Ooi, P.A.C., Lim, G.S. and Khoo, S.G. (1979) Biological control in Malaysia. In: *Proceedings Plant Protection Seminar 1979 Strategies in Plant Protection* (Eds. Lim, T.K. *et al.*), pp. 1–35 Malaysian Plant Protection Society, Kuala Lumpur.

Ooi, P.A.C., Neo, C.E. and Lim G.S. (1981) A compendium of economic entomology in Peninsula Malaysia (1887–1975). Bulletin No. 156 Ministry of Agriculture, Malaysia. 233 pp.

Ooi, P.A.C., Sim, C.H. and Tay, E.B. (1988) Status of the arctiid moth introduced to control Siam weed in Sabah, Malaysia. *Planter, Kuala Lumpur* 64, 298-304.

Simmonds, F.J. (1980) Biological control of *Cordia curassavica* (Boraginaceae) in Malaysia. *Entomophaga* 25, 363–364.

Syed, R.A. (1979) An attempt on biological control of *Eupatorium odoratum* L.f in Sabah, Malaysia. In: *Proceedings of the Sixth Asian-Pacific Weed Science Society Conference* Vol. II, pp. 459–566.

Teoh, C.H. and Ooi, P.A.C. (1986) Towards more effective plant protection in the tropics. *Journal of Plant Protection in the Tropics* 3, 51–72.

Ung, S.H., Ahmad, Y. and Chin, W.H. (1979) Biological control of *Cordia curassavica* (Jacq.) R. & S. in Malaysia by *Schematiza cordiae* Barb. (Coleop: Galerucidae). *Malaysian Agricultural Journal* 52, 154–165.

Waterhouse, D.F. and Norris, K.R. (1989) *Biological Control Pacific Prospects* Supplement 1. Australian Centre for International Agricultural Research, Canberra. 123 pp.

Wood, B.J. (1964) Severe outbreaks of caterpillars on oil palm estates in Malaya induced by the use of residual contact insecticides. In: *Proceedings XII International Congress of Entomology*, 575 pp.

Wood, B.J. (1971) Development of integrated control programs for pests of tropical perennial crops in Malaysia. In: *Biological Control* (Ed. Huflaker, C.B.), pp. 422–457. Plenum Press, New York.

Chapter twenty-eight: **Biotechnology and Pest Management in 2000 AD**

G.S. Khush[1] and G.H. Toenniessen[2]

[1]*International Rice Research Institute, P.O. Box 933, Manila, Philippines*

[2]*The Rockefeller Foundation, 1133 Avenue of the Americas, New York, N.Y. 10036, USA*

Summary

Varietal resistance to diseases and insects has played and will continue to play a major role in pest management programmes. Major advances have been made in developing cultivars with multiple resistance to diseases and insects. In the past, genetic improvement for pest resistance has been achieved mainly through the application of classical Mendelian genetics and conventional plant breeding methods. Plant breeders have relied upon crop germplasm, including wild species and induced mutants, as sources of resistance. However, recent advances in cellular and molecular genetics have led to the development of new tools for producing resistant cultivars. It is now possible to introduce into crop plants novel genes for pest resistance from unrelated plants, animals, or micro-organisms. Tissue culture procedures have helped broaden the gene pools for resistance through production of wide hybrids among distantly related species and genera and through *in vitro* selection of useful mutants. Pest resistance genes, when tagged with isozyme (RFLP) markers, can be moved more rapidly from one cultivar to another. Nucleic acid probes allow the detection of pathogens in breeding materials and aid in the selection process.

Introduction

Varietal resistance to diseases and insects has been a major control strategy in the past and will continue to be a predominant component of the pest management programmes in the future. The dramatic examples of the success of the host resistance programme are the control of Hessian fly (*Mayetiola destructor*, Diptera: Cecidomyiidae) through breeding of Hessian fly-resistant wheats in the United States of America and the control of brown plant hopper of rice in Asia. The use of rust-resistant wheat cultivars throughout Europe, North America, Australia and Asia has been a bulwark against epidemics of this devastating disease. Incorporation of genes for disease and insect resistance into cultivars is the major objective of crop improvement programmes. Numerous cultivars of crop plants with multiple resistance to major diseases and insects have been developed and are now grown worldwide. The cultivation of such cultivars has saved millions of dollars worth of losses caused by pests. Moreover, pest-resistant cultivars have greater yield stability and thus provide greater food security.

Until now, the genetic improvement for pest resistance has been achieved mainly through the application of classical Mendelian genetics and through conventional plant breeding methods. Plant breeders have relied mainly on crop germplasm, including wild species and induced mutants, as sources of resistance. However, recent advances in biotechnology, particularly in cellular and molecular biology, have opened new vistas for developing resistant cultivars. As a result of these breakthroughs, it is now possible to introduce into crop plants novel genes for pest resistance from unrelated plants, animals or micro-organisms. Tissue culture procedures have helped broaden the gene pools for resistance through production of wide hybrids among distantly related species and genera and through *in vitro* selection of useful mutants. Nucleic acid probes allow the detection of pathogens in the breeding materials and thus serve as an aid in the selection process. Similarly, pest resistance genes, when tagged with isozyme or RFLP markers, can be moved more rapidly from one cultivar to another. RFLP markers are also useful in studying the genetic variation in pest populations. This information is useful in host resistance gene deployment strategies.

The following techniques of biotechnology will be widely used in pest management programmes by the year 2000.

- Wide hybridisation for incorporation of alien genes.
- Somaclonal variation and selection of useful mutants.
- Expansion of gene pools through somatic hybridisation.
- Incorporation of novel genes through genetic engineering.
- Isozyme and RFLP markers for tagging resistant genes.
- Nucleic acid probes as diagnostic tools for pathogens.

Wide Hybridisation for Incorporation of Useful Genes

In the past, plant breeders have used mainly the primary gene pools or genetic variability present in the cultivated germplasm, land races, weedy species and closely related wild species. Sometimes, however, the sources of resistance are inadequate or even lacking in the primary gene pools. Notable examples include resistance to sheath blight, grassy stunt virus and stem borers in rice, pink boll worm in cotton, pod borer in pigeonpea, *Ascochyta* blight in chickpea and yellow mosaic in mungbean. In situations like these, plant breeders are forced to utilise distantly related species as sources of resistance.

However, many barriers are encountered in transferring genes from such sources. The first is the abortion of hybrid embryos. Embryo rescue is a very useful tissue culture technique to overcome this barrier. It has been successfully used to produce a series of interspecific and intergeneric hybrids in a number of crop genera such as wheat, barley, rice, legumes, oil seeds, fibre crops, vegetables and horticultural crops (Williams *et al.*, 1987; Khush & Brar, 1988). In rice, the embryo rescue technique was used for transferring genes for resistance to brown plant hopper and whitebacked plant hopper from *O. officinalis* to cultivated rice, *O. sativa* (Jena & Khush, 1990). This technique is now being used for transferring genes from even more distantly related species such as *O. latifolia, O. brachyantha* and *O. granulata*.

Somaclonal Variation and Selection of Useful Mutants

Variation arising through tissue culture in regenerated plants and their progenies is referred to as somaclonal variation. It is an important source of variability in well-adapted genotypes without resorting to hybridisation. Larkin & Scowcroft (1981) reviewed the occurrence of such variation for a series of agronomic traits in a large number of crop species such as wheat, rice, potato, tobacco, tomato, and sugarcane. Some useful pest-resistant somaclones such as sugarcanes resistant to Fiji virus, eyespot and downy mildew (Heinz *et al.*, 1977), maize lines resistant to *Helminthosporium* (Brettell & Ingram, 1979) and potatoes resistant to late blight (Shepard *et al.*, 1980) have been selected.

Exposure of cell cultures to toxic concentrations of selective agents such as culture filtrates and pathotoxins, where applicable, can increase the selection efficiency for useful somaclones. In vitro selection for disease resistance was first demonstrated in 1973 when tobacco plants resistant to wildfire disease *(Pseudomonas tabaci)* were obtained by selecting cell lines tolerant to methionine sulfoximine — an analogue of bacterial toxin for wildfire (Carlson, 1973). Selection of cultured cells tolerant to culture filtrate and/or toxin have been successfully used to produce pest-resistant germplasm. The notable examples include resistance to *Phoma lingam* in Brassica (Sacristan, 1982), *Pseudomonas syringae* cv *tabaci* (Thanutong *et al.*, 1983), *Helminthosporium sacchari* in sugarcane (Larkin & Scowcroft, 1983), *Fusarium oxyporium* in alfalfa (Arcioni *et al.*, 1987) and *Helminthosporium sativum* in wheat and barley (Chawla & Wenzel, 1987). Information on in vitro selection at the cellular level for insect resistance is scarce. Preliminary results on the selection for resistance to southwestern corn borer indicate that the selection for insect tolerance at the callus level may be effective (Williams & Davis, 1985).

While in vitro selection at the cellular level seems to be a promising approach for developing pest-resistant germplasm, there are problems that limit its wide-scale adoption. Lack of appropriate in vitro selection techniques for many diseases and insects, low frequency of plant regeneration in long-term cultures and limited correlation between tolerance for biotic stress at cellular level and whole plant level are just some of the limitations to the wide-scale adoption of this technique.

Expansion of Gene Pools through Somatic Hybridisation

In some cases, crosses between distantly related species cannot be accomplished even through embryo rescue technique. In such cases, somatic hybridisation involving fusion of protoplasts of two species and subsequent plant regeneration can yield somatic hybrids. Such hybrids have 2*n* chromosome complements of both parents and are thus amphiploids. Backcrosses to the cultivated parent yield progenies with a limited amount of chromatin of the wild parent. Backcross progenies can then be screened for useful variations. Many somatic hybrids have been produced between incompatible species (Khush & Brar, 1988), but to date gene tranfers using this technique have not been reported.

Incorporation of Novel Genes through Genetic Engineering

Protocols for the transformation of many crop species have now been established. This allows the introduction of cloned genes from unrelated sources such as other plant species, animals and micro-organisms. Foreign genes have been successfully incorporated into many dicot crop species through *Agrobacterium tumefaciens* as well as through direct DNA transfer methods. Various techniques are now available for transforming monocot crop plants (Uchimiya *et al.*, 1989). Transgenic plants have been produced in more than 30 plant species, and several useful genes for disease and insect resistance have been introduced.

Insect resistance

The protein produced by the bacterium, *Bacillus thuringiensis* is lethal to the caterpillars of lepidopteran insects. When the *Bt* gene responsible for the production of the toxin is transferred to the crop plants through genetic engineering, plants produce their own biocides. *Bt* protein gene has been transferred into tobacco (Vaeck *et al.*, 1987), tomato (Fischhoff *et al.*, 1987; Delannay *et al.*, 1989) and several other crops. The larvae of lepidopteran insects fed on genetically engineered plants were killed within 48 hours, and there was very little evidence of feeding damage to leaves of transformants. Delannay *et al.* (1989) evaluated transgenic tomato plants expressing *Bt* gene for 2*n* years under field conditions. Plants infested with tobacco hornworm showed very limited feeding damage compared to non-transgenic control plants which suffered heavy feeding damage and were almost completely defoliated within two weeks. Significant control of tomato fruitworm and tomato pinworm was also observed. Another approach to insect tolerance is the development of transgenic plants that are capable of producing protease inhibitors. Hilder *et al.* (1987) transferred cowpea trypsin inhibitor (*CpTi*) gene in tobacco. The transgenic plants produced relatively high levels of gene product (more than 1% of the soluble protein) and showed resistance to *Heliothis virescens.* The information from laboratory as well as from field tests of transformed plants indicates that introduction of insect toxin genes into plants is a practical method for providing protection against certain insect pests.

The cystein proteinase inhibitor, oryzacystatin, has been isolated from rice seeds. Both the protein and the gene coding for it have been thoroughly characterised (Abe *et al.*, 1987; Kondo *et al.*, 1989; Chen, 1991). Gut proteases of the rice weevil and the red flour beetle are strongly inhibited by oryzacystatin. Additional inhibitors from other plants, such as the mungbean trypsin inhibitor, potato proteinase inhibitors I and II, arrowhead proteinase inhibitors and towel gourd trypsin inhibitors are being tested against digestive enzymes of the rice insect pests (Chi, 1990) and are likely to be candidate genes for transforming rice and other crops.

Lectins are a group of plant proteins that bind to carbohydrates, including chitin. Some of these have been shown to provide protection against disease and insect attack (Chrispeels & Raikhel, 1991). Acrelin-1, a lectin found in beans, has toxic effects on *Zabrates subfasciatus,* an important bean insect (Osborn *et al.*, 1988). Chrispeels & Raikhel (1991) have suggested that the deleterious effect of chitin binding lectins on insect development is mediated by binding to chitin in the peritrophic membrane that lines the midgut of insects, thus interfering with the uptake of nutrients. Genes for lectin production would be good candidates for transforming crop plants against insect attack.

Disease resistance

Infection of plants by fungal and bacterial pathogens results in an inducible defence response. This can include synthesis and accumulation of phytoalexins, reinforcement of cell walls by deposition of callose, lignin and related phenolics, and an increase in the activity of hydrolytic enzymes such as chitinases and glucanases.

Chitinases and glucanases degrade the major structural polysaccharides of the fungal cell wall. They attack the growing hyphal tip and are thus potent inhibitors of fungal growth (Schlumbaum *et al.*, 1986). Genes coding for such inhibitors and lytic enzymes and for key enzymes in phytoalexin and cell-wall component biosynthetic pathways have been cloned. These are candidate genes for engineering-enhanced disease resistance.

Donors for virus resistance are rare in the germplasm collections. A highly successful strategy against virus infection is termed coat-protein-mediated protection. A chimeric gene is made combining a strong plant promoter with the virus gene for capsid protein. Transgenic plants with capsid protein genes accumulate capsid proteins in plant cells; such plants are highly resistant to infection by the virus where the gene came from (Beachy, 1990). While the mechanism of protection is not yet fully understood, it appears that disassembly of the virus is inhibited (Register & Beachy, 1988). Because little or no virus replication occurs, there is little opportunity for mutants that could overcome the resistance to arise. Incorporation of capsid protein genes has provided protection against infections from viruses in a number of transgenic plants (Table 1). The notable examples of coat-protein-mediated resistance include resistance to tobacco mosaic virus in tobacco and tomato (Abel *et al.*, 1986; Nelson *et al.*, 1988), alfalfa mosaic virus (Loesch-Fries *et al.*, 1987), potato virus X (Hemenway *et al.*, 1988) and cucumber mosaic virus (Cuozzo *et al.*, 1988). The transgenic tomato and tobacco plants expressing coat protein gene had delayed disease development (Tumer *et al.*, 1987). The extent of protection

Table 1. Transgenic plants with novel genes for insect tolerance and disease resistance produced to date

Transgenic plants	Novel gene(s) introduced	Reference
Insect tolerance		
Tomato	*Bt*	Fischhoff *et al.* (1987) Delannay *et al.* (1989)
Tobacco	*Bt*	Vaeck *et al.* (1987)
Tobacco	*CpTi*	Hilder *et al.* (1987)
Disease resistance		
Tobacco	TMV–CP	Beachy *et al.* (1985) Abel *et al.* (1986) Nelson *et al.* (1988)
	AIMV–CP	Van dun *et al.* (1987)
	CMV–CP Antisense CMV–CP	Cuozzo *et al.* (1988)
	PVX–CP	Hemenway *et al.* (1988)
	CMV-satellite RNA	Baulcombe *et al.* (1986)
	TRV–Cp	Van dun & Bol (1988)
Tomato, tobacco	AIMV–CP	Tumer *et al.* (1987)
Potato	PVX–Cp PVY–CP	Hoekema *et al.* (1989) Lawson *et al.* (1990)

Bt, Bacillus thuringiensis; *CpTi,* cowpea trypsin inhibitor; CP, coat protein; TMV, tobacco mosaic virus; AIMV, alfalfa mosaic virus; CMV, cowpea mosaic virus; PVX, potato virus X; PVY, potato virus Y; TRV, tobacco rattle virus.

in transformed plants depends upon the level of expression of coat protein. The coat protein genes of two viruses which cause tungro virus of rice have been cloned (Jones *et al.*, 1990), and efforts are underway to transform rice cultivars with these capsid genes.

Another promising strategy against viral pathogens is the inhibition of cell-to-cell movement of viruses (Hull, 1990). Many viruses encode a 30-kilodalton protein that facilitates such movement through the plasmodesmata (Wolf *et al.*, 1989; Melcher, 1990). A gene whose product suppresses synthesis or inactivates this protein should provide broad-spectrum virus resistance.

Isozyme and RFLP Markers for Tagging Resistance Genes

Evaluation of segregating populations for disease and insect resistance is a laborious and time–consuming process. Quite often, considerable resources in land or greenhouse space have to be allocated for screening. However, if the resistant genes can be tagged with some markers that can be scored in the laboratory, the screening process can be simplified. Fortunately, two types of markers proven useful in breeding for resistance have been identified: biochemical markers, also called isozymes and DNA markers or restriction fragment length polymorphism (RFLP) markers. Certain properties enhance the utility of these markers in plant breeding. These are:

- Genotypes of molecular loci can be determined at whole plant, tissue and cellular levels.
- A relatively large number of naturally occurring alleles are found at molecular markers.
- Alleles of most molecular markers are co-dominant, allowing all possible genotypes to be distinguished in any segregating generation.
- Very few epistatic or pleiotropic effects are observed with molecular markers and a virtually limitless number of segregating markers can be monitored in a single population.

The utility of isozyme markers in breeding was demonstrated in tomatoes. The gene *Mi* for nematode resistance is closely linked with a rare allele of *Aps-1* (acid phosphatase) locus on chromosome 6 (Rick & Fobes, 1974). Many tomato breeding programmes screen segregating progenies for the presence of the *Aps-1* allele instead of using the cumbersome evaluation for nematode resistance.

The only limitation in the use of isozyme markers is that there are not enough of them to cover the entire chromosome map. For tagging a gene with a single marker, a tight linkage (less than 5cM) is necessary. Based on a bracketed marker approach, it would be necessary to have markers throughout the genome at 20-cM intervals. However, only a limited number of isozyme loci are available which cannot cover the entire genome.

Fortunately, RFLP markers are numerous and saturated maps can be prepared. In maize and tomatoes, for example, more than 500 RFLP loci have been established. Similarly, a RFLP map of rice with 450 markers has been prepared. Such comprehensive maps have made it possible to establish tight linkages between RFLP markers and genes for pest resistance, including quantitative trait loci (QTL). Nieunhuis *et al.* (1987) found an association between RFLP loci and QTL affecting expression of resistance to arthropod insects in *Lycopersicon hirsutum,* a wild species of tomato. Association between RFLP loci on three chromosomes and expression of calorimetric absorbance for 2-tridecanone (2TD), a toxic factor for insect resistance was found. A tight linkage was detected between Tm-2a locus for resistance to tobacco mosaic virus in tomato and two RFLP markers, TG79 and TG101 (Young *et al.,* 1988). In tomatoes, Sarfatti *et al.* (1989) were able to tag 1-2 gene for resistance to race 2 of *Fusarium oxysporum* using TG105 marker. In rice, genes for resistance to whitebacked plant hopper (McCouch *et al.,* 1992) and for resistance to blast (Yu *et al.,* 1991) have been tagged with tightly linked RFLP markers. As more genes for resistance are tagged with RFLP loci, the marker-based selection for resistance in segregating populations will become a useful tool for increased selection efficiency.

Nucleic Acid Probes as Diagnostic Tools for Pathogens

The availability of nucleic acid probes has made it possible to screen a large volume of germplasm and breeding materials for the presence of pathogens. The nucleic acid spot hybridisation method is used routinely to detect viruses in potato germplasm. It is reliable and can be completed in a few hours. Efforts are underway to prepare nucleic acid probes for screening rice germplasm for the tungro viruses which are difficult to detect. It is now possible to detect the presence of DNA and RNA virus in plants and in insect vectors by squashing the plant or insect material on to a nylon membrane followed by hybridization with a virus-specific probe. Czosnek & Navot (1988) used the squash blot procedure, a sensitive and simple technique for detection of yellow leaf curl in tomato leaves and also in the viruliferous whiteflies which vector the virus. The squash-blots are very stable, can either be processed immediately, stored for months, or mailed to diagnostic facilities.

Since many viral diseases are spread by insects, it would be feasible to monitor insect populations for infectivity and predict the spread of viral diseases. Thus, the nucleic acid probes offer great promise for use in evaluating germplasm, in screening the breeding materials and in pest management strategies. Therefore, there is an urgent need to develop nucleic acid probes for major fungal, bacterial and viral diseases.

Conclusions

In the past, major advances have been made in developing disease and insect-resistant cultivars by incorporating host resistance genes through the application of Mendelian genetics and conventional breeding methods. Recent advances in biotechnology, particularly in protoplast culture, gene cloning and transformation, offer enormous potential for introducing novel genes for resistance into crop plants. Nucleic acid probes and RFLP markers will be increasingly used for evaluation of germplasm and for increasing the selection efficiency in breeding programmes. The resistant cultivars thus developed will form the backbone of pest management programmes in the year 2000.

References

Abe, K., Emori, Y., Kondo, H., Suzuki, K. and Arai, S. (1987) Molecular cloning of a cysteine proteinase inhibitor of rice (Oryzacystatin): homology with cystatin and transient expression in the ripening process of rice seeds. *J. Biol. Chem.* 262, 16793-16797.

Abel, P.P., Nelson, R.S., De, B., Hoffmann, N., Rogers, S.G., Fraley, R.T. and Beachy, R.N. (1986) Delay of disease development in transgenic plants that express the tobacco mosaic virus coat protein gene. *Science* 232, 738–743.

Arcioni, S., Pezzotti, M. and Damiani, F. (1987) In vitro selection of alfalfa plants resistant to *Fusarium oxysporum* f.sp. *medicaginis*. *Theor. Appl. Genet.* 74,700–705.

Baulcombe, D.C., Saunders, G.R., Bevan, M.W., Mayo, M.A. and Harrison, B.D. (1986) Expression of biologically active viral satellite RNA from the nuclear genome of transformed plants. *Nature* 321,446–449.

Beachy, R.N. (1990) Plant transformation to confer resistance against virus infection. In: *Gene Manipulation in Plant Improvement II* (Ed. Gustafson, J.P.) pp. 305-313. Plenum Press, New York, USA.

Beachy, R.N., Abel, P., Oliver, M.J. De, B., Fraley, R.T., Rogers, S.G. and Horsch, R.B. (1985). In: *Biotechnology in Plant Science-Relevance to Agriculture in the Eighties* (Eds. Zaitlin, M., Day,P., Hollaender, A. Wilson, C.M.), pp. 265–275. Academic Press, New York, USA.

Bretell, R.I.S. and Ingram, D.S. (1979) Tissue culture in the production of novel disease resistant crop plants. *Biol. Rev.* 54, 329–345.

Carlson, P. (1973) Methionine sulfoximine-resistant mutant of tobacco. *Science* 180, 1366–1368.

Chawla, H.S. and Wenzel, G. (1987) In vitro selection of barley and wheat for resistance against *Helminthosporium sativum*. *Theor. Appl. Genet.* 74, 841–845.

Chen, M. (1991) Characterization of insect alpha-amylases, and cloning and expression of oryzacystatin cDNA in *E. coli*. Ph.D. Dissertation, Kansas University, USA.

Chi, C.W. (1990) Three different kinds of proteinase inhibitors as candidates for insecticidal proteins. In: *Molecular and Genetic Approaches to Plant Stress*. pp. T13. 1-T13.3 ICGEB, New Delhi, India.

Chrispeels, J.J. and Raikhel, H.N. (1991) Lectins, lectin genes and their role in plant defense. *Plant Cell* 3, 1–9.

Cuozzo, M., Connell, K.M., Kaniewski, W., Fang, R.X., Chua, N.H. and Tumer, N.E. (1988) Viral protection in transgenic tobacco plants expressing the cucumber mosaic virus coat protein or its antisense RNA. *Bio/Technology* 6, 549–557.

Czosnek, H. and Navot, N. (1988) Virus detection in squash blots of plants and insects: applications in diagnositics, epidemiology and breeding. In: *Biotechnology in Agriculture*. pp. 83-88. Alan R–Liss Inc., USA.

Delannay, X., La Vallee, B.J., Proksch, R.K., Fuchs, R.L., Sims, S.R., Greenplate, J.T., Marrone, P.G., Dodson, R.B., Augustine, J.T., Layton, J.G. and Fischhoff, D.D. (1989) Field performance of transgenic tomato plants expressing the *Bacillus thuringiensis* var. *kurstaki* insect control protein. *Bio/Technology* 7, 1265–1269.

Fischhoff, D.A., Bowdish, K.S., Perlak, F.J., Marrone, P.G., Mc Cormick, S.M., Niedermeyer, J.G., Dean, D.A., Kusano–Kretzmer, K., Mayer, E.J., Rochester, D.E., Rogers, S.G. and Fraley, R.T. (1987) Insect tolerant transgenic tomato plants. *Bio/ Technology* 5,807–813.

Heinz, D.J., Krishnamurthi, M., Nickell, L.G. and Maretzki, A. (1977) Cell, tissue and organ culture in sugarcane improvement. In: *Applied and Fundamental Aspects of Plant Cell Tissue and Organ Culture* (Eds. Reinert, J. and Bajaj, Y.P.S.), pp. 3–17. Springer-Verlag, Berlin.

Hemenway, C., Fang, R.X., Kaniewski, W.K., Chua N.H., and Tumer, N.E. (1988) Analysis of the mechanism of protection in transgenic plants expressing the potato virus X coat protein or its antisense RNA. *EMBO J.* 7,1273–1280.

Hilder, V.A., Gatehouse, A.M.R., Sheerman, S.E., Barker, R.F. and Boulter, D. (1987) A novel mechanism of insect resistance engineered into tobacco. *Nature* 300,160–163.

Hoekema, A., Huisman, M.J., Molendijk, L., van den Elzen, P.J.M. and Cornelissen, B.J.C. (1989) The genetic engineering of two commercial potato cultivars for resistance to potato virus X. *Bio/Technology* 7,273–278.

Hull, R. 1990. Nonconventional resistance to viruses in plants–concepts and risks. In: *Gene Manipulation in Plant Improvement II* (Ed. Gustafson, J.P.) pp. 289–303. Plenum Press, New York, USA.

Jena, K.K. and Khush, G.S. (1990) Introgression of genes from *Oryza officinalis* Well ex Watt to cultivated rice, *O. sativa* L. *Theor. Appl. Genet.* 80,737–745.

Jones, M.I., Dasgupta, J., Cliffe, J., Lee, G., Blakeborough, M., Davies, J.W. and Hull, R. (1990) *The Molecular and Genetic Approaches to Plant Stress*. pp. T2. 1-T2.2. ICGEB, New Delhi, India.

Khush, G.S. and Brar, D.S. (1988) Wide hybridization in plant breeding. In: *Plant Breeding and Genetic Engineering* (Ed. Zakri, A.H.), pp. 141–188. SABRAO, Malaysia.

Kondo, H., Emori, Y., Abe, K., Suzuki, K. and Arai, S. (1989) Cloning and sequence analysis of the genomic DNA fragment encoding Oryzacystatin. *Gene* 81, 259–265.

Larkin, P.J. and Scowcroft, W.R. (1981) Somaclonal variation – a novel source of variability from cell cultures for plant improvement. *Theor. Appl. Genet.* 60, 197–214.

Larkin, P.J. and Scowcroft, W.R. (1983) Somaclonal variation and eyespot toxin tolerance in sugarcane. *Plant Cell, Tissue and Organ Culture* 2,111–121.

Lawson, C., Kaniewski, W., Haley, L., Rozman, R., Newell, C., Sanders, P. and Tumer, N. E. (1990) Engineering resistance to mixed virus infection in a commercial potato cultivar: Resistance to potato virus X and potato virus Y in transgenic Russet Burbank potato. *Bio/Technology* 8,127–134.

Loesch-Fries, L.S., Merlo, T., Zinnen, L. Burhop, Hill K., Krahn, K., Jarvis, N., Nelson, S. and Halk, E. (1987) Expression of alfalfa mosaic virus RNA4 in transgenic plants confers virus resistance. *EMBO J.* 6,1845–1851.

McCouch, S.R., Tanksley, S.D and Khush, G.S. (in press) Tagging genes for disease and insect resistance via linkage to RFLP markers. In: *Rice Genetics II*. International Rice Research Institute, P.O. Box 933, Manila, Philippines.

Melcher, U. (1990) Similarities between putative transport proteins of plant viruses. *J. Gen. Virology* 71, 1009–1018.

Nelson, R.S., McComick, S.M., Delannay, X., Dube, P., Layton, J., Anderson, E J., Kaniewska, M., Proksch, R.K., Horsch, R.B., Rogers, S.G., Fraley, R.T. and Beachy, R.N. (1988) Virus tolerance, plant growth and field performance of transgenic tomato plants expressing coat protein from tobacco mosaic virus. *Bio/Technology* 6, 403–409.

Nieunhuis, J., Halentjaris, T., Slocum, M., Ruggero, B. and Schaefer, A. (1987) Restriction fragment length polymorphism analysis of loci associated with insect resistance in tomato. *Crop Sci.* 27, 797–803.

Osborn, T.C., Alexander, D.C., Sun, S.M., Cardona, C. and Bliss, F. (1988) Insecticidal activity and lectin homology of arcelin seed protein. *Science* 240, 207–210.

Register, J.C. and Beachy, R.N. (1988) Resistance to TMV in transgenic plants results from interference with an early event in infection. *Virology* 166, 524–532.

Rick, C.M. and Fobes, J.F. (1974) Association of an allozyme with nematode resistance. *Rep. Tomato Genet. Coop.* 24, 25.

Sacristan, M.D. (1982) Resistance to *Phoma lingam* of plants regenerated from selected cell and embryogenic cultures of haploid *Brassica napus*. *Theor. Appl. Genet.* 61, 193–200.

Sarfatti, M., Katan, J., Fluhr, R. and Zamir, D. (1989) An RFLP marker in tomato linked to the *Fusarium oxysporum* resistance gene 1–2. *Theor. Appl. Genet.* 78, 755–759.

Schlumbaum, A.F. Mauch, Vogeli, U. and Boller, T. (1986) Potato chitinases are potent inhibitors of fungal growth. *Nature* 324, 365–367.

Shepard, J., Bidney, F.D. and Shahin, E. (1980) Potato protoplasts in crop improvement. *Science* 28, 17–24.

Thanutong, P., Furusawa, I. and Yamamoto, M. (1983) Resistant tobacco plants from protoplast-derived calluses selected for their resistance to *Pseudomonas* and *Alternaria* toxins. *Theor. Appl. Genet.* 6, 209–215.

Tumer, N.E., O'Connell, K.M., Nelson, R.N., Sanders P.R., Beachy, R.N., Fraley, R.T. and Shah, D.M. (1987) Expression of alfalfa mosaic virus coat protein gene confers cross protection in transgenic tobacco and tomato plants. *EMBO J.* 6, 1181–1188.

Uchimiya, H., Handa, T. and Brar, D.S. (1989) Transgenic plants. *J. Biotechnology* 12, 1–20.

Vaeck, M., Reynaerts, A., Hofte, H., Jansens, S., De Beuckeleer, M., Dean, C., Zabeau, M., van Montagu, M. and Leemans, J. (1987) Transgenic plants protected from insect attack. *Nature* 328, 33–37.

Van dun, C.M.P., Bol, J.F. and van Vloten-Doting, L. (1987) Expression of alfalfa mosaic virus and tobacco rattle virus coat protein genes in transgenic tobacco plants. *Virology* 159, 290–305.

Van dun, C.M.P. and Bol, J.F. (1988) Transgenic tobacco plants accumulating tobacco rattle virus coat protein resist infection with tobacco rattle virus and pea early browning virus. *Virology* 167, 649–652.

Williams, W.P. and Davis, F.M. (1985) Southwestern cornborer larval growth on corn callus and its relationship with leaf feeding resistance. *Crop Sci.* 25, 312-319.

Williams, E.G., Maheswaran, G. and Hutchinson, J.F. (1987) Embryo and ovule culture in crop improvement. *Plant Breed. Rev.* 5, 181–236.

Wolf, S., Deom, C.M., Beachy, R.N. and Lucas, W.J. (1989) Movement protein of tobacco mosaic virus modifies plasmodesmatel size exclusion limit. *Science* 246, 377–379.

Young, N.D., Zamir, D., Ganal, M.W. and Tanksley, S.D. (1988) Use of isogenic lines and simultaneous probing to identify DNA markers tightly linked to *Tm-2a* gene in tomato. *Genetics* 120, 579–585.

Yu, Z,H., Mackill, D.J., Bonman, J.M. and Tanksley, S.D. (1991) Tagging genes for blast resistance in rice via linkage to RFLP markers. *Theor. Appl. Genet.* 81, 471–476.

Chapter twenty-nine: **Beyond Mendel's Garden**

G.J. Persley
International Service for National Agricultural Research (ISNAR), PO Box 93375, 2509 AJ The Hague, The Netherlands

Abstract

The paper provides a summary of the major findings of a study on agricultural biotechnology jointly sponsored by the World Bank, the Australian Centre for International Agricultural Research (ACIAR), the Australian International Development Assistance Bureau (AIDAB) and the International Service for National Agricultural Research (ISNAR). The study addressed the following issues:

1) the likely impact of biotechnology on agricultural development;
2) the socio-economic and institutional issues which will affect its successful application;
3) the changing roles of public and private sector research;
4) intellectual property management;
5) biosafety;
6) the role of international development agencies.

The study involved 66 active collaborators, from 18 countries. Thirty-one papers and 14 country studies have been commissioned. The findings have been published in a World Bank technical paper and also in two volumes published by CAB International.

Significant economic impact of modern biotechnology on agricultural production is not expected before the year 2000. After that, it will become an increasingly important component of new technologies for crop production. The development of a notional strategy for biotechnology which involves both the public and private sectors is essential to make effective use of domestic and external resources. The aim of research managers should be to integrate new technologies into existing biological research programmes. New initiatives are required, rather than new institutions.

The likely early applications of new biotechnologies to crop production in the Third World are:

- new diagnostics, which could be made readily available and used widely in many countries;
- genetic mapping of major tropical crops as an aid to conventional breeding programmes;

- plant virus resistance by genetic engineering of the host plant;
- novel biocontrol agents for pest control, to reduce pesticide use.

Several "orphan commodities" have been assessed for their current constraints and likely availability of new technologies to aid their resolution. Substantial progress may be expected in the short term (up to 5 years from now) for rice, in the medium term (5–10 years) for banana/plantain, cassava and coffee and in the long term (10 years or more) for coconut and oil palm.

A major issue that will affect the application of modern biotechnology in agriculture is the management of intellectual property. Each country needs to weigh up the benefits and costs of intellectual property rights in biotechnology and frame its policies accordingly. Model patent systems could be developed by countries who decide to take steps to manage their own intellectual property. Another issue that will affect the application of biotechnology to agriculture is the regulatory climate governing the release of novel products. A safe and efficient regulatory environment is in itself a comparative advantage in biotechnology.

The new features of biotechnology are of sufficient importance to merit the establishment by international development agencies of innovative funding mechanisms, and new systems for the delivery of effective programmes incorporating biotechnology.

Introduction

The current excitement about genetic engineering and its potential impact on agriculture stems from the ability of scientists to manipulate and control genes in new ways. These recent developments need to be seen in a historical perspective, as part of the evolution of the science of genetics. The science of genetics can be considered to have originated in 1866, with Mendel's postulations as to the mode of inheritance of specific characteristics. The evolution of the science of genetics leading to modern biotechnology is illustrated in Fig. 1. The exponential rate at which new knowledge is being gained is shown in Fig. 2. It is this "knowledge explosion" in biology which presents the challenge to those concerned with agricultural research and development.

This paper summarises the major findings of a study on agricultural biotechnology, sponsored by the World Bank, the International Service for National Agricultural Research (ISNAR) and the Australian Government, through the Australian International Development Assistance Bureau (AIDAB), and the Australian Centre for International Agricultural Research (ACIAR). The study addressed the following issues:

- The likely impact of biotechnology on agricultural development.
- The socio-economic and institutional issues which will affect its successful application.

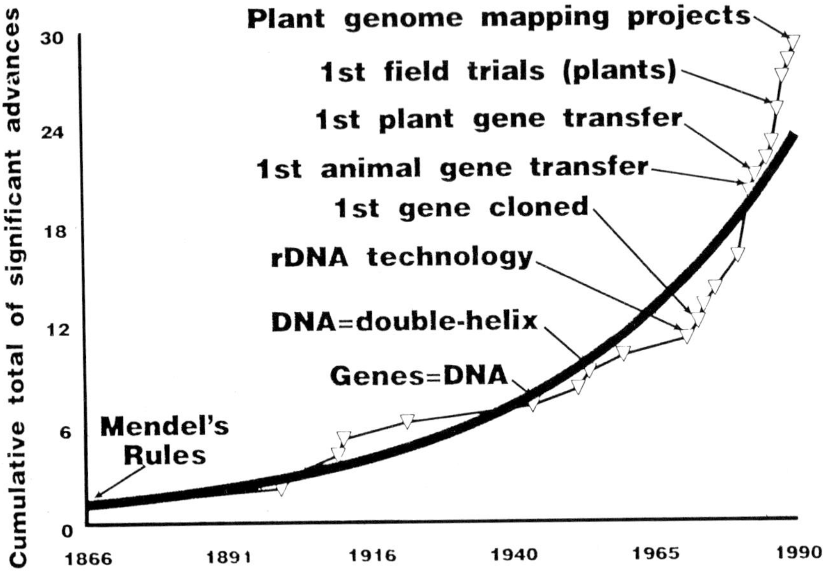

Fig. 1. Biotechnology milestones: 1866–1990

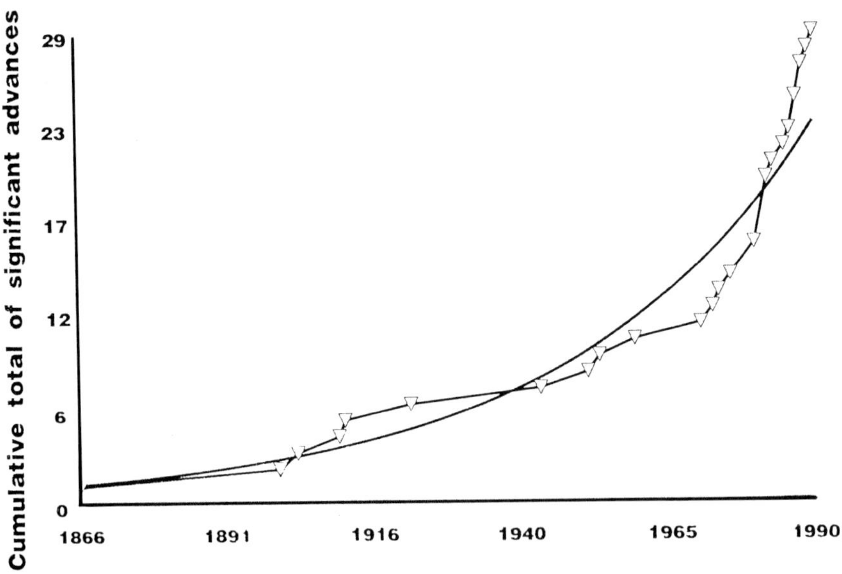

Fig. 2. Technology take-off
Source: Adapted from OTA (1989) and Wyke (1988)

- The changing roles of public and private sector research.
- Intellectual property management.
- Biosafety.
- The role of international development agencies.

The study involved 66 active collaborators, from 18 countries. Thirty-one papers and 14 country studies have been commissioned. The findings have been published in a World Bank technical paper (1991) and in two volumes published by CAB International (Persley, 1990a, b).

Definitions

Biotechnology is defined here as "any technique that uses living organisms, or substances from those organisms, to make or modify a product, to improve plants or animals, or to develop substances for specific uses" (OTA, 1989).

Biotechnology is comprised of a continuum of technologies, ranging from long-established and widely used technologies, for the commercial use of living organisms, through to more strategic research on the genetic engineering of plants and animals (Fig. 3).

Biotechnology includes *traditional biotechnology* covering well-established and widely used technologies for commercially useful operations. These include the biotechnologies presently used in brewing, food fermentation, conventional animal vaccine production and many others. *Modern biotechnology* encompasses the use of more recently available technologies, particularly those based on the use of (i) recombinant DNA technology; (ii) monoclonal antibodies; and (iii) new cell and tissue culture techniques, including novel bioprocessing. *Recombinant DNA technology* is a series of enabling techniques for genetic engineering, which allow the manipulation of the DNA, the essential genetic material in the cell. *Monoclonal antibodies* are specific tools which allow the rapid detection of individual proteins produced by the cells. *New cell and tissue culture techniques* allow the rapid propagation of living cells.

The basic principle of *genetic engineering* is that the genetic material, deoxyribonucleic acid (DNA), can be transferred from a cell of one species to another unrelated species, and made to express itself in the recipient cells. The newly formed DNA in the recipient cells contains its own, naturally occurring genes, plus the new gene. The new, *recombinant DNA* is inserted into the chromosomes of the transgenic plant. The recipient cells are said to be transformed by the arrival of the new genetic information.

Constraints to Agricultural Production

The first step in assessing the usefulness of new technologies to agriculture is to identify the problems which have proved intractable to conventional

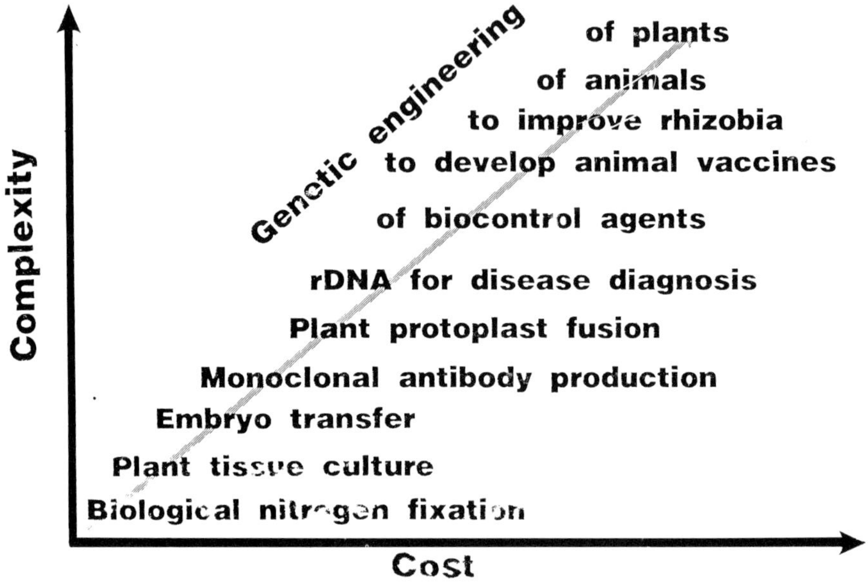

Fig. 3. Classifying biotechnologies

approaches and which may benefit from the application of these new technologies. It is important to combine the skills of agricultural scientists and economists who understand the commodity and its needs, with the skills of molecular biologists who see new ways to approach old problems in deciding what needs to be done. Cell and molecular biologists are the new partners of agricultural scientists, not their replacements.

Technology Assessment

Biotechnology as applied to crop production is concerned with:

- *Agricultural microbiology*, to produce micro-organisms beneficial to crop agriculture.
- *Cell and tissue culture*, including the rapid propagation of useful micro-organisms and plant species.
- *New diagnostics* based on the use of monoclonal antibodies and nucleic acid probes for the diagnosis of plant pests and diseases, and the detection of foreign chemicals in food.
- *Genetic engineering* of plant species to introduce new traits.
- *New genetic mapping techniques* based on the use of restriction fragment length polymorphisms (RFLPs) as an aid to plant breeding programmes.

The likely early applications of new biotechnologies to crop production in the Third World are:

- *New diagnostics*,which could be made readily available and used widely in many countries.
- *Genetic mapping* of major tropical crops as an aid to conventional breeding programmes.
- *Plant virus resistance* by genetic engineering of the host plant.
- *Novel biocontrol agents* for pest control, to reduce pesticide use.

Commodity Analysis

Several commodities have been assessed for their current constraints and likely availability of new technologies to aid their resolution. Substantial progress may be expected in the short term (up to 5 years from now) for potato, rapeseed and rice, in the medium term (5–10 years) for banana/plantain, cassava and coffee, and in the long term (10 years or more) for coconut, oil palm and wheat.

Some of the above commodities could be termed "orphan commodities". These are commodities in which there is little investment in modern biotechnology in industrial countries, either because the commodity is not important in temperate areas, or because there are no likely profits for transnational companies. Yet, they are important food or export commodities in the Third World (e.g. cassava, banana/plantain). More attention should be paid to "orphan commodities" in order to facilitate the application of biotechnology to the resolution of constraints with these crops.

Socio-economic Impact

Significant economic impact of modern biotechnology on agricultural production is not expected before the year 2000. After that, it will become an increasingly important component of new technologies for crop production. This time frame is similar to that for conventional research, in that it takes about ten years to develop a new plant variety, animal vaccine or pharmaceutical.

Biotechnology is likely to change the comparative advantages between countries and between commodities, particularly for export commodities. The application of new technologies to export commodities will improve their comparative position in the international marketplace.

The likely socio-economic effects of biotechnology in the Third World are positive in terms of increasing the productivity of tropical commodities to meet future needs, offering new opportunities for the use of marginal lands and reducing use of agrochemical. The effects are also potentially negative, in that they offer the possibility of producing high-value products in tissue culture, thus displacing crops presently grown for export in the Third World. The potential negative substitution effects should be monitored and remedial strategies developed in which economically damaging effects are identified.

National Strategies

The development of a national strategy for biotechnology which involves both the public and private sector is essential to make effective use of domestic and external resources. Today's modern biotechnology will be tomorrow's conventional technology. The aim of research managers should be to integrate new technologies into existing biological research programmes. As part of a national strategy, there may be some efficiencies in providing some common services from a central laboratory, especially where expensive equipment is required. New initiatives are required, rather than new institutions.

Training

Training is required at several levels:

- Bridging courses for research managers and mature scientists.
- Post-doctoral fellowships.
- Graduate training.
- Undergraduate courses.

There is especially a need for bridging courses to introduce policy makers and research managers to the new possibilities afforded by modern biotechnology. This would enable more informed decisions to be taken on the allocation of human and financial resources.

Post-doctoral fellowships in advanced laboratories are a key component in maintaining contact with the latest advances worldwide. Graduate and undergraduate training underpin the development of capability in the basic biological sciences in a country. Mobilisation of the basic biological sciences usually found in the universities, to solve agricultural problems, will require new policy and institutional arrangements and additional financial resources for most countries.

Public/Private Sector Collaboration

The major change in the funding of agricultural research in industrial countries in the past decade has been the greatly increased role of the private sector, largely in funding research in modern biotechnology. It has been estimated that at least half the current funding in R & D activities in agricultural biotechnology worldwide comes from the private sector. The major reason for the greater role of the private sector is that for many of the new technologies, the process and/or the product is protectable. A company is able to appropriate many of the benefits of its research investments. Private companies are therefore much more likely to invest in modern biotechnology than in more conventional agricultural research.

The trend towards greater private sector investment in agricultural research is established in the major OECD countries. It is an emerging trend in Latin America, parts of Asia and the Middle East. There is little evidence as yet in Africa. Public sector investments are presently the main source of financing for biotechnology in most developing countries.

There is a need for greater public/private sector collaboration in relating to biotechnology and its application to problems in the Third World. This will require:

- continued public sector investments from domestic and external resources;
- public/private sector partnerships;
- innovative funding mechanisms;
- involvement of both local private sector companies and transnational companies.

Patent Issues

The major issue that will affect the application of modern biotechnology in agriculture is the management of intellectual property. The lack of patent protection is a major disincentive for private sector investments in biotechnology in the Third World, both by local private sector companies and transnational companies.

The advantages of the provision of intellectual property rights are to encourage the development of local research capacity, and greater in-country investments in biotechnology. The major disadvantage is that it involves giving proprietary protection to living organisms which some consider to be part of the common heritage of mankind.

Each country needs to weigh up the benefits and costs of intellectual property rights in biotechnology and frame its policies accordingly. Model systems are being developed by countries which have decided to take steps to manage their own intellectual property.

Biosafety

A major issue that will affect the application of biotechnology to agriculture is the regulatory climate governing the release of novel products. A safe and efficient regulatory environment is in itself a comparative advantage in biotechnology. Action has been taken in several countries to ensure environmental safety and public health in biotechnology applications in agriculture.

Over the past few years, risk assessment procedures have been developed (mainly in the USA) to enable the assessment of instances of high, medium and low levels of risk associated with the release of genetically engineered organisms. Guidelines have been developed in several countries to govern national regulatory systems.

National regulatory systems are based on a tiered system of responsibility involving:

- an institutional biosafety committee;
- a national biosafety committee;
- project biosafety reviews.

World Bank Investments

The options by which the World Bank could provide further support for agricultural biotechnology are:

- *Additional loans* for specific biotechnology components within national agricultural research projects.
- *New joint ventures* to assist the participation of local private sector companies in the commercial development of biotechnology.
- *Additional, targeted grant funds* to the international agricultural research centres to increase their capacity in biotechnology, and their ability to assist national agricultural research systems in acquiring and adapting new technologies.

International Development Agencies

There are three main policy options for international development agencies interested in facilitating the application of biotechnology to agriculture. These are:

1) *Public sector support*: Provide support for biotechnology in association with other support for agricultural research at public sector institutions and universities. This includes the provision of training opportunities and research facilities. This would require little change from present policy and practice,
2) *Public/private sector support:* Provide support for biotechnology at public sector institutions in conjunction with collaborative research with advanced laboratories in either the public or private sector. Private companies may choose to participate in such collaborative activities either on a contract

research basis, or because participation in the project may confer commercial advantages that are not linked directly to the project itself.

3) *Commercial joint ventures:* Facilitate the establishment of commercially viable joint ventures between public and/or private sector agencies in developing industrialised countries. These new partnerships need to include both the research and development to produce new products and processes, and the business systems necessary for the delivery of novel products to farmers.

Agricultural Biotechnology Services

From the extensive consultations undertaken in the study, it is clear that there are common concerns for policy makers, research managers and scientists in many countries. There is a need in many countries for ready access to impartial advice and information as to how best to integrate modern biology (biotechnology) into existing research efforts and agricultural development programmes. This information should be available prior to decisions on whether to seek substantial external financing of new programmes. The information provided needs to focus on ways to identify current problems where biotechnology may be helpful, appropriate institutional and management arrangements and the necessary policy framework, including regulatory requirements and intellectual property management. ISNAR is seeking to incorporate some of these services within its advisory services for national agricultural research systems.

There is also a need for international development agencies such as the World Bank to strengthen their in-house capacity to appraise biotechnology requests. This may be done by ensuring existing staff have ready access to current information and sources of expertise on new biotechnology developments and the policy and management issues related to modern biotechnology.

Conclusion

The new features of biotechnology are of sufficient importance to merit the establishment by international development agencies of innovative funding mechanisms and new systems for the delivery of effective programmes incorporating biotechnology. The design of such programmes requires a policy dialogue among the international development agencies, and individual countries, as well as among the major players in modern biotechnology in the public and private sectors. It requires especially the international development agencies to devise new ways of doing business.

References

OTA (1989) US Congress Office of Technology Assessment, *New Developments in Biotechnology: Patenting life*. Special Report OTA-BA-370. US Government Printing Office, Washington DC, 196pp.

Persley, G.J. (1990a) *Beyond Mendel's Garden: Biotechnology in the service of world agriculture*. CAB International, Wallingford, UK, 155pp.

Persley, G.J. (ed.) (1990b) *Agricultural Biotechnology: Opportunities for international development*. CAB International, Wallingford, UK, 481pp.

World Bank (1991) Agricultural Biotechnology: The Next Green Revolution? *World Bank Technical Paper No. 133*, World Bank, Washington DC, 51pp.

Wyke, A. (1988) The genetic alternative: a summary of biotechnology, *The Economist*, 30 April.

Chapter thirty:
Crop Protection and Information Technology in the Year 2000

P.R. Scott and J.H. Gilmore
CAB International, Wallingford, Oxfordshire OX10 8DE, UK

Summary

The effects of advances in information technology (IT) over 20 years on the working methods of research scientists are outlined. Advances in the capability of personal computers are emphasised. Remarkable computing power is now available to both developed and developing countries, if funds permit. Examples of new opportunities to use IT in pest management are reviewed. CD-ROM technology allows compact storage of huge arrays of data; for example, the entire set of bibliographic abstracts in crop protection compiled by CAB International from 1913 to the present, approximately 800,000 abstracts, can be accommodated on two disks. Other developments include: multi-media electronic compendia of crop protection information; expert systems, especially in pest diagnosis; genetic cataloguing; compilations of plant quarantine information; pest risk analysis. Personal computers and CD-ROMs are seen as playing a central role in pest management by the year 2000.

Introduction

Information technology and biotechnology represent two of the most exciting developments for pest management in the year 2000. Both are full of promise, as yet largely unfulfilled, and in some respects unpredictable. Previous papers have examined the prospects for biotechnology in pest management. In this paper we move to information technology.

In 1989, CABI, FAO and CTA organised the International Crop Protection Information Workshop in the UK. In the published proceedings (Harris & Scott, 1989), a recurring theme is the desire for more effective electronic information systems for crop protection, especially in developing countries. Analysis of the responses to a questionnaire, published in the proceedings, showed that while most respondents considered the availability of bibliographic abstracts adequate in paper form, even in developing countries, only 14% were satisfied with their availability in electronic form. At the Workshop, there was a strong sense of expectation that information technology still had so much to deliver in support of pest management, and this is reflected in the published Recommendations. The other recurring theme of the Workshop was a sense of frustration that our ability to amass information for pest management tends to be much greater than our ability to communicate it.

Crop Protection and Pest Management

The basic approaches to pest management are unlikely to change, though the balance among them may shift:

- Quarantine
- Sound husbandry
- Genetic resistance
- Biological control
- Pesticides.

We place quarantine first on this occasion, partly because exclusion should be the first consideration in pest control, but particularly because information technology has a special role to play in this context, as discussed below. We also emphasise genetic resistance.

Information Technology

The Pace of Change

The basic parameters are changing rapidly. To illustrate this, we can look at two extremes, noting a remarkable development in just two decades.

In 1971, a research scientist working at a very well endowed institute in a developed country would be entirely dependent for published research information on the resources of his institute library. Furthermore, he would only have access to printed materials, or possibly microfilm, in the form of primary publications and secondary indexes. The task, for example, of carrying out a literature survey, even in a specialist discipline, would be extremely time-consuming and very imperfect.

For data processing or compiling statistics, he might have access to an institutional mainframe computer which would have limited power, would be slow, and would probably be difficult to use. The computer software available would be restricted in capability, and the scientist might even need to write his own software or call on computing staff to prepare software to satisfy his requirements.

He would probably have to write his research papers longhand, or at best in rough form on a typewriter and then rely on another person to prepare the final form, with a number of correction cycles. Communication with other scientists would be limited to mail or telephone contact and attendance at meetings.

At the other extreme, the scientist in a poor institute in a developing country in 1971 would have virtually no library facilities, certainly no access to computers, very rudimentary facilities for publishing, and almost no possibility of communication with other scientists outside his immediate region. Furthermore, other than through massive injections of funding, he would have no prospect of improving this situation.

Twenty years later, in 1991, the "rich" scientist will have at his disposal, through international communications and online services, access to a vast store of information in virtually every area of relevance to his work. He will also have, on his desk, one or more CD-ROM (Compact Disk-Read Only Memory) databases of the highest relevance. He will have for his personal use a computer workstation with the power to carry out any data processing other than that which might require the use of "super-computers". This will include powerful software which he will be able to tailor for his specific requirements without the need for expertise in computer programming.

The system will also provide him with facilities for directly preparing his research papers to a very high publishing standard, and printing these or transmitting them direct to publishers in electronic form. His computer will be connected to telecommunication networks giving contact with his colleagues virtually anywhere in the world. This scenario is by no means fanciful: it can already be seen in many scientific research institutions and we are not very far from the time when such facilities will be regarded as standard tools for research scientists.

The remarkable difference between the environment of 1971 and that of 1991 is that virtually all of these resources can be provided to scientists anywhere in the world. Leaving aside the question of cost for a moment, it is technically possible for the developing-country scientist to have facilities comparable with the scientist in the developed world. The major exception is that many developing countries are still poorly served by telecommunications, and where they are available the costs of use may be prohibitive. Of course, the costs of equipment, software and the information resources are significant factors but these costs continue to decrease and the "entry level" can be relatively low. Although it is unlikely that any scientists in developing countries would at present have anything approaching the kind of capability described here, there is no doubt that the use of microcomputers is spreading very rapidly and there is already a substantial base of equipment and experience from which the enhanced facilities can be developed.

The Central Significance of the Personal Computer

The underlying technology is the personal computer (PC). The pace of development of the PC has probably been unparalleled by any other technology, and there are few areas of computing which cannot now be satisfied by PCs. In fact, with the development of more powerful processors, very high volume storage devices, and the advent of networking, it is quite conceivable that other forms of computing will become obsolescent except those requiring so-called super-computing capabilities, such as very complex mathematical modelling and meteorological data processing.

This will be significant in providing access to large information databases, such as the bibliographic databases. Access to these in the past has been restricted to institutes which could afford the high capital costs of acquiring and maintaining mini or mainframe computer systems and specialised software. It is now feasible for smaller organisations to provide information services using microcomputer networks and low-cost software systems. Furthermore, we can expect to see a new generation of delivery mechanisms for very large databases such that, for example, a whole database such as *CAB ABSTRACTS* could be supplied as a "black box" with software, and simply plugged into a PC network for institutional use with none of the complexity or cost of large computer maintenance support.

Information Technology and Pest Management

We can only look at a few examples of the opportunities that these developments in information technology offer to pest management. CD-ROM is probably the most important.

CD-ROM Technology

CD-ROM is now well established as a powerful and compact delivery mechanism for databases. The growth in CD-ROM availability has been explosive, led by the producers of bibliographic databases. Each disk can store 600 million characters — for example the entire *CAB ABSTRACTS* database over three years, or at least 400,000 abstracts. Other bibliographic databases such as *AGRICOLA, AGRIS* and *BIOSIS* previews are also available on CD-ROM. The technology is reliable and inexpensive and remarkably simple to learn. No special computer expertise is needed, and hardly any understanding of the principles of information retrieval beyond a conception of what is being searched for.

Despite the remarkable capacity of CD-ROM for information, it does have a limit, and three years of data may not be enough. CAB International, in partnership with the Natural Resources Institute, is therefore developing specialist CD-ROMs for crop protection and pest management. Figure 1 shows the complete set of *Review of Agricultural Entomology* from 1913 to 1990 — about 350,000 abstracts. They would fill about three-quarters of one CD-ROM disk. What is actually planned is a project to combine abstracts on entomology, plant pathology, nematology and weed science on two disks, one for pre-1973 abstracts and one for 1973 onwards — some 800,000 abstracts in all.

Much of the literature of crop protection retains its value over time, for example that on basic pest biology, climatic requirements, host range, geographic distribution, natural-enemy relationships, pest resistance, biodiversity, taxonomy and nomenclature. An important part of this information was gathered before 1973. Therefore, high priority is attached to including the earlier disk in the project.

Once the bibliographic record has been captured on CD-ROM, CAB International plans to link it with a multi-media CD-ROM compendium of information on important organisms, for example quarantine pests. The compendium will include pest data sheets, distribution maps, illustrations,

Fig. 1. The complete set of *Review of Agricultural Entomology* from 1913 to 1990, containing approximately 350,000 bibliographic abstracts. The text would fill three-quarters of one CD-ROM disk.

taxonomic data, and perhaps appropriate expert systems. Such a compendium will offer an entirely new approach to accessing and integrating a range of types of information about pest organisms.

Expert Systems

Another branch of information technology that shows great promise for pest management is that known as expert systems. Figure 2 shows a small part of the decision tree that a human expert uses in making recommendations about whether to spray insecticides to control brown plant hopper on rice. If decision making can be reduced to a set of logical rules, then a computer can be equipped with software (an expert system) to emulate the human expertise. The use of expert-systems in pest management is reviewed by Lindsey & Novak (1989), Heong (1989) and Scott (1991).

Another important example of the use of expert systems is in diagnosis. CAB International has developed a prototype expert system for diagnosis of cocoa pests and diseases, based on observations of symptoms and damage.

Diagnosis in a taxonomic framework lends itself particularly well to computerisation, and represents a major advance over the traditional printed dichotomous key. INTKEY, for example, uses its database of information about the characteristics by which a group of organisms is classified to conduct a question and answer session with the user, through which it narrows down the diagnostic possibilities until one or a small group of taxa remains (Watson *et al.*, 1989). Unlike a printed key, the system can interact with the user, avoiding questions to which the answer is not available, or suggesting an optimal path to diagnosis. INTKEY is a powerful system, capable of handling information about many characters for large groups of organisms, e.g. all the plant viruses in the VIDE database (Boswell *et al.*, 1986). It is being enhanced by the addition of a more user-friendly front end with illustrations (Whitten, Chapter 2, this volume).

CAB International's CABIKEY is an interactive diagnostic key system with an emphasis on user-friendly presentation. It includes coloured illustrations of diagnostic differences between taxa and of the whole organisms to which a diagnosis points. Recent advances in screen displays have made a major contribution, in this and other contexts, not only to the accessibility of graphical information but also in the development of user-friendly interfaces. The system readily accepts a user's own data, so that a key can be developed for any group. CABIKEY is being applied particularly to insects, for example in the identification of fruitfly species, or in diagnosis to generic level within classes such as Coleoptera (CABI, 1991).

Genetic Cataloguing

Breeding plants for resistance to pests and diseases is one of the oldest and

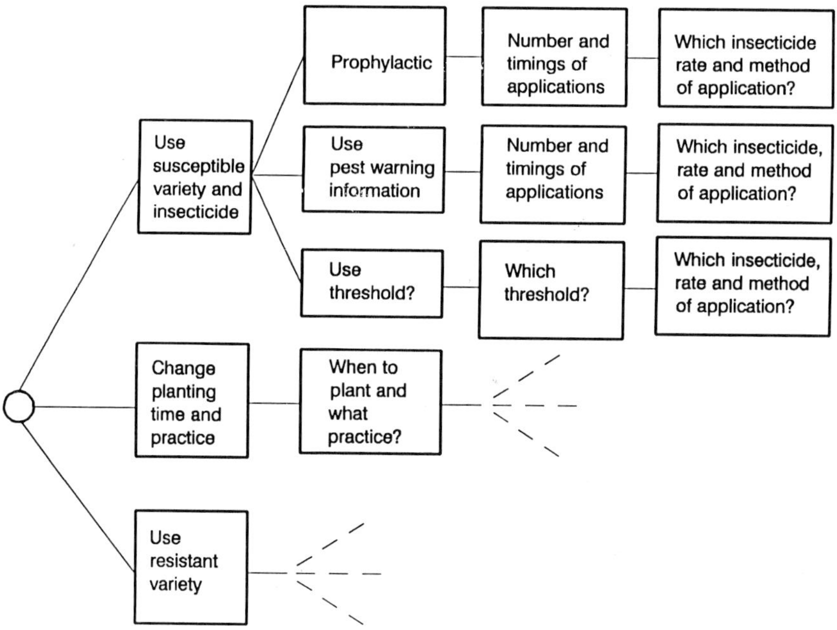

Fig. 2. Portion of a strategic decision tree used in making recommendations for control of brown plant hopper in rice, suitable as a basis for an expert system (after Heong, 1989)

soundest approaches to crop protection. It harnesses the crop's own reproductive capacity to deliver the pest management tool to every daughter plant of the pest-resistant parent. Genetic manipulation of plants has the potential to offer many new possibilities in this arena.

Information technology can contribute to the cataloguing of genetic information, without which mistakes are easily made. Excessive use of a few accessible resistance genes leads to the risk of genetic vulnerability. In 1970 a disastrous epidemic of southern corn leaf blight, caused by the fungal pathogen *Cochliobolus heterostrophus,* brought havoc to the maize industry in the USA. It was the result of the spread of a new race of the fungus and its highly specific virulence to maize varieties with a cytoplasmic gene for male sterility. This gene, in the "Texas" cytoplasm, had been introduced into more than three-quarters of the national crop because of its great value in the development of hybrid varieties. The susceptibility it would bring to the leaf blight was unsuspected. This episode led to the publication of an influential monograph on *Genetic Vulnerability of Major Crops* (National Academy of Sciences, 1972).

Since then, maize breeders have had the benefit of an extensive database, now on CD-ROM, maintained by the International Maize and Wheat Improvement Center (CIMMYT), cataloguing the genetic and phenotypic characters of huge numbers of maize varieties and accessions. Information technology has an important role to play in cataloguing the genetic diversity of plants, especially for the benefit of plant breeders trying to improve resistance to pests and diseases.

As we enter the era of genetically manipulated plants, it is difficult to avoid another phase of over-using a few convenient genes. The *Bt* insecticidal toxin genes are an obvious example, and may lead again to genetic vulnerability. Maintaining an easily searched electronic record of genetic diversity will be a major contribution of information technology to plant improvement.

This is of course an element of the resource of biodiversity discussed in other papers. It is worth noting in passing that information technology has found another application in this context. Those seeking to conserve biodiversity face the need to set priorities. Biodiversity at a given location may simply be measured in terms of number of species or as an index combining number of species with abundance. It is preferable to assess taxonomic diversity; for example Vane-Wright *et al.* (1991) have introduced an index based on the information content of cladistic classifications. Information technology has a significant role to play in handling large arrays of taxonomic data for such purposes.

Quarantine Information

The International Crop Protection Information Workshop (Harris & Scott, 1989) discussed many other applications of information technology to pest management. For this chapter, we turn finally to one of the central themes of that Workshop, plant quarantine. A list of 26 recommendations was published in the report of the Working Group on Plant Quarantine Information; seven of the most important are listed in Table 1.

This is an aspect of pest management to which the simplest application of information technology can contribute greatly. CAB International and the European and Mediterranean Plant Protection Organisation (EPPO) are the joint contractors to the European Communities to compile data sheets on each of the quarantine pests for Europe. The data sheets conform to a standard pattern for quarantine organisms specified by the Food and Agriculture Organisation of the United Nations, as indicated in Table 2. The headings of the data sheets correspond to fields of a database, allowing for simple cataloguing and retrieval of information. The intention is to link this database with another containing detailed distribution maps for each organism. The electronic medium greatly facilitates updating, which is essential for the quarantine application.

Electronic cataloguing of quarantine information lends itself to regional treatment. CAB International and EPPO have started with Europe, but this is only the beginning of what could be done for other regions. In time, a global plant quarantine database could be developed. The factual information of the

Table 1. Partial list of information needs for plant quarantine, recommended by the International Crop Protection Information Workshop (Harris & Scott, 1989)

Pest data sheets
Pest distribution maps
Host lists
Quarantine pest databases
Pest risk assessment
Pest forecasting systems
Computerised international quarantine information network

pest data sheets could be extended to quarantine regulations and treatments, and the whole could be packaged in such a way that it becomes a quarantine expert system, providing guidance to quarantine staff and others on the limitations and risks of particular importations of plant material. Furthermore, the data would become part of the multi-media compendium concept discussed above, for application beyond the confines of plant quarantine itself.

A greater challenge for the longer term is to use information technology to link the established facts of biology and distribution for a particular pest with the factual record of the environment in regions where it is not yet found, to arrive at a prediction of the likelihood if its becoming damaging there if it were introduced. This process, known as pest risk assessment, is attracting much attention. Common-sense interpretation of the opportunities to establish in known environments is the first requirement, for pests whose biology is understood. This is made easier by the provision of ready access to databases of the biological and environmental information. Climatic considerations are particularly important, and here it is possible to enlist the help of computers to make a quantitative match between the known climatic requirements of a pest (or its requirements inferred from its existing distribution) and worldwide climate data to predict where the pest has the potential to establish. This approach has been developed by Kohlmann *et al.* (1988), and particularly by Sutherst & Maywald (1985) with the CLIMEX computer modelling system.

Conclusion

We believe the opportunities for the application of information technology to improved pest management systems are of great importance. Scott (1991) discusses some of these in greater detail. In this paper we have selected a few that appear likely to lead to particularly significant advances by the year 2000.

Table 2. Headings of standardised plant quarantine pest data sheet, as used by CAB International and the European and Mediterranean Plant Protection Organisation

- Identity
  - Names
  - Synonyms
  - Taxonomic Position
  - Common names
  - Notes on taxonomy and nomenclature
- Host
- Geographical distribution
  - Europe
  - Asia
  - Africa
  - North America
  - Central America and Caribbean
  - South America
  - Oceania
- Biology
- Detection and identification
- Means of movement/dispersal
- Pest significance
  - Economic impact
  - Control
  - Quarantine risk
- Phytosanitary measures
- Bibliography

The PC is almost certain to be at the centre of developments. The CD-ROM is at present the obvious medium for large-scale data delivery. There is no technical limitation to the widespread use of PC and CD-ROM technology in developing as well as in developed countries. Databases, expert systems and visual-display technology are three areas where developments are occurring rapidly. The pace of technical change is indeed so rapid that even the year 2000 is really too far ahead for realistic prediction.

References

Boswell, K.F., Dallwitz, M.J., Gibbs, A.J. and Watson, L. (1986) The VIDE (Virus Identification Data Exchange) Project: a data bank for plant viruses. *Review of Plant Pathology* 65, 221–231.

CABI (1991) Identification of fruit flies. *CAB International: 1990 in review*, p. 24. CAB International, Wallingford.

Harris, K.M. and Scott, P.R. (eds.) (1989) *Crop Protection Information: an international perspective.* CAB International, Wallingford.

Heong, K.L. (1989) Sources of rice crop protection information. In: *Crop Protection Information: an international perspective* (Eds. Harris, K.M. and Scott, P.R.), pp. 51–66. CAB International, Wallingford.

Kohlmann, B., Nix, H. and Shaw, D.D. (1988) Environmental predictions and distributional limits of chromosomal taxa in the Australian grasshopper *Caledia captiva. Oecologia* 75, 483–493.

Lindsey, G. and Novak, K. (1989) Developments in information dissemination techniques: directions for international agricultural research centres. In: *Crop Protection Information: an international perspective* (Eds. Harris, K.M. and Scott, P.R.), pp. 27–48. CAB International, Wallingford.

National Academy of Sciences (1972) *Genetic Vulnerability of Major Crops.* National Academy of Sciences, Washington DC.

Scott, P.R. (1991) The universal issue: information transfer. In: *The Biodiversity of Microorganisms and Invertebrates: Its role in sustainable agriculture* (Ed. Hawksworth, D.L.), pp. 245–266. CAB International, Wallingford.

Sutherst, R.W. and Maywald, G.F. (1985) A computerised system for matching climates in ecology. *Agriculture, Ecosystems and Environment* 13, 281–299.

Vane-Wright, R.I., Humphries, C.J. and Williams, P.H. (1991) What to protect? - Systematics and the agony of choice. *Biological Conservation* 55, 235–254.

Watson, L., Gibbs Russell, G.E. and Dallwitz, M.J. (1989) Grass genera of southern Africa: interactive identification and information retrieval from an automated data bank. *South African Journal of Botany* 55, 452–463.

Chapter thirty-one: *New Technologies Discussion*

Chairman: Prof D.J. Greenland

Dr H.Y. Yeang (RRIM) You spoke of the application of embryo rescue for wide crosses. Presumably, the F1 would by itself be of limited immediate use and several generations of backcrosses would be required to fix the gene of interest into the cultivar of commercial interest? You have given examples of crops whose embryo rescue has been successful. These are cereals and other plants such as tomatoes that have short breeding cycles. In Malaysia, tree species have much longer breeding cycles. Do you therefore see embryo rescue as workable in these tree crops? Would you say that direct gene transfer would be more efficient in these dicot crop species?
Dr Khush The embryo rescue technique has greater applicability to annual crops where the life cycle is short. Generally, you have to make two or three backcrosses where embryo rescue techniques are to be used. Obviously, this technique cannot be applied to long life perennial tree crops. For such crops, techniques of genetic engineering are suitable for transferring genes from unrelated sources.
Dr Peter Dart RRIM has made a good start in this direction, by demonstrating that they have a strain of *Agrobacterium* which infects rubber and can transfer genes to rubber. There are other examples of gene transfer on tree crops such as poplars and walnut. So the prospect is there.
Dr R.J. Williams (CIBA-Ceigy, Switzerland) Do you have any idea of the fermentation capacity needed for the production of fungal insecticides for effective control of the desert locust?
Dr J. Waage This is hard to estimate at this early stage in the project, but the intention is that any myco-insecticide product would be produced by natural programmes relatively near the site of use (for reasons of storage and transport). Small-scale production plants for these same fungi produce kilogram quantities on a weekly basis for the control of sugarcane grasshopper in sugarcane.
Dr Max Whitten You gave some nice examples of identifying major genes for disease and insect resistance in relation to crop plants and various crossing techniques for transferring these genes to crop plants. How often do you find such resistance under major or quantitative gene control? How soon do you see plant breeders having the ability to identify and isolate the gene for resistance and transfer it directly by recombinant-DNA techniques into crop plants such as rice?

Dr Khush The chances of detecting major genes bearing quantitative resistance depend upon the screening techniques. If the disease and insect pressure during screening is very thigh, then the donors with major genes are identified and donors with moderate resistance are eliminated. However, if only mild selection pressure is applied, donors with quantitative resistance can be identified. There is great likelihood of identifying genes for disease and insect resistance especially if such genes can be tagged with RFLP markers and then cloned through chromosome working. Once such genes are cloned, they can be introduced into any crop variety provided transformation techniques are available.

Norman Hj Kamarudin (PORIM) You have mentioned classical biological control over the last few decades. The most recent example is the importation of *Liothrips mikaniae* for the control of *Mikania micrantha.* Could you elaborate on the current progress in establishing the biocontrol agent in the field?

Dr Peter A.C. Ooi Currently we are attempting to establish *Liothrips mikaniae* in the field. There are some problems with general predators removing the thrips. However, we hope that establishment will occur when large numbers of insects released.

Dr O. Mochida Are there any good examples of natural enemies which are bred to be resistant to pesticides through biotechnology methods?

Dr J. Waage Yes, substantial use is made today in orchards and field crops of predatory mites which have been selected for resistance to acaricides. This permits use of acaricides against pest mites, when necessary, without disrupting the biological control provided by predators. Perhaps Dr Dart may know of sophisticated biotechnology methods.

Dr P. Dart I know of no case where you can engineer a resistance to an insecticide, because resistance is almost certainly polygenic. The principle of resistance is of course present in bacteria, where genes can be moved.

Mr Jose Soriano Biological control from the perspective of an average farmer may seem like a very radical approach compared to his more conventional or traditional control methods. Could you share with us some constraints you have experienced in Malaysia in the process of establishing the three trends you mentioned earlier?

Dr Peter A.C. Ooi Initially we have to demonstrate successful biological control. This is akin to demonstrating predation to farmers by feeding pests to specific predators. Once we have success, we can build up confidence in farmers and policy makers who will then support future biological control programmes.

Dr K. Gallagher You mentioned that banning aerial spraying of pesticides resulted in pressure for investment in microbials. Legislation and policy will be important for promoting all new technologies in pest management. Does CAB International have a special programme on policy legislation, and intellectual property?

Dr J. Waage Although this is an issue in which I personally take a close interest, we do not at present have such a policy unit to provide the public and decision makers with the necessary information and confidence. Thus progress on biological control will continue to depend largely on opportunities created by political decisions to reduce pesticide use. The technical scientific support needs to be there, and scientists must be prepared to devote time and effort to persuading decision makers. Thus we create our own markets for biological control products.

Dr Lakchai Meenakanit A recently reported cross between a coconut and an oil palm proved infertile. How can such a hybrid be made to produce fruit?

Dr P. Dart Biotechnology can assist by transferring particular genes perhaps for disease or pest resistance from such a "marriage". Already trees such as poplar and walnut have been transformed by foreign genes and the trees regenerated. Transformation by a crown gall forming *Agrobacterium tumefaciens* of rubber has been demonstrated at RRIM. So the prospects for genetic engineering of either of these cultivars in the future is good. Laboratories are currently working on regeneration of parts from tissue culture from both these species.

Dr Khush There are limits to the crosses you can make and expect to prove useful. In the case you mention, you cannot do anything with the hybrid. If you have a single trait which you wish to transfer, it is best to use genetic engineering.

Dr Greenland In developing countries there will be an enormous need for training in biotechnology. Would Dr Dart care to comment?

Dr P. Dart I know of no courses dealing with problems between experts, and implementation of their technologies in developing countries. I am aware of the difficulties: there is a gap between those comfortable with terms and technologies and others with different skills who do not know how to tap in to these experts. It is a problem which also exists in developed countries. The dialogue between the two sides is enormously valuable. The research institutes themselves have to learn about these technologies. A national policy is needed to make institutes in developing countries familiar with such advanced technologies. Each country needs to decide which of its institutes is to be chosen to undertake a sufficiently large training programme. We need to develop networking with developed countries, and to train people to take part in that dialogue. The IRRI/Rockefeller Foundation partnership shows how people can co-operate. It is a process which requires goodwill on both sides. The basic issue is to recognise that people are assuming learning and training roles as equals.

Dr Khush We already have a training course in rice technology, tissue culture, hybridisation, etc., at IRRI. We envisage continuing to run this course for some years.

Abbreviations and Acronyms

ACIAR	Australian Centre for International Agricultural Research
a.i.	active ingredient
AIDAB	Australian International Development Assistance Bureau
ANRPC	Association of Natural Rubber Producing Countries
ARC	Agricultural Research Committee
ASEAN	Association of Southeast Asian Nations
ASEAN PLANTI	Plant Quarantine Centre and Training Institute
AUIBMPP	All-Union Institute of Biological Methods of Plant Protection
AVRDC	Asian Vegetable Research and Development Centre
Bt	*Bacillus thuringiensis*
BMC	Behaviour-modifying chemicals
BPH	Brown plant hopper (*Nilaparvata lugens*)
BSLR	Broad spectrum long residual (insectides)
CABI	CAB International
CD-ROM	Compact disk-read only memory
CIMMYT	International Maize and Wheat Improvement Centre
CLF	Corynespora leaf fall
CMV	Cucumber mosaic virus
CPB	Cocoa pod borer
CSIRO	Commonwealth Scientific and Industrial Research Organisation
CVMV	Chilli veinal mottle virus
DBM	Diamondback moth
DOAS	Department of Agriculture Sabah
EBDC	Ethylenebisdithio-carbamate
EEC	European Economic Community
EMBO	European Molecular Biology Organisation
EPPO	European and Mediterranean Plant Protection Organisation
ETL	Economic threshold level
ETU	Ethylenethiourea
FAO	Food and Agriculture Organisation
FRAC	Fungicide Resistance Action Committee
FRIM	Forest Research Institute of Malaysia
GDP	Gross domestic product
GIFAP	International Group of National Associations of Agrochemical Manufacturers
HRAC	Herbicide Resistance Action Committee
IGR	Insect growth regulators
IIBC	International Institute of Biological Control
IOPRM	International Organisation for Pest Resistance Management

IPM	Integrated pest management
IRAC	Insecticide Resistance Action Committee
IRM	Integrated resistance management
IRPA	Intensification of research in priority areas
IRRDB	International Rubber Research Development Board
IRRI	International Rice Research Institute
ISNAR	International Service for National Agricultural Research
IT	Information technology
JH	Juvenile hormone
JHA	Juvenile hormone analogue
MAPPS	Malaysian Plant Protection Society
MARDI	Malaysian Agricultural Research and Development Institute
MOPGC	Malaysian Oil Palm Growers Council
MRL	Marginal residue limit
NGO	Non-government organisation
NPV	Nuclear polyhedrosis virus
P & D	Pests and diseases
PC	Personal computer
PEG	Pyrethroids Efficiency Group
PLANTI ASEAN	Plant Quarantine Centre and Training Institute
PORIM	Palm Oil Research Institute of Malaysia
PORLA	Palm Oil Registration and Licensing Authority
p.p.m.	Parts per million
PQ	Plant quarantine
QTL	Quantitative trait loci
R & D	Research and Development
RFLP	Restriction fragment length polymorphism
RNA	Ribonucleic acid
RRIM	Rubber research Institute of Malaysia
SABRAO	Society for the Advancement of Breeding Researches in Asia and Oceania
SALB	South American leaf blight
SLF	Secondary leaf fall
T & V	Training and Visitation
TEPP	Tetraethyl pyrophosphate
ULV	Ultra-low volume
UNDP	United Nations Development Program
UNEP	United Nations Environmental Program
UPAM	United Planting Association of Malaysia
UPRD	United Plantations Research Department
VAM	Vesicular–arbuscular mycorrhizal
VIDE	Virus Identification Data Exchange
VIZR	All-Union Research Institute for Plant Protection (USSR)
VPM	Universiti Pertanian Malaysia
VSD	Vascular Streak Dieback
WHO	World Health Organisation

Index

NOTE: Common and Latin names are used throughout these proceedings. Users are advised to look under both headings as they are not always cross-referenced.